NONSMOOTH MECHANICS AND ANALYSIS

Theoretical and Numerical Advances

Advances in Mechanics and Mathematics

VOLUME 12

NONSMOOTH MECHANICS AND ANALYSIS

Theoretical and Numerical Advances

Edited by

P. ALART
Université Montpellier II, Montpellier, France

O. MAISONNEUVE
Université Montpellier II, Montpellier, France

R.T. ROCKAFELLAR
University of Washington, Seattle, Washington, U.S.A.

Library of Congress Control Number: 2005932761

ISBN-10: 0-387-29196-2 e-ISBN 0-387-29195-4

ISBN-13: 978-0387-29196-3

Printed on acid-free paper.

Printed in the United States of America.

9 8 7 6 5 4 3 2 1

springeronline.com

Contents

Preface

In the course of the last fifty years, developments in nonsmooth analysis and nonsmooth mechanics have often been closely linked. The present book acts as an illustration of this. Its objective is two-fold. It is of course intended to help to diffuse the recent results obtained by various renowned specialists. But there is an equal desire to pay homage to Jean Jacques Moreau, who is undoubtedly the most emblematic figure in the correlated, not to say dual, advances in these two fields.

Jean Jacques Moreau appears as a rightful heir to the founders of differential calculus and mechanics through the depth of his thinking in the field of nonsmooth mechanics and the size of his contribution to the development of nonsmooth analysis. His interest in mechanics has focused on a wide variety of subjects: singularities in fluid flows, the initiation of cavitation, plasticity, and the statics and dynamics of granular media. The 'Ariadne's thread' running throughout is the notion of unilateral constraint. Allied to this is his investment in mathematics in the fields of convex analysis, calculus of variations and differential measures. When considering these contributions, regardless of their nature, one cannot fail to be struck by their clarity, discerning originality and elegance. Precision and rigor of thinking, clarity and elegance of style are the distinctive features of his work.

In 2003, Jean Jacques Moreau's colleagues decided to celebrate his 80th birthday by organizing a symposium in Montpellier, the city in which he has spent the major part of his university life and where he is still very active scientifically in the Laboratoire de Mécanique et de Génie Civil. In the appendix, the reader will find a list of those who participated in the symposium, entitled "Theoretical and Numerical Nonsmooth Mechanics". This book is one of the outcomes of the symposium. It does not represent the acts, stricto sensu, although it does reflect its structure in five parts. Each of these corresponds to a field of mathematics or mechanics in which Jean Jacques Moreau has made remarkable contributions. The works that follow are those of eminent specialists, united by the fact that they have all, at some time, appreciated and utilized these contributions.

The diversity of the topics presented in the following pages may come as a surprise. In a way this reflects the open and mobile mind of the man they pay tribute to; it should allow the reader to appreciate, at least in part, the importance of the scientific wake produced. Specialists in one or a number of the subjects dealt with will enjoy discovering new results in their own fields. But they will also discover subjects outside their usual domain, linked through certain aspects and likely to open up new horizons. Not surprisingly, the first part is devoted to Convex and Nonsmooth Analysis, two modern and powerful approaches to the study of theories like optimization, calculus of variations, optimal control, differential inclusions, and Hamilton-Jacobi equations, in particular. Some new aspects are presented regarding the evolution of these theories. When a convex function is propagated by a Hamilton-Jacobi partial differential equation involving a concave-convex hamiltonian, the associated proximal mapping is shown to locally exhibit Lipschitz dependence on time. Models of transportation in a city are presented, which in connection with Monge-Kantorovich, embrace optimal transportation networks, pricing policies and the design of the city. The following chapter deals with various types of convex hulls involving a differential inclusion with given boundary value. Next, we find proof of a formula giving the Legendre-Fenchel transformation of a convex composite function in terms of the transformation of its components. Last is a synthetic survey of balayage involving bistochastic operators, using inequalities that go back to Hardy, Littlewood and Pólya.

Some very different applications of nonsmooth analysis in mechanics and thermomechanics are tackled in the second part. The existence and uniqueness of steady state solutions are investigated for thermoelastic contact problems, where a variable contact heat flow resistance is considered as well as frictional heating. A necessary and sufficient condition is obtained, with a direct mechanical interpretation, for statical admissibility of loads in unilateral structural analysis. In the case of plasticity, an overall presentation of shakedown theorems in the framework of generalized standard models is derived by min-max duality and some new results are proposed for the kinematic safety coefficient. This is followed by a review of recent studies and applications carried out on the different forms of a model coupling adhesion and friction, based on introducing the adhesion intensity variable. From a general point of view, through various examples in thermomechanics, it is shown how Clausius-Duhem inequality may be productive, by identifying quantities that are related as constitutive laws and by suggesting useful experiments. The reader will also find a thorough investigation, in two dimensions, of the unilateral crack identification problem in elastodynamics. This problem may be quite complicated, nonsmooth and nonconvex. The last two papers in this part concern, firstly, a penalty approximation of the Painlevé problem with convergence to a solution and, secondly, a block relaxation solutions algorithm, in the case of a stress-based static formulation of a discrete contact problem with friction.

Drop formation and turbulence are among the most fascinating topics in fluid mechanics. In the third part a new approach of coarse-grained approximation of velocity during turbulence follows a history of drop formation focusing on the periodic rediscovery of non linear features.

The dynamics of a collection of rigid bodies interacting via contact, dry friction and multiple collisions is a vast subject, to which Jean Jacques Moreau has devoted his attention over the last twenty years. The mathematical formulation of the equations of motion, in the sense of differential inclusions, within an implicit scheme, generates efficient algorithms that are generically referred to as the "contact dynamics method" in the field of granular media. The application of such algorithms, as an alternative to smoothing numerical strategies, has rapidly asserted itself in very different applications. The fourth part deals both with new developments in the mathematical classification of numerical algorithms and applications to the field of granular materials.

The last part, Topics in Nonsmooth Science, provides evidence of the wide possibilities of applying the concepts developed in convex and nonsmooth analysis. Often used in mechanics, these have been applied and extended in other general frameworks. The sweeping process, useful for constructive proofs and numerical solutions, may be applied to generalized dynamical systems and to higher order differential equations. The existence may be proved of solutions for evolutionary dynamical systems, involving differential measures, by constructive algorithms. Such algorithms are used to reach stationary solutions by taking advantage of friction. Finally, the duality theory, so pertinent in the convex context, is still useful in nonconvex programming.

The editors hope that, as a result, this book will attract a wide public. They would like to thank the "Département des Sciences de l'Ingénieur"of the "Centre National de la Recherche Scientifique" for its financial support, and David Dureisseix for his efficient help for finalizing the layout of this book.

P. Alart, O. Maisonneuve and R. T. Rockfellar

A short biography of Jean Jacques MOREAU

Jean Jacques Moreau was born on 31 July 1923, in BLAYE (GIRONDE). "Agrégé" in Mathematics and Doctor of Mathematics (University of Paris), he began his career as a researcher at the Centre National de la Recherche Scientifique (CNRS) before being appointed as Professor of Mathematical Methods in Physics at Poitiers University, and then Professor of General Mechanics at Montpellier University II, where he has spent most of his career. Today he is Emeritus Professor in the Laboratoire de Mécanique et Génie Civil, joint research unit at Montpellier University II - CNRS.

The central theme of his research is nonsmooth mechanics, a field whose applications concern for example contacts between rigid or deformable bodies, friction, plastic deformation of materials, wakes in fluid flows, and cavitation... The helicity invariant in the dynamics of ideal fluids, discovered by Jean Jacques Moreau in 1962, provides a starting point for the consideration of certain problems arising in fluid dynamics. His mathematical knowledge equipped him to develop theoretical tools adapted to these subjects and these have become standard practice in nonsmooth mechanics. Since the sixties, this activity has led him to important contributions in the construction of nonsmooth analysis, a mathematical field that is likewise of interest to specialists in optimisation, operational research and economics. He thus founded the Convex Analysis Group in the 1970s, at the Mathematics Institute at Montpellier University II, which has continued, under a succession of titles, to produce outstanding contributions.

Since the end of the 1980s, Jean Jacques Moreau has focused more closely on the numerical aspects of the subjects he has been studying. He notably devised novel calculation techniques for the statics or dynamics of collections of very numerous bodies. The direct applications concern, on one hand, the dynamics of masonry works subjected to seismic effects and, on the other, the largely interdisciplinary field of the mechanics of granular media. His computer simulations have allowed him to make substantial personal contributions to this mechanics, while his numerical techniques have found applications in seismic engineering and rail engineering (TGV, train à grande vitesse, ballast behavior).

Jean Jacques Moreau has been awarded a number of prizes by the Science Academy, including the Grand Prix Joanidès. He spent a year as guest researcher at the Mathematics Research Centre at Montreal University, and has been invited abroad on numerous occasions by the top research teams in his field. He is author, co-author and editor of several advanced works on contact mechanics and more generally, on nonsmooth mechanics, and has also published a two-volume course in mechanics that has greatly influenced the teaching of this discipline. For numerous academics Jean Jacques Moreau has been and truly remains a Master of Mechanics.

I

CONVEX AND NONSMOOTH ANALYSIS

Chapter 1

MOREAU'S PROXIMAL MAPPINGS AND CONVEXITY IN HAMILTON-JACOBI THEORY

R. Tyrrell Rockafellar *

Department of Mathematics, box 352420, University of Washington, Seattle, WA 98195-4350, USA

rtr@math.washington.edu

Abstract Proximal mappings, which generalize projection mappings, were introduced by Moreau and shown to be valuable in understanding the subgradient properties of convex functions. Proximal mappings subsequently turned out to be important also in numerical methods of optimization and the solution of nonlinear partial differential equations and variational inequalities. Here it is shown that, when a convex function is propagated through time by a generalized Hamilton-Jacobi partial differential equation with a Hamiltonian that is concave in the state and convex in the co-state, the associated proximal mapping exhibits locally Lipschitz dependence on time. Furthermore, the subgradient mapping associated of the value function associated with this mapping is graphically Lipschitzian.

Keywords: Proximal mappings, Hamilton-Jacobi theory, convex analysis, subgradients, Lipschitz properties, graphically Lipschitzian mappings.

1. Introduction

In some of his earliest work in convex analysis, J.J. Moreau introduced in (Moreau, 1962, Moreau, 1965), the *proximal* mapping P associated with a lower semicontinuous, proper, convex function f on a Hilbert

*Research supported by the U.S. National Science Foundation under grant DMS–104055

space $\mathcal{H}$, namely

$$P(z) = \text{argmin}_x \Big\{ f(x) + \tfrac{1}{2}||x - z||^2 \Big\}. \tag{1.1}$$

It has many remarkable properties. Moreau showed that P is everywhere single-valued as a mapping from $\mathcal{H}$ into $\mathcal{H}$, and moreover is nonexpansive:

$$||P(z') - P(z)|| \leq ||z' - z|| \text{ for all } z, z'. \tag{1.2}$$

In this respect P resembles a projection mapping, and indeed when f is the indicator of a convex set C, P is the projection mapping onto C. He also discovered an interesting duality. The proximal mapping associated with the convex function f^* conjugate to f, which we can denote by

$$Q(z) = \text{argmin}_x \Big\{ f^*(y) + \tfrac{1}{2}||y - z||^2 \Big\}, \tag{1.3}$$

which likewise is nonexpansive of course, satisfies

$$Q = I - P, \qquad P = I - Q. \tag{1.4}$$

In fact the mappings P and Q serve to parameterize the generally set-valued subgradient mapping ∂f associated with f:

$$y \in \partial f(x) \iff (x, y) = (P(z), Q(z)) \text{ for some } z, \tag{1.5}$$

this z being determined uniquely through (1.4) by $z = x + y$. Another important feature is that the *envelope* function associated with f, namely

$$E(z) = \min_x \Big\{ f(x) + \tfrac{1}{2}||x - z||^2 \Big\}, \tag{1.6}$$

is a finite convex function on $\mathcal{H}$ which is Fréchet differentiable with gradient mapping

$$\nabla E(z) = Q(z). \tag{1.7}$$

Our objective in this article is to tie Moreau's proximal mappings and envelopes into the Hamilton-Jacobi theory associated with convex optimization over absolutely continuous arcs $\xi : [0, t] \to \mathbb{R}^n$; for this setting we henceforth will have $\mathcal{H} = \mathbb{R}^n$. Let the space of such arcs be denoted by $\mathcal{A}_n^1[0, t]$.

The optimization problems in question concern the functions f_t on $\mathbb{R}^n$ defined by $f_0 = f$ and

$$f_t(x) = \min_{\xi \in \mathcal{A}_n^1[0,t], \xi(t)=x} \Big\{ f(\xi(0)) + \int_0^t L(\xi(\tau), \dot{\xi}(\tau)) d\tau \Big\} \text{ for } t > 0, \tag{1.8}$$

which represent the propagation of f forward in time t under the "dynamics" of a Lagrangian function L.

A pair of recent articles (Rockafellar and Wolenski, 2001, Rockafellar and Wolenski, 2001a), has explored this in the case where L satisfies the following assumptions, which we also make here:

(A1) The function L is convex, proper and lsc on $I\!R^n \times I\!R^n$.

(A2) The set $F(x) := \operatorname{dom} L(x, \cdot)$ is nonempty for all x, and there is a constant ρ such that $\operatorname{dist}(0, F(x)) \leq \rho(1 + ||x||)$ for all x.

(A3) There are constants α and β and a coercive, proper, nondecreasing function θ on $[0, \infty)$ such that $L(x, v) \geq \theta(\max\{0, ||v|| - \alpha||x||\}) - \beta||x||$ for all x and v.

The convexity of $L(x, v)$ with respect to (x, v) in (A1), instead of just with respect to v, is called *full convexity*. It opens the way to broad use of the tools of convex analysis. The properties in (A2) and (A3) are dual to each other in a sense brought out in (Rockafellar and Wolenski, 2001) and provide coercivity and other needed features of the integral functional.

It was shown in (Rockafellar and Wolenski, 2001, Theorem 2.1) that, under these assumptions, f_t is, for every t, a *lower semicontinuous, proper, convex function on $I\!R^n$ which depends epi-continuously on t* (i.e., the set-valued mapping $t \mapsto \operatorname{epi} f_t$ is continuous with respect to $t \in [0, \infty)$ in the sense of Kuratowsk-Painlevé set convergence (Rockafellar and Wets, 1997)). The topic we wish to address here is how, in that case, the associated proximal mappings

$$P_t(z) = \operatorname{argmin}_x \left\{ f_t(x) + \tfrac{1}{2}||x - z||^2 \right\}, \tag{1.9}$$

with $P_0 = P$, and envelope functions

$$E_t(z) = \min_x \left\{ f_t(x) + \tfrac{1}{2}||x - z||^2 \right\}, \tag{1.10}$$

with $E_0 = E$, behave in their dependence on t. It will be useful for that purpose to employ the notation

$$\bar{P}(t, z) = P_t(z), \qquad \bar{E}(t, z) = E_t(z). \tag{1.11}$$

Some aspects of this dependence can be deduced readily from the epi-continuity of f_t in t, for example the continuity of $\bar{P}(t, z)$ and $\bar{E}(t, z)$ with respect to $t \in [0, \infty)$; cf. (Rockafellar and Wets, 1997, 7.37, 7.38). From that, it follows through the nonexpansivity of $\bar{P}(t, z)$ in z and the finite convexity of $\bar{E}(t, z)$ in z, that both $\bar{P}(t, z)$ and $\bar{E}(t, z)$ are continuous with respect to $(t, z) \in [0, \infty) \times I\!R^n$.

At the end of our paper (Rockafellar, 2004) in an application of other results about variational problems with full convexity, we were able to show more recently that $\bar{E}(t,z)$ is in fact *continuously differentiable with respect to* (t,z), not just with respect to z, as would already be a consequence of the convexity and differentiability of E_t, noted above. But this property does not, by itself, translate into any extra feature of the dependence of $\bar{P}(t,z)$ on t, beyond the continuity we already have at our disposal.

The following new result which we contribute here thus reaches a new level, moreover one where $\bar{P}$ and $\bar{E}$ are again on a par with each other.

Theorem 1. *Under (A1), (A2) and (A3), both* $\bar{P}(t,z)$ *and* $\nabla\bar{E}(t,z)$ *are locally Lipschitz continuous with respect to* (t,z). *Thus,* $\bar{E}$ *is a function of class* $\mathcal{C}^{1+}$.

Our proof will rely on the Hamilton-Jacobi theory in (Rockafellar and Wolenski, 2001a) for the forward propagation expressed by (1.8). It concerns the characterization of the function

$$\bar{f}(t,x) = f_t(x) \text{ for } (t,x) \in [0,\infty) \times \mathbb{R}^n \tag{1.12}$$

in terms of a generalized "method of characteristics" in subgradient format.

2. Hamilton-Jacobi framework

The Hamiltonian function H that corresponds to the Lagrangian function L is obtained by passing from the convex function $L(x,\cdot)$ to its conjugate:

$$H(x,y) := \sup_v \Big\{ \langle v,y\rangle - L(x,v) \Big\}. \tag{2.1}$$

Because of the lower semicontinuity in (A1) and the properness of $L(x,\cdot)$ implied by (A2), the reciprocal formula holds that

$$L(x,v) = \sup_y \Big\{ \langle v,y\rangle - H(x,y) \Big\}, \tag{2.2}$$

so L and H are completely dual to each other. It was established in (Rockafellar and Wolenski, 2001) that a function $H : \mathbb{R}^n \times \mathbb{R}^n \to \overline{\mathbb{R}}$ is the Hamiltonian for a Lagrangian L fulfilling (A1), (A2) and (A3) if and only if it satisfies

(H1) $H(x,y)$ is concave in x, convex in y, and everywhere finite,

(H2) There are constants α and β and a finite, convex function φ such that

$$H(x,y) \le \varphi(y) + (\alpha||y|| + \beta)||x|| \text{ for all } x,\ y.$$

(H3) There are constants γ and δ and a finite, concave function ψ such that

$$H(x,y) \geq \psi(x) - (\gamma||x|| + \delta)||y|| \text{ for all } x,\ y.$$

The convexity-concavity of H in (H1) is a well known counterpart to the full convexity of L under the "partial conjugacy" in (2.1) and (2.2), cf. (Rockafellar, 1970). It implies in particular that H is locally Lipschitz continuous; cf. (Rockafellar, 1970, §35). The growth conditions in (H2) and (H3) are dual to (A3) and (A2), respectively. This duality underscores the refined nature of (A2) and (A3); they are tightly intertwined. They have also been singled out because of the role they can play in control theory of fully convex type. For example, L satisfies (A1), (A2), (A3), when it has the form

$$L(x,v) = g(x) + \min_u\{\, h(u) \mid Ax + Bu = v\} \tag{2.3}$$

for matrices $A \in \mathbb{R}^{n\times n}$, $B \in \mathbb{R}^{n\times m}$, a finite convex function g and a lower semicontinuous, proper convex function h that is coercive, or equivalently, has finite conjugate h^*. Then $f_t(x)$ is the minimum in the problem of minimizing

$$f(\xi(0)) + \int_0^t \{g(\xi(\tau)) + h(\omega(\tau))\}dt$$

over all summable control functions $\omega : [0,t] \to \mathbb{R}^m$ such that

$$\dot{\xi}(\tau) = A\xi(\tau) + B\omega(\tau) \text{ for a.e. } \tau, \quad \xi(t) = x.$$

The corresponding Hamiltonian in this case is

$$H(x,y) = \langle Ax, y\rangle - g(x) + h^*(y). \tag{2.4}$$

In the control context, backward propagation from a terminal time would be more natural than forward propagation from time 0, but it is elementary to reformulate from one to the other. Forward propagation is more convenient mathematically for the formulas that can be developed.

For any finite, concave-convex function H on $\mathbb{R}^n \times \mathbb{R}^m$, there is an associated *Hamiltonian dynamical system*, which can be written as the differential inclusion

$$\dot{\xi}(\tau) \in \partial_y H(\xi(\tau),\eta(\tau)), \quad -\dot{\eta}(\tau) \in \partial_x H(\xi(\tau),\eta(\tau)) \text{ for a.e. } \tau, \tag{2.5}$$

where ∂_y refers to subgradients in the convex sense in the y argument, and ∂_x refers to subgradients in the concave sense in the x argument.

In principle, the candidates ξ and η for a solution over an interval $[0,t]$ could just belong to $\mathcal{A}_n^1[0,t]$, but the local Lipschitz continuity of H, and the local boundedness it entails for the subgradient mappings that are involved (Rockafellar, 1970, §35), guarantee that ξ and η belong to $\mathcal{A}_n^\infty[0,t]$, i.e., that they are Lipschitz continuous.

Dynamics of the kind in (2.5) were first introduced in (Rockafellar, 1970) for their role in capturing optimality in variational problems with fully convex Lagrangians. In the present circumstances where (A1), (A2) and (A3) hold, it has been established in (Rockafellar and Wolenski, 2001) that

$$\xi \text{ solves (1.8)} \iff \begin{cases} \xi(t) = x \text{ and } (\xi,\eta) \text{ solves (2.5)} \\ \text{for some } \eta \text{ with } \eta(0) \in \partial f(\xi(0)) \,. \end{cases} \tag{2.6}$$

Again, ∂f refers to subgradients of the convex function f in the traditions of convex analysis. Another powerful property obtained in (Rockafellar and Wolenski, 2001), which helps in characterizing the functions f_t and therefore $\bar{f}$, is that

$$y \in \partial f_t(x) \iff \begin{cases} (\xi(t),\eta(t)) = (x,y) \text{ for some } (\xi,\eta) \\ \text{satisfying (2.5) with } \eta(0) \in \partial f(\xi(0)). \end{cases} \tag{2.7}$$

Yet another property from (Rockafellar and Wolenski, 2001), which we can take advantage of here, is that, for any $(x_0,y_0) \in I\!R^n \times I\!R^n$, the Hamiltonian system has at least one trajectory pair (ξ,η) that starts from (x_0,y_0) and continues forever, i.e., for the entire time interval $[0,\infty)$. This implies further that any trajectory up to a certain time t can be continued indefinitely beyond t. Such trajectories need not be unique, however.

Subgradients of the value function $\bar{f}$ in (1.12) must be considered as well. The complication there is that $\bar{f}(t,x)$ is only convex with respect to x, not (t,x). Subgradient theory beyond convex analysis is therefore essential. In this respect, we use $\partial\bar{f}$ to denote subgradients with respect to (t,x) in the broader sense of variational analysis laid out, for instance, in (Rockafellar and Wets, 1997). These avoid the convex hull operation in the definition utilized earlier by Clarke and are merely "limiting subgradients" in that context.

The key result from (Rockafellar and Wolenski, 2001) concerning subgradients $\partial\bar{f}$, which we will need to utilize later, reveals that, for $t > 0$,

$$(s,y) \in \partial\bar{f}(t,x) \iff y \in \partial f_t(x) \text{ and } s = -H(x,y). \tag{2.8}$$

Observe that the implication "$\Rightarrow$" in (2.8) says that $\bar{f}$ satisfies a subgradient version of Hamilton-Jacobi partial differential equation for H

and the initial condition $\bar{f}(0,x) = f(x)$. It becomes the classical version when $\bar{f}$ is continuously differentiable, so that $\partial\bar{f}(t,x)$ reduces to the singleton $\nabla\bar{f}(t,x)$. This subgradient version turns out, in consequence of other developments in this setting, to agree with the "viscosity" version of the Hamilton-Jacobi equation, but is not covered by the uniqueness results that have so far been achieved in that setting. The uniqueness of $\bar{f}$ as a solution, under our conditions (H1), (H2), (H3), and the initial function f follows, nonetheless, from independent arguments in variational analysis; cf. (Galbraith, 1999, Galbraith).

By virtue of its implication "$\Leftarrow$" in our context of potential nonsmoothness, (2.8) furnishes more than just a generalized Hamilton-Jacobi equation. Most importantly, it can be combined with (2.7) to see that

$$(s,y) \in \partial\bar{f}(t,x) \iff \begin{cases} \exists(\xi,\eta) \text{ satisfying (2.6) over } [0,t] \\ \text{such that } (\xi(t),\eta(t)) = (x,y) \\ \text{and } -H(\xi(t),\eta(t)) = s. \end{cases} \tag{2.9}$$

This constitutes a generalized "method of characteristics" of remarkable completeness, and in a global pattern not dreamed of in classical Hamilton-Jacobi theory, where everything depends essentially on the implicit function theorem with its local character. Instead of relying on such classical underpinnings, the characterization in (2.9) is based on convex analysis and extensive appeals to duality.

Proof of Theorem 1. We concentrate first on the claims about $\bar{P}$, which we already know to have the property that

$$||\bar{P}(t,z') - \bar{P}(t,z)|| \leq ||z'-z|| \text{ for all } z,z' \in I\!R^n,\ t \in [0,\infty). \tag{2.10}$$

To confirm the local Lipschitz continuity of $\bar{P}(t,z)$ with respect to (t,z), it will be enough, on this basis, to demonstrate local Lipschitz continuity in t with a constant that is locally uniform in z. Therefore, we fix any $t^* \in [0,\infty)$ and $z^* \in I\!R^n$, and take

$$x^* = \bar{P}(t^*,z^*), \quad y^* = \bar{Q}(t^*,z^*), \tag{2.11}$$

where

$$\bar{Q}(t,z) = Q_t(z) \text{ for } Q_t = I - P_t. \tag{2.12}$$

Fix any $(t^*,z^*) \in [0,\infty) \times I\!R^n$ along with a compact neighborhood $T_0 \times Z_0$ of this pair. The mapping

$$M : (t,z) \to \left(\bar{P}(t,z), \bar{Q}(t,z)\right) \in I\!R^{2n}, \tag{2.13}$$

which we already know is continuous, takes $T_0 \times Z_0$ into a compact set $M(T_0,Z_0) \subset I\!R^{2n}$. Utilizing the fact that H is locally Lipschitz continuous on $\mathbb{R}^{2n}$, we can select compact subsets U_0 and U_1 of $\mathbb{R}^n$ such that $M(T_0, Z_0) \subset U_1 \subset \operatorname{int} U_0$ and furthermore

$$(x,y) \in U_0,\ u \in \partial_x H(x,y),\ v \in \partial_y H(x,y) \implies \begin{cases} ||u|| \le \kappa, \\ ||v|| \le \kappa. \end{cases}$$

Trajectories (ξ, η) to the Hamiltonian system in (2.5) are then necessarily Lipschitz continuous with constant κ over time intervals during which they stay inside U_0. It is possible next, therefore, to choose an interval neighborhood T_1 of t^* within T_0 such that any Hamiltonian trajectory (ξ, η) over T_1 that touches U_1 remains entirely in U_0 (and thus has the indicated Lipschitz property). Finally, we can choose a neighborhood $T \times Z$ of (t^*, z^*) within $T_1 \times Z_0$, with T_1 being an interval, such that $M(T, Z) \subset U_1$.

With these preparations completed, consider any $z \in Z$ and any interval $[t, t'] \subset T$, with $t < t'$. Let $(x, y) = M(t, z)$, so that

$$x = \bar{P}(t, z) = P_t(z), \qquad y = \bar{Q}(t, z) = Q_t(z),$$

and consequently

$$y \in \partial f_t(x), \qquad x + y = z,$$

from the basic properties of proximal mappings. Also, $(x, y) \in U_1$. By (2.7), there is a Hamiltonian trajectory (ξ, η) over $[0, t]$ with $(\xi(t), \eta(t)) = (x, y)$. It can be continued over $[t, t']$. Our selection of $[t, t']$ ensures that, during that time interval, both ξ and η are Lipschitz continuous with constant κ.

We also have $\eta(\tau) \in \partial f_t(\xi(\tau))$; this follows by applying (2.7) to the interval $[0, \tau]$ in place of $[0, t]$. Let $\zeta(\tau) = \xi(\tau) + \eta(\tau)$ for $\tau \in [t, t']$. Then $\zeta(t) = z$ and ζ is Lipschitz continuous with constant 2κ. Moreover

$$\xi(\tau) = P_\tau(\zeta(\tau)) = \bar{P}(\tau, \zeta(\tau)), \qquad \eta(\tau) = Q_\tau(\zeta(\tau)) = \bar{Q}(\tau, \zeta(\tau)),$$

again according to Moreau's theory of proximal mappings. Now, by writing

$$\bar{P}(t', z) - \bar{P}(t, z) = [\bar{P}(t', \zeta(t)) - \bar{P}(t', \zeta(t'))] + [\bar{P}(t', \zeta(t')) - \bar{P}(t, \zeta(t))],$$

where $||P(t', \zeta(t)) - P(t', \zeta(t'))|| \le ||\zeta(t) - \zeta(t')||$ and, on the other hand, $P(t, \zeta(t)) = \xi(t)$ and $P(t', \zeta(t')) = \xi(t')$, we are able to estimate that

$$||\bar{P}(t, z) - \bar{P}(t, z)|| \ \le\ ||\zeta(t') - \zeta(t)|| + ||\xi(t') - \xi(t)|| \ \le\ 3\kappa|t' - t|.$$

Because this holds for all $z \in Z$ and $[t, t'] \subset T$, we have the locally uniform Lipschitz continuity property that was required for $\bar{P}$ in its time argument.

Note that the local Lipschitz continuity of $\bar{P}$ implies the same property for $\bar{Q}$, inasmuch as $\bar{Q}(t,z) = z - \bar{P}(t,z)$.

Turning now to the claims about $\bar{E}$, we observe, to begin with, that since $\nabla E_t = Q_t$ from proximal mapping theory, we have $\nabla_z \bar{E}(t,z) = \bar{Q}(t,z)$. A complementary fact, coming from (Rockafellar, 2004, Theorem 4 and Corollary), is that

$$\frac{\partial \bar{E}}{\partial t}(t,z) = -H(x,y) \quad \text{for } (x,y) = (\bar{P}(t,z), \bar{Q}(t,z)).$$

In these terms we have

$$\nabla \bar{E}(t,z) = \big(-H(\bar{P}(t,z), \bar{Q}(t,z)), \bar{Q}(t,z)\big).$$

Since H is locally Lipschitz continuous, and both $\bar{P}$ and $\bar{Q}$ are locally Lipschitz continuous, as just verified, we conclude that $\nabla \bar{E}$ is locally Lipschitz continuous, as claimed, too. □

3. Subgradient graphical Lipschitz property

The facts in Theorem 1 lead to a further insight into the subgradients of the function $\bar{f}$. To explain it, we recall the concept of a set-valued mapping $S : I\!R^p \to I\!R^q$ being *graphically Lipschitzian of dimension d* around a point $(\bar{u}, \bar{v})$ in its graph. This means that there is some neighborhood of $(\bar{u}, \bar{v})$ in which, under a smooth change of coordinates, the graph of S can be identified with that of a Lipschitz continuous mapping on a d-dimensional parameter space.

The subgradient mappings $\partial f : I\!R^n \rightrightarrows I\!R^n$ associated with lower semicontinuous, proper, convex functions f on $I\!R^n$, like here, are prime examples of graphically Lipschitzian mappings. Indeed, this property is provided by Moreau's theory of proximal mappings. In passing from the x, y, "coordinates" in which the relation $y \in \partial f(x)$ is developed, to the z, w, "coordinates" specified by $z = x + y$ and $w = x - y$, we get just the kind of representation demanded, since the graph of ∂f can be viewed parameterically in terms of the pairs $(P(z), Q(z))$ as z ranges over $I\!R^n$; cf. (1.5). Thus, ∂f is graphically Lipschitzian of dimension n, everywhere.

Is there an extension of this property to the mapping $\partial \bar{f}$ from $(0,\infty) \times I\!R^n$ to $I\!R \times I\!R^n$? The next theorem says yes.

Theorem 2. *Under (A1), (A2) and (A3), the subgradient mapping $\partial \bar{f}$ is everywhere graphically Lipschitzian of dimension $n+1$.*

Proof. We get this out of (2.8) and the parameterization properties developed in the proof of Theorem 1. These tell us that the representation

$$\big(-H(\bar{P}(t,z), \bar{Q}(t,z)), \bar{Q}(t,z)\big) \in \partial \bar{f}\big(t, \bar{P}(t,z)\big)$$

fully covers the graph of $\partial \bar{f}$ in a one-to-one manner relative to $(0, \infty) \times \mathbb{R}^n$ as (t, z) ranges over $(0, \infty) \times \mathbb{R}^n$. This is an $n + 1$-dimensional parameterization in which the mappings are locally Lipschitz continuous, so the assertion of the theorem is fully justified. □

References

J.J. Moreau, Fonctions convexes duales et points proximaux dans un espace hilbertien. *Comptes Rendus de l'Académie des Sciences de Paris* **255** (1962), 2897–2899.

J.J. Moreau, Proximité et dualité dans un espace hilbertien. *Bulletin de la Société Mathématique de France* **93** (1965), 273–299.

R. T. Rockafellar, P. R. Wolenski, Convexity in Hamilton-Jacobi theory I: dynamics and duality. *SIAM Journal on Control and Optimization* **39** (2001), 1323–1350.

R. T. Rockafellar, P. R. Wolenski, Convexity in Hamilton-Jacobi theory II: envelope representations. *SIAM Journal on Control and Optimization* **39** (2001), 1351–1372.

R. T. Rockafellar, *Convex Analysis*, Princeton University Press, 1970.

R. T. Rockafellar, R. J-B Wets, *Variational Analysis*, Springer-Verlag, Berlin, 1997.

R. T. Rockafellar, Hamilton-Jacobi theory and parametric analysis in fully convex problems of optimal control. *Journal of Global Optimization*, to appear.

R. T. Rockafellar, Generalized Hamiltonian equations for convex problems of Lagrange. *Pacific Journal of Mathematics* **33** (1970), 411–428.

G. Galbraith, *Applications of Variational Analysis to Optimal Trajectories and Nonsmooth Hamilton-Jacobi Theory*, Ph.D. thesis, University of Washington, Seattle (1999).

G. Galbraith, The role of cosmically Lipschitz mappings in nonsmooth Hamilton-Jacobi theory (preprint).

Chapter 2

THREE OPTIMIZATION PROBLEMS IN MASS TRANSPORTATION THEORY

Giuseppe Buttazzo
Dipartimento di Matematica, Università di Pisa, Via Buonarroti 2, I-56127 Pisa, Italy
buttazzo@dm.unipi.it

Abstract We give a model for the description of an urban transportation network and we consider the related optimization problem which consists in finding the design of the network which has the best transportation performances. This will be done by introducing, for every admissible network, a suitable metric space with a distance that inserted into the Monge-Kantorovich cost functional provides the criterion to be optimized. Together with the optimal design of an urban transportation network, other kinds of optimization problems related to mass transportation can be considered. In particular we will illustrate some models for the optimal design of a city, and for the optimal pricing policy on a given transportation network.

Keywords: Monge problem, transport map, optimal pricing policies

1. Introduction

In this paper we present some models of optimization problems in mass transportation theory; they are related to the optimal design of urban structures or to the optimal management of structures that already exist. The models we present are very simple and do not pretend to give a careful description of the urban realities; however, adding more parameters to fit more realistic situations, will certainly increase the computational difficulties but does not seem to modify the theoretical scheme in an essential way. Thus we remain in the simplest framework, since our goal is to stress the fact that mass transportation theory is the right tool to attack this kind of problems.

The problems we will present are of three kinds; all of them require the use of the Wasserstein distances W_p between two probabilities f^+ and f^-, that we will introduce in the next section. Let us illustrate here shortly the problems that we are going to study later in more details.

Optimal transportation networks. In a given urban area Ω, with two given probabilities f^+ and f^-, which respectively represent the density of residents and the density of services, a transportation network has to be designed in an optimal way. A cost functional has to be introduced through a suitable Wasserstein distance between f^+ and f^-, which takes into account the cost of residents to move outside the network (by their own means) and on the network (for instance by paying a ticket). The admissible class of networks where the minimization will be performed consists of all closed connected one-dimensional subsets of Ω with prescribed total length.

Optimal pricing policies. With the same framework as above (Ω, f^+, f^- given) we also consider the network as prescribed. The unknown is here the ticket pricing policy the manager of the network has to choose, and the goal is to maximize the total income. Of course, a too low ticket price policy will not be optimal, but also a too high ticket price policy will push customers to use their own transportation means, decreasing in this way the total income of the company.

Optimal design of an urban area. In this case the urban area Ω is still considered as prescribed, whereas f^+ and f^- are the unknowns of the problem that have to be determined in an optimal way taking into account the following facts:

- there is a transportation cost for moving from the residential areas to the services poles;
- people desire not to live in areas where the density of population is too high;
- services need to be concentrated as much as possible, in order to increase efficiency and decrease management costs.

2. The Wasserstein distances

Mass transportation theory goes back to Gaspard Monge (1781) when he presented a model in a paper on *Académie des Sciences de Paris*. The elementary work to move a particle x into $T(x)$, as in Figure 2.1, is given by $|x - T(x)|$, so that the total work is

$$\int_{\text{remblais}} |x - T(x)| \, dx \, .$$

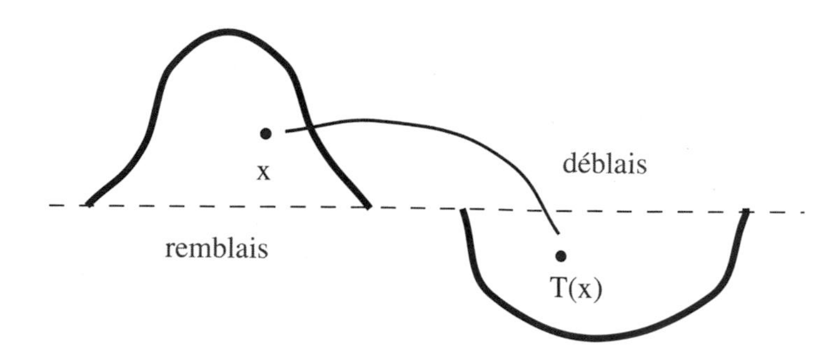

Figure 2.1. The Monge problem.

A map T is called *admissible transport map* if it maps "remblais" into "déblais". The Monge problem is then

$$\min\left\{\int_{\text{remblais}} |x - T(x)|\,dx \ :\ T \text{ admissible}\right\}.$$

It is convenient to consider the Monge problem in the framework of metric spaces:

- (X, d) is a metric space;
- f^+, f^- are two probabilities on X (f^+ represents the "remblais", f^- the "déblais");
- T is an admissible transport map if it maps f^+ onto f^-, that is $T^\# f^+ = f^-$.

The Monge problem is then

$$\min\left\{\int_X d\big(x, T(x)\big)\,df^+(x) \ :\ T \text{ admissible}\right\}.$$

The question about the existence of an optimal transport map T_{opt} for the Monge problem above is very delicate and does not belong to the purposes of the present paper (we refer the interested reader to the several papers available in the literature). Since we want to consider f^+ and f^- as general probabilities, it is convenient to reformulate the problem in a relaxed form (due to Kantorovich (Kantorovich, 1942, Kantorovich, 1948)): instead of transport maps we consider measures γ on $X \times X$ (called *transport plans*); γ is said an admissible transport plan if

$$\pi_1^\# \gamma = f^+, \qquad \pi_2^\# \gamma = f^-$$

where π_1 and π_2 respectively denote the projections of $X \times X$ on the first and second factors. In this way, the Monge-Kantorovich problem becomes:

$$\min\Big\{\int_{X\times X} d(x,y)\,d\gamma(x,y)\ :\ \gamma \text{ admissible}\Big\}.$$

THEOREM 2.1 *There exists an optimal transport plan γ_{opt}; in the Euclidean case γ_{opt} is actually a transport map T_{opt} whenever f^+ and f^- are in L^1.*

We denote by $MK(f^+,f^-,d)$ the minimum value in the Monge-Kantorovich problem above. This defines the Wasserstein distance (of exponent 1) by

$$W_1(f^+,f^-,d) = MK(f^+,f^-,d)$$

where the metric space (X,d) is considered as fixed. The Wasserstein distances of exponent $p > 1$ are defined in a similar way:

$$W_p(f^+,f^-,d) = \min\Big\{\Big(\int_{X\times X} d^p(x,y)\,d\gamma(x,y)\Big)^{1/p}\ :\ \gamma \text{ admissible}\Big\}.$$

When X is a compact metric space all the distances W_p are topologically equivalent, and the topology generated by them coincides with the weak* topology on the probabilities on X.

3. Optimal transportation networks

We consider here a model for the optimal planning of an urban transportation network, see (Brancolini and Buttazzo, 2003). Suppose that the following objects are given:

- a compact regular domain Ω of $\mathbb{R}^N$ $(N \geq 2)$; it represents the geographical region or urban area we are dealing with;
- a nonnegative measure f^+ on Ω; it represents the density of residents in the urban area Ω;
- a nonnegative measure f^- on Ω; it represents the density of services in the urban area Ω.

We assume that f^+ and f^- have the same mass, that we normalize to 1; so f^+ and f^- are supposed to be probability measures. The main unknown of the problem is the transportation network Σ that has to be designed in an optimal way to transport the residents f^+ into the services f^-. The goal is to introduce a cost functional $F(\Sigma)$ and to minimize it on a class of admissible choices. We assume that Σ varies among all closed connected 1-dimensional subsets of Ω with total length bounded by a given constant L. Thus the admissible class where Σ varies is

$$\mathcal{A}_L = \{\Sigma \subset \Omega, \text{ closed, connected, } \mathcal{H}^1(\Sigma) \leq L\}. \tag{2.1}$$

In order to introduce the optimization problem we associate to every "admissible urban network" Σ a suitable "point-to-point cost function" d_Σ which takes into account the costs for residents to move by their own means as well as by using the network. The cost functional will be then

$$F(\Sigma) = W_p(f^+, f^-, d_\Sigma) \tag{2.2}$$

for some fixed $p \geq 1$, so that the optimization problem we deal with is

$$\min\{F(\Sigma) \ : \ \Sigma \in \mathcal{A}_L\}. \tag{2.3}$$

It remains to introduce the function d_Σ (that in the realistic situations will be a semi-distance on Ω). To do that, we consider:

- a continuous and nondecreasing function $A : [0, +\infty[\to [0, +\infty[$ with $A(0) = 0$, which measures the cost for residents of traveling by their own means;

- a lower semicontinuous and nondecreasing function $B : [0, +\infty[\to [0, +\infty[$ with $B(0) = 0$, which measures the cost for residents of traveling by using the network.

More precisely, $A(t)$ represents the cost for a resident to cover a length t by his own means (walking, time consumption, car fuel, ...), whereas $B(t)$ represents the cost to cover a length t by using the transportation network (ticket, time consumption, ...). The assumptions made on the pricing policy function B allow us to consider the usual cases below: the flat rate policy of Figure 2.2 (a) as well as the multiple-zones policy of Figure 2.2 (b).

Therefore, the function d_Σ is defined by:

$$d_\Sigma(x, y) = \inf\{A(\mathcal{H}^1(\phi \setminus \Sigma)) + B(\mathcal{H}^1(\phi \cap \Sigma)) \ : \ \phi \in \mathcal{C}_{x,y}\}, \tag{2.4}$$

where $\mathcal{C}_{x,y}$ denotes the class of all curves in Ω connecting x to y.

THEOREM 2.2 *The optimization problem (2.3) admits at least a solution* Σ_{opt}.

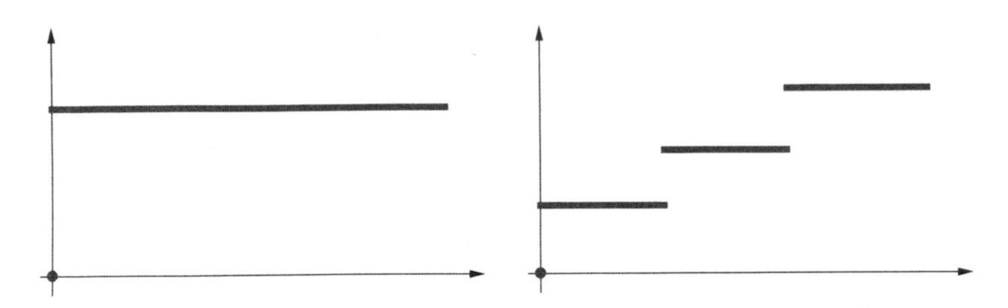

Figure 2.2. (a) flat rate policy (b) multiple-zones policy.

Once the existence of Σ_{opt} is established, several interesting questions arise:

- study the regularity properties of Σ_{opt}, under reasonable regularity assumptions on the data f^+ and f^-;
- study the geometrical necessary conditions of optimality that Σ_{opt} has to fulfill (nonexistence of closed loops, bifurcation points, distance from the boundary $\partial\Omega$, ...);
- perform an asymptotic analysis of the optimization problem (2.3) as $L \to 0$ and as $L \to +\infty$.

Most of the questions above are still open in the general framework covered by the existence Theorem 2.2 above. However, some partial results are available in particular situations; we refer the interested reader to the several recent papers on the subject, see for instance (Brancolini and Buttazzo, 2003, Buttazzo et al., 2002, Buttazzo and Stepanov, Mosconi and Tilli, 2003).

4. Optimal pricing policies

With the notation above, we consider the urban area Ω and the measures f^+, f^- as fixed, as well as the transportation network Σ. The unknown is in this case the pricing policy function B that the manager of the network has to choose among all lower semicontinuous monotone nondecreasing functions B, with $B(0) = 0$. The goal is to maximize the total income, which of course depends on the policy B chosen, so it can be seen as a functional $F(B)$.

The function B can be seen as a control variable and the corresponding Kantorovich transport plan γ_B as a state variable, which solves the minimum problem

$$\min\left\{\int_{\Omega\times\Omega} d_B^p(x,y)\,d\gamma(x,y)\ :\ \gamma \text{ admissible}\right\} \tag{2.5}$$

where p is the Wasserstein exponent and d_B is the cost function

$$d_B(x,y) = \inf\left\{A\big(\mathcal{H}^1(\phi \setminus \Sigma)\big) + B\big(\mathcal{H}^1(\phi \cap \Sigma)\big) \ : \ \phi \in \mathcal{C}_{x,y}\right\}. \qquad (2.6)$$

The quantity $d_B(x,y)$ can be seen as the total minimal cost a customer has to pay to go from a point x to a point y, using the best path ϕ. This cost is divided in two parts: a part $A\big(\mathcal{H}^1(\phi \setminus \Sigma)\big)$ due to the use of his own means, and a part $i_B(x,y) = B\big(\mathcal{H}^1(\phi \cap \Sigma)\big)$ due to the ticket to pay for using the transportation network. The only condition we assume to make the problem well posed is that, in case several paths ϕ realize the minimum in (2.6), the customer chooses the one with minimal own means cost (and so with maximal network cost). The total income is then

$$F(B) = \int_{\Omega\times\Omega} i_B(x,y)\, d\gamma_B(x,y), \qquad (2.7)$$

so that the optimization problem we consider is:

$$\max\left\{F(B) \ : \ B \text{ l.s.c., nondecreasing, } B(0) = 0\right\}. \qquad (2.8)$$

The following result has been proved in (Buttazzo et al.).

THEOREM 2.3 *There exists an optimal pricing policy B_{opt} solving the maximal income problem (2.8).*

Also in this case some necessary conditions of optimality can be obtained. It may happen that several functions B_{opt} solve the maximal income problem (2.8); in this case, as a canonical representative, we choose the smallest one, with respect to the usual order between functions. It is possible to show that it is still a solution of problem (2.8). In particular, the function B_{opt} turns out to be continuous, and its Lipschitz constant can be bounded by the one of A. We refer to (Buttazzo et al.) for all details as well as for the proofs above.

5. Optimal design of an urban area

We consider the following model for the optimal planning of an urban area, see (Buttazzo and Santambrogio, 2003).

- The domain Ω (the geographical region or urban area), a regular compact subset of $\mathbb{R}^N$, is prescribed;
- the probability measure f^+ on Ω (the density of residents) is unknown;
- the probability measure f^- on Ω (the density of services) is unknown.

Here the distance d in Ω is taken for simplicity as the Euclidean one, but with a similar procedure one could also study the cases in which the distance is induced by a transportation network Σ, as in the previous sections. The unknowns of the problem are f^+ and f^-, that have to be determined in an optimal way taking into account the following facts:

- residents have to pay a transportation cost for moving from the residential areas to the services poles;
- residents like to live in areas where the density of population is not too high;
- services need to be concentrated as much as possible, in order to increase efficiency and decrease management costs.

The transportation cost will be described through a Monge-Kantorovich mass transportation model; it is indeed given by a p-Wasserstein distance ($p \geq 1$) $W_p(f^+, f^-)$.

The *total unhappiness* of residents due to high density of population will be described by a penalization functional, of the form

$$H(f^+) = \begin{cases} \int_\Omega h(u)\,dx & \text{if } f^+ = u\,dx \\ +\infty & \text{otherwise,} \end{cases}$$

where h is assumed to be convex and superlinear (i.e. $h(t)/t \to +\infty$ as $t \to +\infty$). The increasing and diverging function $h(t)/t$ then represents the *unhappiness* to live in an area with population density t.

Finally, there is a third term $G(f^-)$ which penalizes sparse services. We force f^- to be a sum of Dirac masses and we consider $G(f^-)$ as a functional defined on measures, of the form studied by Bouchitté and Buttazzo in (Bouchitté and Buttazzo, 1990, Bouchitté and Buttazzo, 1992, Bouchitté and Buttazzo, 1992):

$$G(f^-) = \begin{cases} \sum_n g(a_n) & \text{if } f^- = \sum_n a_n \delta_{x_n} \\ +\infty & \text{otherwise,} \end{cases}$$

where g is concave and with infinite slope at the origin ((i.e. $g(t)/t \to +\infty$ as $t \to 0^+$). Every single term $g(a_n)$ in the sum above represents the cost for building and managing a service pole of dimension a_n, located at the point $x_n \in \Omega$.

We have then the optimization problem

$$\min\left\{W_p(f^+, f^-) + H(f^+) + G(f^-) \ :\ f^+, f^- \text{ probabilities on } \Omega\right\}. \quad (2.9)$$

THEOREM 2.4 *There exists an optimal pair (f^+, f^-) solving the problem above.*

Also in this case we obtain some necessary conditions of optimality. In particular, if Ω is sufficiently large, the optimal structure of the city consists of a finite number of disjoint subcities: circular residential areas with a service pole at their center.

Acknowledgments

This work is part of the European Research Training Network *"Homogenization and Multiple Scales" (HMS2000)* under contract HPRN-2000-00109.

References

L. Ambrosio, *Lecture notes on optimal transport problems.* In "Mathematical Aspects of Evolving Interfaces", Madeira 2–9 June 2000, Lecture Notes in Mathematics **1812**, Springer-Verlag, Berlin (2003), 1–52.

L. Ambrosio, A. Pratelli, *Existence and stability results in the L^1 theory of optimal transportation.* In "Optimal Transportation and Applications", Martina Franca 2–8 September 2001, Lecture Notes in Mathematics **1813**, Springer-Verlag, Berlin (2003), 123–160.

G. Bouchitté, G. Buttazzo, *New lower semicontinuity results for nonconvex functionals defined on measures.* Nonlinear Anal., **15** (1990), 679–692.

G. Bouchitté, G. Buttazzo, *Integral representation of nonconvex functionals defined on measures.* Ann. Inst. H. Poincaré Anal. Non Linéaire, **9** (1992), 101–117.

G. Bouchitté, G. Buttazzo, *Relaxation for a class of nonconvex functionals defined on measures.* Ann. Inst. H. Poincaré Anal. Non Linéaire, **10** (1993), 345–361.

G. Bouchitté, G. Buttazzo, *Characterization of optimal shapes and masses through Monge-Kantorovich equation.* J. Eur. Math. Soc., **3** (2001), 139–168.

G. Bouchitté, G. Buttazzo, P. Seppecher, *Shape optimization solutions via Monge-Kantorovich equation.* C. R. Acad. Sci. Paris, **324-I** (1997), 1185–1191.

A. Brancolini, G. Buttazzo, *Optimal networks for mass transportation problems.* Preprint Dipartimento di Matematica Università di Pisa, Pisa (2003).

G. Buttazzo, A. Davini, I. Fragalà, F. Macià, *Optimal Riemannian distances preventing mass transfer.* J. Reine Angew. Math., (to appear).

G. Buttazzo, L. De Pascale, *Optimal shapes and masses, and optimal transportation problems.* In "Optimal Transportation and Applications", Martina Franca 2–8 September 2001, Lecture Notes in Mathematics **1813**, Springer-Verlag, Berlin (2003), 11–52.

G. Buttazzo, E. Oudet, E. Stepanov, *Optimal transportation problems with free Dirichlet regions.* In "Variational Methods for Discontinuous Structures", Cernobbio 2001, Progress in Non-Linear Differential Equations **51**, Birkhäuser Verlag, Basel (2002), 41–65.

G. Buttazzo, A. Pratelli, E. Stepanov, *Optimal pricing policies for public transportation networks.* Paper in preparation.

G. Buttazzo, F. Santambrogio, *A model for the optimal planning of an urban area.* Preprint Dipartimento di Matematica Università di Pisa, Pisa (2003).

G. Buttazzo, E. Stepanov, *On regularity of transport density in the Monge-Kantorovich problem.* SIAM J. Control Optim., **42** (3) (2003), 1044-1055.

G. Buttazzo, E. Stepanov, *Transport density in Monge-Kantorovich problems with Dirichlet conditions.* Preprint Dipartimento di Matematica Università di Pisa, Pisa (2001).

G. Buttazzo, E. Stepanov, *Optimal transportation networks as free Dirichlet regions for the Monge-Kantorovich problem.* Ann. Scuola Norm. Sup. Pisa Cl. Sci., (to appear).

G. Carlier, I. Ekeland, *Optimal transportation and the structure of cities.* Preprint 2003, available at `http://www.math.ubc.ca`.

L. De Pascale, A. Pratelli, *Regularity properties for Monge transport density and for solutions of some shape optimization problem.* Calc. Var., **14** (2002), 249–274.

L. De Pascale, L. C. Evans, A. Pratelli, *Integral estimates for transport densities.* Bull. London Math. Soc., (to appear).

L. C. Evans, *Partial differential equations and Monge-Kantorovich mass transfer.* Current Developments in Mathematics, Int. Press, Boston (1999), 65–126.

L. C. Evans, W. Gangbo, *Differential Equations Methods for the Monge-Kantorovich Mass Transfer Problem.* Mem. Amer. Math. Soc. **137**, Providence (1999).

M. Feldman, R. J. McCann, *Monge's transport problem on a Riemannian manifold.* Trans. Amer. Math. Soc., **354** (2002), 1667–1697.

W. Gangbo, R. J. McCann, *The geometry of optimal transportation.* Acta Math., **177** (1996), 113–161.

L. V. Kantorovich, *On the transfer of masses.* Dokl. Akad. Nauk. SSSR, **37** (1942), 227–229.

L. V. Kantorovich, *On a problem of Monge.* Uspekhi Mat. Nauk., **3** (1948), 225–226.

G. Monge, *Mémoire sur la Théorie des Déblais et des Remblais.* Histoire de l'Académie Royale des Sciences de Paris, avec les Mémoires de Mathématique et de Physique pour la même année, (1781), 666–704.

S. J. N. Mosconi, P. Tilli, *Γ-convergence for the irrigation problem.* Preprint Scuola Normale Superiore, Pisa (2003), available at `http://cvgmt.sns.it`.

S. T. Rachev, L. Rüschendorf, *Mass transportation problems. Vol. I Theory, Vol. II Applications.* Probability and its Applications, Springer-Verlag, Berlin (1998).

C. Villani, *Topics in Optimal Transportation.* Graduate Studies in Mathematics, Amer. Math. Soc. (2003).

Chapter 3

SOME GEOMETRICAL AND ALGEBRAIC PROPERTIES OF VARIOUS TYPES OF CONVEX HULLS

Bernard Dacorogna

EPFL-DMA, CH 1015 Lausanne, Switzerland

bernard.dacorogna@epfl.ch

Abstract This article is dedicated to Jean Jacques Moreau for his 80th birthday. He is a master of convex analysis.

When dealing with differential inclusions of the form

$$Du(x) \in E, \text{ a.e. in } \Omega$$

together with some boundary data $u = \varphi$ on $\partial\Omega$, one is lead to consider several types of convex hulls of sets. It is the aim of the present article to discuss these matters.

Keywords: quasiconvexity, polyconvexity, rank one convexity, convex hulls

1. Introduction

We start in this introduction with the analytical motivations for studying some extensions of the notion of convex hull of a given set. In the remaining part of the article we will however only discuss geometrical and algebraic aspects of these notions. Therefore the reader only interested on these aspects can completely skip the introduction, since we will not use any of the notions discussed now.

We let $\Omega \subset \mathbb{R}^n$ be a bounded open set, $u : \Omega \subset \mathbb{R}^n \to \mathbb{R}^m$ and therefore the gradient matrix Du belongs to $\mathbb{R}^{m\times n}$ and finally we let $E \subset \mathbb{R}^{m\times n}$ be a compact set.

We are interested in solving the following Dirichlet problem

$$(D) \qquad \begin{cases} Du(x) \in E, \text{ a.e. } x \in \Omega \\ u(x) = \varphi(x), \ x \in \partial\Omega \end{cases}$$

where $\varphi : \overline{\Omega} \to \mathbb{R}^m$ is a given map.

In the scalar case ($n = 1$ or $m = 1$) a sufficient condition for solving the problem is

$$D\varphi(x) \in E \cup \operatorname{int}\operatorname{co} E, \text{ a.e. in } \Omega \tag{3.1}$$

where $\operatorname{int}\operatorname{co} E$ stands for the interior of the convex hull of E. This fact was observed by several authors, with different proofs and different levels of generality; notably in (Bressan and Flores, 1994, Cellina, 1993, Dacorogna and Marcellini, 1996, Dacorogna and Marcellini, 1997, Dacorogna and Marcellini, 1999a, De Blasi and Pianigiani, 1999) or (Friesecke, 1994). It should be noted that this sufficient condition is very close from the necessary one, which, when properly formulated, is

$$D\varphi(x) \in \overline{\operatorname{co}} E, \text{ a.e. in } \Omega, \tag{3.2}$$

where $\overline{\operatorname{co}} E$ denotes the closure of the convex hull of E.

When turning to the vectorial case ($n, m \geq 2$) the problem becomes considerably harder and conditions (3.1) and (3.2) are not anymore appropriate. One needs to introduce several extensions of the notion of convex hull, namely the *polyconvex*, *quasiconvex* and *rank one convex hulls*. We will define these notions precisely in the next section, but let us quote first an existence theorem involving the quasiconvex hull, $\overline{\operatorname{Qco}} E$.

We start with the following definition introduced by Dacorogna-Marcellini in (Dacorogna and Marcellini, 1999) (cf. also (Dacorogna and Marcellini, 1999a) and (Dacorogna and Pisante)), which is the key condition to get existence of solutions.

Definition 3.1 (Relaxation property) *Let $E \subset \mathbb{R}^{m \times n}$. We say that $\overline{\operatorname{Qco}} E$ has the relaxation property if for every bounded open set $\Omega \subset \mathbb{R}^n$, for every affine function u_ξ satisfying*

$$Du_\xi(x) = \xi \in \operatorname{int} \overline{\operatorname{Qco}} E,$$

there exists a sequence $\{u_\nu\}$ of piecewise affine functions in $\overline{\Omega}$, so that

$$u_\nu \in u_\xi + W_0^{1,\infty}(\Omega; \mathbb{R}^m), \; Du_\nu(x) \in \operatorname{int} \overline{\operatorname{Qco}} E, \text{ a.e. in } \Omega$$

$$u_\nu \overset{*}{\rightharpoonup} u_\xi \text{ in } W^{1,\infty}, \quad \int_\Omega \operatorname{dist}(Du_\nu(x); E)\, dx \to 0 \text{ as } \nu \to \infty.$$

The main existence theorem is then

Theorem 3.2 *Let $\Omega \subset \mathbb{R}^n$ be open. Let $E \subset \mathbb{R}^{m \times n}$ be such that E is compact. Assume that $\overline{\operatorname{Qco}} E$ has the relaxation property. Let φ be* *piecewise* $C^1\left(\overline{\Omega};\mathbb{R}^m\right)$ *and verifying*

$$D\varphi(x) \in E \cup \operatorname{int}\overline{\operatorname{Qco}E},\ a.e.\ in\ \Omega.$$

Then there exists (a dense set of) $u \in \varphi + W_0^{1,\infty}(\Omega;\mathbb{R}^m)$ *such that*

$$Du(x) \in E,\ a.e.\ in\ \Omega.$$

REMARK 3.3 *The theorem was first proved by Dacorogna-Marcellini in (Dacorogna and Marcellini, 1999) (cf. also Theorem 6.3 in (Dacorogna and Marcellini, 1999a)) under an extra further hypothesis. This hypothesis was later removed by Sychev in (Sychev, 2001) (see also Müller-Sychev (Müller and Sychev, 2001)), using the method of convex integration of Gromov as revisited by Müller-Sverak (Müller and Sverak, 1996), and Kirchheim (Kirchheim, 2001). As stated it has been recently proved by Dacorogna-Pisante (Dacorogna and Pisante).*

2. The different types of convex hulls

We now discuss the central notions of our article, for more background we refer to Dacorogna-Marcellini (Dacorogna and Marcellini, 1999a).

We will discuss the different notions of hulls by dual considerations on functions; we therefore start with the following definitions.

DEFINITION 3.4 *(i) A function* $f : \mathbb{R}^{m\times n} \to \overline{\mathbb{R}} = \mathbb{R} \cup \{+\infty\}$ *is said to be polyconvex if*

$$f\left(\sum_{i=1}^{\tau+1} t_i\xi_i\right) \leq \sum_{i=1}^{\tau+1} t_i f(\xi_i)$$

whenever $t_i \geq 0$ *and*

$$\sum_{i=1}^{\tau+1} t_i = 1,\ T\left(\sum_{i=1}^{\tau+1} t_i\xi_i\right) = \sum_{i=1}^{\tau+1} t_i T(\xi_i)$$

where for a matrix $\xi \in \mathbb{R}^{m\times n}$ *we let*

$$T(\xi) = (\xi, adj_2\xi, \ldots, adj_{m\wedge n}\xi)$$

where $adj_s\xi$ *stands for the matrix of all* $s \times s$ *subdeterminants of the matrix* ξ, $1 \leq s \leq m \wedge n = \min\{m,n\}$ *and where*

$$\tau = \tau(m,n) = \sum_{s=1}^{m\wedge n} \binom{m}{s}\binom{n}{s} \text{ and } \binom{m}{s} = \frac{m!}{s!\,(m-s)!}.$$

(ii) A Borel measurable function $f : \mathbb{R}^{m\times n} \to \mathbb{R}$ *is said to be quasiconvex if*

$$\int_U f(\xi + D\varphi(x))\,dx \geq f(\xi)\ \mathrm{meas}(U)$$

for every bounded domain $U \subset \mathbb{R}^n$, $\xi \in \mathbb{R}^{m\times n}$, *and* $\varphi \in W_0^{1,\infty}(U;\mathbb{R}^m)$.

(iii) A function $f : \mathbb{R}^{m\times n} \to \overline{\mathbb{R}} = \mathbb{R} \cup \{+\infty\}$ *is said to be rank one convex if*

$$f(t\xi_1 + (1-t)\xi_2) \leq t\,f(\xi_1) + (1-t)\ f(\xi_2)$$

for every ξ_1, ξ_2 *with rank* $\{\xi_1 - \xi_2\} = 1$ *and every* $t \in [0,1]$.

(iv) The different envelopes of a given function f *are defined as*

$$\begin{aligned} Cf &= \sup\{g \leq f : g \text{ convex}\}, \\ Pf &= \sup\{g \leq f : g \text{ polyconvex}\}, \\ Qf &= \sup\{g \leq f : g \text{ quasiconvex}\}, \\ Rf &= \sup\{g \leq f : g \text{ rank one convex}\}. \end{aligned}$$

We are now in a position to define the main notions of the article.

DEFINITION 3.5 *We let, for* $E \subset \mathbb{R}^{m\times n}$,

$$\begin{aligned} \overline{\mathcal{F}}_E &= \left\{f : \mathbb{R}^{m\times n} \to \overline{\mathbb{R}} = \mathbb{R}\cup\{+\infty\} : f|_E \leq 0\right\} \\ \mathcal{F}_E &= \left\{f : \mathbb{R}^{m\times n} \to \mathbb{R} : f|_E \leq 0\right\}. \end{aligned}$$

We then have respectively, the convex, polyconvex, rank one convex, rank one convex finite and (closure of the) quasiconvex hull defined by

$$\begin{aligned} \mathrm{co}\,E &= \left\{\xi \in \mathbb{R}^{m\times n} : f(\xi) \leq 0, \text{ for every convex } f \in \overline{\mathcal{F}}_E\right\} \\ \mathrm{Pco}\,E &= \left\{\xi \in \mathbb{R}^{m\times n} : f(\xi) \leq 0, \text{ for every polyconvex } f \in \overline{\mathcal{F}}_E\right\} \\ \mathrm{Rco}\,E &= \left\{\xi \in \mathbb{R}^{m\times n} : f(\xi) \leq 0, \text{ for every rank one convex } f \in \overline{\mathcal{F}}_E\right\} \\ \mathrm{Rco}_f E &= \left\{\xi \in \mathbb{R}^{m\times n} : f(\xi) \leq 0, \text{ for every rank one convex } f \in \mathcal{F}_E\right\} \\ \overline{\mathrm{Qco}}E &= \left\{\xi \in \mathbb{R}^{m\times n} : f(\xi) \leq 0, \text{ for every quasiconvex } f \in \mathcal{F}_E\right\}. \end{aligned}$$

REMARK 3.6 *The definition of rank one convex hull* $\mathrm{Rco}\,E$ *that we adopted is called, by some authors, lamination convex hull of* E*; while the same authors call* $\mathrm{Rco}_f E$ *the rank one convex hull of* E.

We start by pointing out several important facts.

1) The definition of convex hull is equivalent to the classical one, i.e. the smallest convex set that contains E. In fact it is enough to consider

only one function: the convex envelope, $C\chi_E$ $(= \chi_{\operatorname{co} E})$, of the indicator function of the set E, namely

$$\chi_E(\xi) = \begin{cases} 0 & \text{if } \xi \in E \\ \\ +\infty & \text{if } \xi \notin E. \end{cases}$$

2) If we replace $\overline{\mathcal{F}}_E$ by $\mathcal{F}_E$ in the definition of $\operatorname{co} E$, we get $\overline{\operatorname{co}}E$ the closure of the convex hull.

3) Similar considerations apply to the polyconvex hull, in particular it is sufficient to consider the polyconvex envelope, $P\chi_E$, of the indicator function of the set E.

4) If the set E is compact then so are $\operatorname{co} E$ and $\operatorname{Pco} E$.

5) We also have as a consequence of Carathéodory theorem

PROPOSITION 3.7 *Let $E \subset \mathbb{R}^{m \times n}$, then the following representation holds*

$$\operatorname{co} E = \left\{ \xi \in \mathbb{R}^{m \times n} : \xi = \sum_{i=1}^{mn+1} t_i \xi_i,\ \xi_i \in E,\ t_i \geq 0 \text{ with } \sum_{i=1}^{mn+1} t_i = 1 \right\}$$

$$\operatorname{Pco} E = \left\{ \xi \in \mathbb{R}^{m \times n} : T(\xi) = \sum_{i=1}^{\tau+1} t_i T(\xi_i),\ \xi_i \in E,\ t_i \geq 0 \text{ with} \sum_{i=1}^{\tau+1} t_i = 1 \right\}$$

6) Matters are however very different with the other definitions, but let us first start with some resemblances.

PROPOSITION 3.8 *Let $E \subset \mathbb{R}^{m \times n}$ and set $R_0 \operatorname{co} E = E$, and let for $i \in \mathbb{N}$*

$$\begin{aligned} R_{i+1} \operatorname{co} E = \{ & \xi \in \mathbb{R}^{m \times n} : \xi = t\xi_1 + (1-t)\xi_2, \\ & \xi_1, \xi_2 \in R_i \operatorname{co} E,\ \text{rank}\{\xi_1 - \xi_2\} = 1,\ t \in [0,1]\}. \end{aligned}$$

Then

$$\operatorname{Rco} E = \underset{i \in \mathbb{N}}{\cup} R_i \operatorname{co} E.$$

7) The rank one convex hull is equivalently defined through the rank one convex envelope, $R\chi_E$, of the indicator function of the set E and it is the smallest rank one convex set that contains E.

We now turn our attention to some differences of behavior among these notions.

8) Contrary to $\operatorname{co} E$ and $\operatorname{Pco} E$, if the set E is compact then $\operatorname{Rco} E$ is not necessarily compact as was pointed out by Kolar (Kolar, 2003).

9) Contrary to $\operatorname{co} E$ and $\operatorname{Pco} E$, the set $\operatorname{Rco}_f E$ is, in general (see below for an example), strictly larger than the closure of $\operatorname{Rco} E$, i.e.,

$$\overline{\operatorname{Rco} E} \subsetneqq \operatorname{Rco}_f E.$$

10) There is no good definition of the quasiconvex hull if we replace $\mathcal{F}_E$ by $\overline{\mathcal{F}}_E$, since quasiconvex functions with values in $\overline{\mathbb{R}} = \mathbb{R} \cup \{+\infty\}$ are not yet well understood. In particular if $E = \{\xi_1, \xi_2\} \subset \mathbb{R}^{m \times n}$, then $Q\chi_E = \chi_E$, independently of the fact that $\xi_1 - \xi_2$ is of rank one or not.

11) For any set $E \subset \mathbb{R}^{m \times n}$ we have

$$E \subset \operatorname{Rco} E \subset \operatorname{Pco} E \subset \operatorname{co} E$$

$$\overline{E} \subset \overline{\operatorname{Rco} E} \subset \operatorname{Rco}_f E \subset \overline{\operatorname{Qco} E} \subset \overline{\operatorname{Pco} E} \subset \overline{\operatorname{co} E}.$$

We now discuss two examples that might shed some light on some differences between these new hulls and the convex one. In both examples we will consider the case $m = n = 2$ and denote by $\mathbb{R}_d^{2\times 2}$ the set of 2×2 diagonal matrices, we will write any such matrix as a vector of $\mathbb{R}^2$.

EXAMPLE 3.9 *The first example is by now classical and is due to Tartar and a very similar one by Casadio (cf. Example 2 page 116 in (Dacorogna, 1989)). It shows the difference between* $\operatorname{Rco}_f E$ *and* $\operatorname{Rco} E$. *Let* $E = \{\xi_1, \xi_2, \xi_3, \xi_4\} \subset \mathbb{R}_d^{2\times 2}$ *be defined by* $\xi_1 = (2,1)$, $\xi_2 = (1,-2)$, $\xi_3 = (-2,-1)$, $\xi_4 = (-1,2)$.

It is easy to see that since $\operatorname{rank}\{\xi_i - \xi_j\} = 2$ *for every* $i \neq j$, *then* $E = \operatorname{Rco} E$. *However*

$$\begin{aligned}\operatorname{Rco}_f E = &\left\{\xi \in \mathbb{R}_d^{2\times 2} : \xi = (x,y) \in [-1,1]^2\right\} \\ &\cup \left\{\xi \in \mathbb{R}_d^{2\times 2} : \xi = (x,1),\ x \in [1,2]\right\} \\ &\cup \left\{\xi \in \mathbb{R}_d^{2\times 2} : \xi = (1,y),\ y \in [-2,-1]\right\} \\ &\cup \left\{\xi \in \mathbb{R}_d^{2\times 2} : \xi = (x,-1),\ x \in [-2,-1]\right\} \\ &\cup \left\{\xi \in \mathbb{R}_d^{2\times 2} : \xi = (-1,y),\ y \in [1,2]\right\}.\end{aligned}$$

The second example exhibits another peculiarity of the rank one convex hull that is very different from the convex one. It is also a phenomenon that one wants to avoid when applying the results to the analytical problem discussed in the Introduction.

EXAMPLE 3.10 *Let* $E = \{\xi_1, \xi_2, \xi_3, \xi_4, \xi_5, \xi_6\} \subset \mathbb{R}_d^{2\times 2}$ *be defined by* $\xi_1 = (1,0)$, $\xi_2 = (1,-1)$, $\xi_3 = (0,-1)$, $\xi_4 = (-1,0)$, $\xi_5 = (-1,1)$, $\xi_6 = (0,1)$. *It is easy to find that*

$$\operatorname{Rco} E = \{\xi : \xi = (x,y) \in [0,1] \times [-1,0]\} \cup \{\xi : \xi = (x,y) \in [-1,0] \times [0,1]\}$$

and its interior (relative to $\mathbb{R}_d^{2\times 2}$) is given by

$$\operatorname{int}\operatorname{Rco} E = \{\xi : \xi = (x,y) \in (0,1)\times(-1,0)\} \cup \\ \{\xi : \xi = (x,y) \in (-1,0)\times(0,1)\}.$$

However there is no way of finding a set E_δ with the following "approximation property" (cf. (Dacorogna and Marcellini, 1999a) and (Dacorogna and Pisante) for more details concerning the use of this property):

(1) $E_\delta \subset \operatorname{Rco} E_\delta \subset \operatorname{int}\operatorname{Rco} E$ *for every* $\delta > 0$;

(2) for every $\epsilon > 0$ *there exists* $\delta_0 = \delta_0(\epsilon) > 0$ *such that* $\operatorname{dist}(\eta; E) \le \epsilon$ *for every* $\eta \in E_\delta$ *and* $\delta \in [0, \delta_0]$;

(3) if $\eta \in \operatorname{int}\operatorname{Rco} E$ *then* $\eta \in \operatorname{Rco} E_\delta$ *for every* $\delta > 0$ *sufficiently small.*

In fact $\operatorname{Rco} E_\delta$ *will be reduced, at best, to four segments and will have empty interior (and condition (3) will be violated).*

3. The singular values

One of the most general example of such hulls concern sets that involve singular values. Let us first recall that the singular values of a given matrix $\xi \in \mathbb{R}^{n\times n}$, denoted by $0 \le \lambda_1(\xi) \le ... \le \lambda_n(\xi)$, are the eigenvalues of $(\xi\xi^t)^{1/2}$.

We will consider three types of sets, letting $0 < \gamma_1 \le ... \le \gamma_n$ and $\alpha \le \beta$ with $\alpha \ne 0$,

$$E = \left\{\xi \in \mathbb{R}^{n\times n} : \lambda_i(\xi) = \gamma_i,\ i = 1, ..., n\right\}$$

$$E_\alpha = \left\{\xi \in \mathbb{R}^{n\times n} : \lambda_i(\xi) = \gamma_i,\ i = 1, ..., n,\ \det\xi = \alpha\right\}$$

$$E_{\alpha,\beta} = \left\{\xi \in \mathbb{R}^{n\times n} : \lambda_i(\xi) = \gamma_i,\ i = 2, ..., n,\ \det\xi \in \{\alpha,\beta\}\right\}$$

where, since $|\det\xi| = \prod_{i=1}^n \lambda_i(\xi)$ and the singular values are ordered as $0 \le \lambda_1(\xi) \le ... \le \lambda_n(\xi)$, we should respectively impose in the second and third cases that

$$\prod_{i=1}^n \gamma_i = |\alpha|$$

$$\gamma_2 \prod_{i=2}^n \gamma_i \ge \max\{|\alpha|, |\beta|\}.$$

Note that the third case contains the other ones as particular cases. Indeed the first one is deduced from the last one by setting $\beta = -\alpha$ and

$$\gamma_1 = \beta\left[\prod_{i=2}^n \gamma_i\right]^{-1}$$

while the second one is obtained by setting $\beta = \alpha$ and

$$\gamma_1 = |\alpha| \left[\prod_{i=2}^{n} \gamma_i \right]^{-1}$$

in the third case.

Our result (cf. (Dacorogna and Marcellini, 1999a, Dacorogna and Tanteri, 2002) for the two first ones and for the third case: Dacorogna-Ribeiro (Dacorogna and Ribeiro)) is then

THEOREM 3.11 *Under the above conditions and notations the following set of identities holds*

$$\operatorname{co} E = \left\{ \xi \in \mathbb{R}^{n \times n} : \sum_{i=\nu}^{n} \lambda_i(\xi) \leq \sum_{i=\nu}^{n} \gamma_i, \ \nu = 1, ..., n \right\}$$

$$\operatorname{Pco} E = \overline{\operatorname{Qco}} E = \operatorname{Rco} E = \left\{ \xi \in \mathbb{R}^{n \times n} : \prod_{i=\nu}^{n} \lambda_i(\xi) \leq \prod_{i=\nu}^{n} \gamma_i, \ \nu = 1, ..., n \right\}.$$

$$\operatorname{Pco} E_\alpha = \operatorname{Rco} E_\alpha =$$
$$\left\{ \xi \in \mathbb{R}^{n \times n} : \prod_{i=\nu}^{n} \lambda_i(\xi) \leq \prod_{i=\nu}^{n} \gamma_i, \ \nu = 2, ..., n, \ \det \xi = \alpha \right\}.$$

$$\operatorname{Pco} E_{\alpha,\beta} = \operatorname{Rco} E_{\alpha,\beta} =$$
$$\left\{ \xi \in \mathbb{R}^{n \times n} : \prod_{i=\nu}^{n} \lambda_i(\xi) \leq \prod_{i=\nu}^{n} \gamma_i, \ \nu = 2, ..., n, \ \det \xi \in [\alpha, \beta] \right\}.$$

As it was pointed out by Buliga (Buliga) there is a surprising formal connection, still not well understood, between the above theorem (with $\alpha = \beta$) and some classical results of H. Weyl, A. Horn and C.J. Thompson (see (Horn and Johnson, 1985), (Horn and Johnson, 1991) page 171 or (Marshall and Olkin, 1979)). The result states that if we denote, as above, the singular values of a given matrix $\xi \in \mathbb{R}^{n \times n}$ by $0 \leq \lambda_1(\xi) \leq ... \leq \lambda_n(\xi)$ and its eigenvalues, which are complex in general, by $\mu_1(\xi), ..., \mu_n(\xi)$ and if we order them by their modulus $(0 \leq |\mu_1(\xi)| \leq ... \leq |\mu_n(\xi)|)$ then the following result holds

$$\prod_{i=\nu}^{n} |\mu_i(\xi)| \leq \prod_{i=\nu}^{n} \lambda_i(\xi), \ \nu = 2, ..., n$$
$$\prod_{i=1}^{n} |\mu_i(\xi)| = \prod_{i=1}^{n} \lambda_i(\xi)$$

for any matrix $\xi \in \mathbb{R}^{n \times n}$.

Acknowledgments

We have benefitted from the financial support of the Fonds National Suisse (grant 21-61390.00)

References

Bressan A. and Flores F., On total differential inclusions; Rend. Sem. Mat. Univ. Padova, 92 (1994), 9-16.

Buliga M., Majorisation with applications in elasticity; to appear.

Cellina A., On minima of a functional of the gradient: sufficient conditions, Nonlinear Anal. Theory Methods Appl., 20 (1993), 343-347.

Croce G., A differential inclusion: the case of an isotropic set, to appear.

Dacorogna B., *Direct methods in the calculus of variations*, Applied Math. Sciences, 78, Springer, Berlin (1989).

Dacorogna B. and Marcellini P., Théorèmes d'existence dans le cas scalaire et vectoriel pour les équations de Hamilton-Jacobi, C.R. Acad. Sci. Paris, 322 (1996), 237-240.

Dacorogna B. and Marcellini P., General existence theorems for Hamilton-Jacobi equations in the scalar and vectorial case, Acta Mathematica, 178 (1997), 1-37.

Dacorogna B. and Marcellini P., On the solvability of implicit nonlinear systems in the vectorial case; in the AMS Series of Contemporary Mathematics, edited by G.Q. Chen and E. DiBenedetto, (1999), 89-113.

Dacorogna B. and Marcellini P., *Implicit partial differential equations*; Birkhäuser, Boston, (1999).

Dacorogna B. and Pisante G., A general existence theorem for differential inclusions in the vector valued case; to appear in Abstract and Applied Analysis.

Dacorogna B. and Ribeiro A.M., Existence of solutions for some implicit pdes and applications to variational integrals involving quasiaffine functions; to appear.

Dacorogna B. and Tanteri C., Implicit partial differential equations and the constraints of nonlinear elasticity; J. Math. Pures Appl. 81 (2002), 311-341.

De Blasi F.S. and Pianigiani G., On the Dirichlet problem for Hamilton-Jacobi equations. A Baire category approach; Nonlinear Differential Equations Appl. 6 (1999), 13–34.

Friesecke G., A necessary and sufficient condition for non attainment and formation of microstructure almost everywhere in scalar variational problems, Proc. Royal Soc. Edinburgh, 124A (1994), 437-471.

Horn R.A. and Johnson C.A.: *Matrix Analysis*; Cambridge University Press (1985).

Horn R.A. and Johnson C.A.: *Topics in Matrix Analysis*; Cambridge University Press (1991).

Kirchheim B., Deformations with finitely many gradients and stability of quasiconvex hulls; C. R. Acad. Sci. Paris, 332 (2001), 289–294.

Kolar J., Non compact lamination convex hulls; Ann. I. H. Poincaré, AN 20 (2003), 391-403.

Marshall A. W. and Olkin I., *Inequalities: theory of majorisation and its applications*; Academic Press (1979).

Müller S. and Sverak V., Attainment results for the two-well problem by convex integration; ed. Jost J., International Press, (1996), 239-251.

Müller S. and Sychev M., Optimal existence theorems for nonhomogeneous differential inclusions; J. Funct. Anal. 181 (2001), 447–475.

Sychev M., Comparing two methods of resolving homogeneous differential inclusions; Calc. Var. Partial Differential Equations 13 (2001), 213–229.

Chapter 4

A NOTE ON THE LEGENDRE-FENCHEL TRANSFORM OF CONVEX COMPOSITE FUNCTIONS

Jean-Baptiste Hiriart-Urruty

Université Paul Sabatier, 118, route de Narbonne, 31062 Toulouse Cedex 4, France

jbhu@cict.fr

Dedicated to J.J. Moreau on the occasion of his 80th birthday.

Abstract In this note we present a short and clear-cut proof of the formula giving the Legendre-Fenchel transform of a convex composite function $g \circ (f_1, \cdots, f_m)$ in terms of the transform of g and those of the f_i's.

Keywords: Convex functions; Composite functions; Legendre-Fenchel transform.

1. Introduction

The Legendre-Fenchel transform (or conjugate) of a function $\varphi : X \to \mathbb{R} \cup \{+\infty\}$ is a function defined on the topological dual space $X^\star$ of X as

$$p \in X^\star \longmapsto \varphi^*(p) := \sup_{x \in X} \left[\langle p, x\rangle - \varphi(x)\right].$$

In Convex analysis, the transformation $\varphi \rightsquigarrow \varphi^*$ plays a role analogous to that of Fourier's or Laplace's transform in other places in Analysis. In particular, one cannot avoid it in analysing a variational problem, more specifically the so-called dual version of it. That explains why the Legendre-Fenchel transform occupies a key place in any work on Convex analysis.

The purpose of the present note is to analyse the formula giving the Legendre-Fenchel transform of the convex composite function $g \circ (f_1, \cdots, f_m)$, with g and all the f_i convex, in terms of g^* and the $f_i{}^*$. Actually such a formula is not new, even if it does not appear explicitly in books but only in some specialized research papers. We intend here to derive such a result in a short and clear-cut way, using only a "pocket theorem" from Convex analysis; in particular, we shall not appeal to any result on convex mappings taking values in ordered vector spaces, as is usually done in the literature.

Before going further, some comments on the historical development of the convex analysis of $g \circ (f_1, \cdots, f_m)$ are in order. First of all, the setting: the f_i's are convex functions on some general vector space X and g is an increasing convex function on $\mathbb{R}^m$ (increasing means here that $g(y) \leq g(z)$ whenever $y_i \leq z_i$ for all i); the resulting composite function $g \circ (f_1, \cdots, f_m)$ is convex on X. Then, how things evolved:

- Expressing the subdifferential of $g \circ (f_1, \cdots, f_m)$ in terms of that of g and those of the $f_i's$ was the first work carried out in the convex analysis of $g \circ (f_1, \cdots, f_m)$, as early as in the years $1965 - 1970$. The goal was made easier to achieve by the fact that one knew the formula aimed at (by extending to subdifferentials the so-called chain rule in Differential calculus). The objective of obtaining the subdifferential of the convex composite function $g \circ F$, with a vector-valued convex operator F, was pursued by several authors in various ways, see (Combari et al., 1994, Combari et al., 1996, Zalinescu, 2002) and references therein for recent contributions.

- To our best knowledge, the first attempt to derive $[g \circ (f_1, \cdots, f_m)]^*$ in terms of g^* and the $f_i^{*\prime}s$ is due to Kutateladze in his note (Kutateladze, 1977) and full-fledged paper (Kutateladze, 1979). The working context was that of convex operators taking values in ordered vector spaces, and this was also the case in most of the subsequent papers on the subject. Not only was the case of real-valued $f_i's$ somehow hidden in the main results in these papers (Theorem 3.7.1 in (Kutateladze, 1979), Proposition 4.11 (ii) in (Combari et al., 1994), Theorem 3.4 (ii) in (Combari et al., 1996), Theorem 2.8.10 in (Zalinescu, 2002)), but more importantly, the theorems were derived after some heavy preparatory work: on subdifferential calculus rules for vector-valued mappings in (Kutateladze, 1979), on perturbation functions in (Combari et al., 1996), on ε-subdifferentials in (Zalinescu, 2002). All these aspects are summarized in section 2.8 (especially bibliographical notes) of (Zalinescu, 2002).

In the setting we are considering in the present paper, the expected formula for the Legendre-Fenchel conjugate of $g \circ (f_1, \cdots, f_m)$ is as follows: for all $p \in X^\star$,

$$[g \circ (f_1, \cdots, f_m)]^*(p) = \min_{\alpha_i \geq 0} \left[g^*(\alpha_1, \cdots, \alpha_m) + \left(\sum_{i=1}^m \alpha_i f_i \right)^* (p) \right]. \tag{4.1}$$

This formula was proved in a simple way when $m = 1$ in (Hiriart-Urruty and Lemarechal, 1993), Chapter X, Section 2.5; we shall follow here the same approach as there, using only a standard result in Convex analysis, the one giving the Legendre-Fenchel conjugate of a sum of convex functions. However the formula (4.1) is not always informative, take for example $g(y_1, \cdots, y_m) := \sum_{i=1}^m y_i$, a situation in which (4.1) does not say anything new; we therefore shall go a step further in the expression of $[g \circ (f_1, \cdots, f_m)]^*(p)$ by developing $\left(\sum_{i=1}^m \alpha_i f_i \right)^* (p)$; hence the final formula (4.7) below is obtained.

We end with some illustrations enhancing the versatility of the proved formula.

2. The Legendre-Fenchel transform of $g \circ (f_1, \cdots, f_m)$

We begin by recalling some basic notations and results from Convex analysis.

Let X be a real Banach space; by $X^\star$ we denote the topological dual space of X, and $(p, x) \in X^\star \times X \longmapsto \langle p, x \rangle$ stands for the duality pairing. The Legendre-Fenchel transform (or conjugate) of a function $\varphi : X \to \mathbb{R} \cup \{+\infty\}$ is defined on $X^\star$ as

$$p \in X^\star \longmapsto \varphi^*(p) := \sup_{x \in X} \left[\langle p, x \rangle - \varphi(x) \right]. \tag{4.2}$$

Clearly only those x in $dom\, \varphi := \{x \in X \mid \varphi(x) < +\infty\}$ are relevant in the calculation of the supremum in (4.2).

As a particular example of φ, consider the indicator function of a nonempty set C in X, that is

$$i_C(x) := 0 \text{ if } x \in C, \quad +\infty \text{ if not}; \tag{4.3}$$

then i_C^* is the so-called support function of C, that is

$$i_C^* =: \sigma_C : p \in X^\star \longmapsto \sigma_C(p) = \sup_{x \in C} \langle p, x \rangle. \tag{4.4}$$

When $\alpha > 0$, there is no ambiguity in defining $\alpha\varphi$ and the resulting conjugacy result is: $(\alpha\varphi)^*(p) = \alpha\varphi^*(\frac{p}{\alpha})$. As for $\alpha = 0$, one should be more careful: we set $(0\varphi)(x) = 0$ if $x \in dom\, \varphi$, $+\infty$ if not; in other words $0\varphi = i_{dom\varphi}$. The corresponding conjugacy result is $(0\varphi)^* = \sigma_{dom\, \varphi}$, a fact coherent with the following result:

$$\alpha\varphi^*(\frac{p}{\alpha}) \longrightarrow_{\alpha\to 0^+} \sigma_{dom\, \varphi}(p)$$

(at least for convex lower-semicontinuous φ).

We denote by $\Gamma_0(X)$ the set of functions $\varphi : X \longrightarrow \mathbb{R} \cup \{+\infty\}$ which are convex, lower-semicontinuous and not identically equal to $+\infty$ on X. The next theorem is the key result we shall rely on in our proofs; it is a classical one in Convex analysis (see (Laurent, 1972), Théorème 6.5.8 for example).

Theorem 1. *Let $f_1, \cdots, f_k \in \Gamma_0(X)$. Suppose there is a point in $\bigcap_{i=1}^{k} dom\, f_i$ at which $f_1, \cdots, f_{k-1}$ are continuous. Then, for all $p \in X^\star$:*

$$(f_1 + \cdots + f_k)^*(p) = \min_{p_1+\cdots+p_k=p} [f_1^*(p_1) + \cdots + f_k^*(p_k)]\,. \qquad (4.5)$$

The context of our work is the following one:

- $f_1, \cdots, f_m \in \Gamma_0(X)$;
- $g \in \Gamma_0(\mathbb{R}^m)$ and is increasing on $\mathbb{R}^m$, i.e. $g(y) \leq g(z)$ whenever $y_i \leq z_i$ for all $i = 1, \cdots, m$.

The composite function $g \circ (f_1, \cdots, f_m)$ is defined on X as follows:

$$[g \circ (f_1, \cdots, f_m)](x) = \begin{cases} g[f_1(x), \cdots, f_m(x)] & \text{if } f_i(x) < +\infty \text{ for all } i, \\ +\infty & \text{if not.} \end{cases}$$

The resulting function $g \circ (f_1, \cdots, f_m)$ is now convex on X. The minimal assumption to secure that $g \circ (f_1, \cdots, f_m)$ does not identically equal $+\infty$ is: there is a point $x_0 \in \bigcap_{i=1}^{m} dom\, f_i$ such that $(f_1(x_0), \cdots, f_m(x_0)) \in dom\, g$. We shall actually assume a little more to derive our main result below.

Theorem 2. *With the assumptions listed above on g and the f_i's, we suppose:*

$$(\mathcal{H}) \left\{ \begin{array}{c} \textit{There is a point } x_0 \in \bigcap_{i=1}^{m} \operatorname{dom} f_i \textit{ such that} \\ (f_1(x_0), \cdots, f_m(x_0)) \textit{ lies in the interior of } \operatorname{dom} g. \end{array} \right.$$

Then, for all $p \in X^\star$,

$$[g \circ (f_1, \cdots, f_m)]^*(p) = \min_{\alpha_1 \geq 0, \cdots, \alpha_m \geq 0} \left[g^*(\alpha_1, \cdots, \alpha_m) + \left(\sum_{i=1}^{m} \alpha_i f_i \right)^* (p) \right] \quad (4.6)$$

(with $0\, f_i = i_{dom\, f_i}$).

If moreover there is a point in $\bigcap_{i=1}^{m} \operatorname{dom} f_i$ at which $f_1, \cdots, f_{m-1}$ are continuous, then for all $p \in X^\star$,

$$[g \circ (f_1, \cdots, f_m)]^*(p) = \min_{\substack{\alpha_1 \geq 0, \cdots, \alpha_m \geq 0 \\ p_1 + \cdots + p_m = p}} \left[g^*(\alpha_1, \cdots, \alpha_m) + \sum_{i=1}^{m} \alpha_i f_i^*(\frac{p_i}{\alpha_i}) \right] \quad (4.7)$$

(where we interpret $0\, f_i^*(\frac{p_i}{0}) = \sigma_{dom\, f_i}(p_i)$).

Proof. By definition, given $p \in X^\star$,

$$\begin{aligned} -[g \circ (f_1, \cdots, f_m)]^*(p) &= \\ &= \inf_{x \in X} \{[g \circ (f_1, \cdots, f_m)](x) - \langle p, x \rangle\} \\ &= \inf_{\substack{f_i(x) < +\infty \\ \text{for all } i}} \{g[f_1(x), \cdots, f_m(x)] - \langle p, x \rangle\} \\ &= \inf_{x \in X, (y_1, \cdots, y_m) \in \mathbb{R}^m} [g(y_1, \cdots, y_m) \mid f_i(x) \leq y_i \text{ for all } i] \quad (4.8) \end{aligned}$$

because g is assumed increasing on $\mathbb{R}^m$.

Let now F_1 and F_2 be defined on $X \times \mathbb{R}^m$ as follows:
For $(x, y_1, \cdots, y_m) \in X \times \mathbb{R}^m$,

$$F_1(x, y_1, \cdots, y_m) := -\langle p, x \rangle + g(y_1, \cdots, y_m)$$

$$F_2(x, y_1, \cdots, y_m) := i_{epi\, f_1}(x, y_1) + \cdots + i_{epi\, f_m}(x, y_m)$$

where $epi\, f_i$ denotes the epigraph of f_i, that is the set of $(x, y_i) \in X \times \mathbb{R}$ such that $f_i(x) \leq y_i$.

Thus, (4.8) can be written as

$$- [g \circ (f_1, \cdots, f_m)]^*(p) = \inf_{(x, y_1, \cdots, y_m) \in X \times \mathbb{R}^m} [F_1(x, y_1, \cdots, y_m) + F_2(x, y_1, \cdots, y_m)] . \quad (4.9)$$

We then have to compute the conjugate (at 0) of a sum of functions, however in a favorable context since:

$F_1 \in \Gamma_0(X \times \mathbb{R})$, $F_2 \in \Gamma_0(X \times \mathbb{R})$, $dom\, F_1 = X \times dom\, g$.

According to the assumption $(\mathcal{H})$ we made, there is a point $(x_0, f_1(x_0), \cdots, f_m(x_0)) \in dom\, F_2$ at which F_1 is continuous (indeed g is continuous at $(f_1(x_0), \cdots, f_m(x_0))$ whenever $(f_1(x_0), \cdots, f_m(x_0))$ lies in the interior of $dom\, g$). We therefore are in a situation where Theorem 1 applies:

$$\begin{aligned} [g \circ (f_1, \cdots, f_m)]^*(p) &= (F_1 + F_2)^*(0) \quad \text{[from (4.9)]} \\ &= \min_{\substack{s \in X^\star \\ (\alpha_1, \cdots, \alpha_m) \in \mathbb{R}^m}} [F_1^*(-s, \alpha_1, \cdots, \alpha_m) + F_2^*(s, -\alpha_1, \cdots, -\alpha_m)] . \end{aligned} \quad (4.10)$$

The computation of the above two conjugate functions (F_1^* and F_2^*) is easy and gives:

$$F_1^*(-s, \alpha_1, \cdots, \alpha_m) = \begin{cases} g^*(\alpha_1, \cdots, \alpha_m) \text{ if } s = p, \\ +\infty \text{ if not ;} \end{cases}$$

$$\begin{aligned} F_2^*(s, \beta_1, \cdots, \beta_m) &= \sup_{\substack{f_i(x) \leq y_i \\ \text{for all } i}} (\langle s, x \rangle + \sum_{i=1}^m \beta_i y_i) \\ &= +\infty \text{ if at least one } \beta_i \text{ is } > 0, \\ &= \sup_{\substack{x \in dom\, f_i \\ \text{for all } i}} [\langle s, x \rangle + \sum_{i=1}^m \beta_i\, f_i(x)] \text{ if } \beta_i \leq 0 \text{ for all } i. \end{aligned}$$

Consequently, the minimum in (4.10) is taken over $s = p$ and those $(\alpha_1, \cdots, \alpha_m) \in \mathbb{R}^m$ whose components are all nonnegative. Whence the formula (4.6) follows.

If there is a point $\tilde{x} \in \bigcap_{i=1}^{m} dom\, f_i$ at which $f_1, \cdots, f_{m-1}$ are continuous, then $\tilde{x} \in \bigcap_{i=1}^{m} dom(\alpha_i\, f_i)$ and $\alpha_1\, f_1, \cdots, \alpha_{m-1}\, f_{m-1}$ are continuous at $\tilde{x}$ (if $\alpha_{i_0} = 0$, $\alpha_{i_0}\, f_{i_0} = i_{dom\, f_{i_0}}$ is indeed continuous at $\tilde{x}$, since $\tilde{x}$ necessarily lies in the interior of $dom\, f_{i_0}$). We apply Theorem 1 again:

$$(\sum_{i=1}^{m} \alpha_i\, f_i)^*(p) = \min_{p_1+\cdots+p_m=p} \left[\sum_{i=1}^{m} \alpha_i\, f_i{}^*(\frac{p}{\alpha_i}) \right].$$

Plugging this into (4.6) yields the formula (4.7). ■

3. By way of illustrations

Since g is assumed increasing in our setting, it is easy to prove that $dom\, g^*$ is indeed included in $(\mathbb{R}_+)^m$, thus the restriction in (4.6) or (4.7) to those α_i which are positive is not surprising. However one may wonder if the optimal α_i's in these formulas are strictly positive or not. The answer is no, some of the $\alpha_i' s$ may be null, possibly all of them. As a general rule, one can say more about the optimal $\alpha_i' s$ only in particular instances.

Consider for example $p \in X^\star$ such that $[g \circ (f_1, \cdots, f_m)]^*(p) < +\infty$; then, if $\sigma_{\bigcap_{i=1}^{m} dom\, f_i}(p) = +\infty$ or if $g^*(0, \cdots, 0) = +\infty$, one is sure that some of the optimal $\alpha_i' s$ in (6) or (7) are strictly positive (in either case, having all the α_i null leads to impossible equalities (4.6) and (4.7)). In some other situations however, one is sure that all the optimal $\alpha_i' s$ are strictly positive; see some examples below.

3.1 The sum of the k largest values

Suppose that the increasing $g \in \Gamma_0(\mathbb{R}^m)$ is positively homogeneous; it is then the support function of a closed convex set $C \subset (\mathbb{R}_+)^m$ (namely the subdifferential of g at the origin). The Legendre-Fenchel transform g^* of g is the indicator function of C so that (4.6) and (4.7) are simplified into:

$$[g \circ (f_1, \cdots, f_m)]^*(p) = \min_{\alpha_1, \cdots, \alpha_m \in C} \left[(\sum_{i=1}^{m} \alpha_i\, f_i)^*(p) \right] \tag{4.11}$$

$$= \min_{\substack{\alpha_1, \cdots, \alpha_m \in C \\ p_1+\cdots+p_m=p}} \left[\sum_{i=1}^{m} \alpha_i\, f_i^*(\frac{p_i}{\alpha_i}) \right]. \tag{4.12}$$

A first example, with a bounded C, is $g(y_1, \cdots, y_m) = \sum_{i=1}^{m} y_i^+$ (where y_i^+ stands for the positive part of y_i). Here g is the support function of $C = [0,1]^m$ and (11)-(12) give an expression of $\left(\sum_{i=1}^{m} f_i^+\right)^*(p)$.

A more interesting example of g as the support function of C, still with a bounded C, is the following one: for an integer $k \in \{1, \cdots, m\}$, let $g_k(y_1, \cdots, y_m) :=$ the sum of the k largest values among the y_i's.

It is not difficult to realize that g_k is the support function of the compact convex polyhedron

$$C_k := \{(\alpha_1, \cdots, \alpha_m) \in [0,1]^m \mid \alpha_1 + \cdots + \alpha_m = k\}. \tag{4.13}$$

Since $dom\, g_k = \mathbb{R}^m$, the corollary below readily follows from Theorem 2.

Corollary 3. *Let $f_1, \cdots, f_m \in \Gamma_0(X)$, let φ_k be defined on X as*

$$\varphi_k(x) := \text{ the sum of the } k \text{ largest values among the } f_i(x)'s.$$

We suppose there is a point in $\bigcap_{i=1}^{m} dom\, f_i$ at which $f_1, \cdots, f_{m-1}$ are continuous. Then, for all $p \in X^\star$

$$\varphi_k^*(p) = \min_{\substack{\alpha_1, \cdots, \alpha_m \in C_k \\ p_1 + \cdots + p_m = p}} \left[\sum_{i=1}^{m} \alpha_i\, f_i^*(\frac{p_i}{\alpha_i})\right]. \tag{4.14}$$

The two "extreme" cases, $k = 1$ or $k = m$, are interesting to consider.

If $k = 1$, $\varphi_1(x) = \max_{i=1,\cdots,m} f_i(x)$, C_1 is the so-called unit-simplex in $\mathbb{R}^m$, and (4.14) reduces to a well-known formula on $(\max_i f_i)^*(p)$:

$$(\max_i f_i)^*(p) = \min_{\substack{\alpha_1 \geq 0, \cdots, \alpha_m \geq 0 \\ \alpha_1 + \cdots + \alpha_m = 1 \\ p_1 + \cdots + p_m = p}} \left[\sum_{i=1}^{m} \alpha_i\, f_i^*(\frac{p_i}{\alpha_i})\right]. \tag{4.15}$$

If $k = m$, $\varphi_m(x) = \sum_{i=1}^{m} f_i(x)$, C_m is the singleton $\{(1, \cdots, 1)\}$, and Corollary 3 takes us back to Theorem 1.

3.2 Smoothing the max function

A way of smoothing the nondifferentiable function $\max_i(y_1, \cdots, y_m)$ is via the so-called log-exponential function.

Given $\varepsilon > 0$, let $\Theta_\varepsilon : \mathbb{R}^m \longrightarrow \mathbb{R}$ be defined as following:

$$(y_1, \cdots, y_m) \in \mathbb{R}^m \longmapsto \Theta_\varepsilon(y_1, \cdots, y_m) := \varepsilon \, log \, (e^{\frac{y_1}{\varepsilon}} + \cdots + e^{\frac{y_m}{\varepsilon}}).$$

Such a function is studied in full detail in (Rockafellar and Wets, 1998) for example: Θ_ε is a (finite-valued) increasing convex function on $\mathbb{R}^m$, whose Legendre-Fenchel transform is given below (Rockafellar and Wets, 1998), p. 482:

$$\Theta_\varepsilon^*(\alpha_1, \cdots, \alpha_m) = \begin{cases} \varepsilon \sum_{i=1}^m \alpha_i \, log \, \alpha_i & \text{if } \alpha_1 \geq 0, \cdots, \alpha_m \geq 0 \\ & \text{and } \alpha_1 + \cdots + \alpha_m = 1, \\ +\infty & \text{if not} \end{cases} \quad (4.16)$$

(the entropy function multiplied by ε).

When $f_1, \cdots, f_m \in \Gamma_0(X)$, the nonsmooth function $\max_{i=1,\cdots,m} f_i$ can be approximated by the function $\Theta_\varepsilon \circ (f_1, \cdots, f_m)$, smooth whenever all the f_i's are smooth.

Here again $dom \, \Theta_\varepsilon = \mathbb{R}^m$, so that the next corollary is an immediate application of Theorem 2.

Corollary 4. *Let $f_1, \cdots, f_m \in \Gamma_0(X)$; we assume there is a point in $\bigcap_{i=1}^m dom \, f_i$ at which $f_1, \cdots, f_{m-1}$ are continuous.*

Then, for all $p \in X^\star$,

$$[\Theta_\varepsilon \circ (f_1, \cdots, f_m)]^*(p) = \min_{\substack{\alpha_1 \geq 0, \cdots, \alpha_m \geq 0 \\ \alpha_1 + \cdots + \alpha_m = 1 \\ p_1 + \cdots + p_m = p}} \left[\varepsilon \sum_{i=1}^m \alpha_i \, log \, \alpha_i + \sum_{i=1}^m \alpha_i \, f_i^*(\frac{p_i}{\alpha_i}) \right]. \quad (4.17)$$

Compare (4.17) with (4.15): since the entropy function $\varepsilon \sum_{i=1}^m \alpha_i \, log \, \alpha_i$ is negative and bounded from below by $-\varepsilon \, log \, m$ (achieved at $\alpha_1 = \cdots = \alpha_m = \frac{1}{m}$), we have:

$$(\max_i f_i)^* - \varepsilon \, log \, m \leq [\Theta_\varepsilon \circ (f_1, \cdots, f_m)]^* \leq (\max_i f_i)^*. \quad (4.18)$$

3.3 Optimality conditions in Convex minimization

Let K be a closed convex cone in $\mathbb{R}^m$, let $f_1, \cdots, f_m$ be convex functions on X and

$$S := \{x \in X \mid (f_1(x), \cdots, f_m(x)) \in K\}. \tag{4.19}$$

As an example, suppose $K = (\mathbb{R}_-)^m$: S is then a constraint set represented by inequalities in Convex minimization.

The indicator function i_S of S is nothing else than $i_K \circ (f_1, \cdots, f_m)$. We are again in the context considered in section 3.1 with $g = i_K$ the support function of the polar cone K° of K. We can thus express the support function of S, that is the Legendre-Fenchel transform of the composite function $i_K \circ (f_1, \cdots, f_m)$ in terms of the $f_i^{*\prime}s$.

Corollary 5. *We assume the following on the $f_i^{\prime}s$ and K:*

- *All the f_i: $X \longrightarrow \mathbb{R}$ are convex and continuous on X (as it is usually the case in applications);*
- *There is a point $x_0 \in X$ such that $(f_1(x_0), \cdots, f_m(x_0))$ lies in the interior of K (this is Slater's constraint qualification condition);*
- *K is a closed convex cone of $\mathbb{R}^m$ containing $(\mathbb{R}_-)^m$.*

Then, for all $0 \neq p \in X^\star$ such that $\sigma_S(p) < +\infty$,

$$\sigma_S(p) = \min_{\substack{0 \neq (\alpha_1, \cdots, \alpha_m) \in K^\circ \\ p_1 + \cdots + p_m = p}} \left[\sum_{i=1}^{m} \alpha_i \, f_i^*(\frac{p_i}{\alpha_i}) \right]. \tag{4.20}$$

Proof. The function $i_K \in \Gamma_0(X)$ is increasing because of the assumption $(\mathbb{R}_-)^m \subset K$. We have assumed there is a point $x_0 \in X$ such that $(f_1(x_0), \cdots, f_m(x_0))$ lies in the interior of the domain of i_K. Thus, by applying Theorem 2, we obtain:

$$\sigma_S(p) = [i_K \circ (f_1, \cdots, f_m)]^*(p) = \min_{\substack{(\alpha_1, \cdots, \alpha_m) \in K^\circ \\ p_1 + \cdots + p_m = p}} \left[\sum_{i=1}^{m} \alpha_i \, f_i^*(\frac{p_i}{\alpha_i}) \right].$$

Note that $K^\circ \subset (\mathbb{R}_+)^m$ but, since $\sigma_{dom\, f_i}(p_i) = +\infty$ for $p_i \neq 0$, all the optimal α_i are not null simultaneously. ■

To pursue our illustration further, consider the following minimization problem:

$$(\mathcal{P}) \quad \text{Minimize } f_0(x) \text{ over } S,$$

where S is described as in (4.19).

We suppose:

- $f_0 \in \Gamma_0(X)$ is continuous at some point of S;
- $v_{opt} := \inf_S f_0 > -\infty$ (f_0 is bounded from below on S);
- $v_{opt} > \inf_X f_0$ (($\mathcal{P}$) is genuinely a constrained problem);
- Assumptions on the $f_i's$ and K made in Corollary 5.

Corollary 6. *Under the assumptions listed above, we have:*

$$-v_{opt} = \min_{\substack{0 \neq (\alpha_1, \cdots, \alpha_m) \in K^\circ \\ p_1 + \cdots + p_m = p \neq 0}} \left[f_0^*(-p) + \sum_{i=1}^{m} \alpha_i f_i^*\left(\frac{p_i}{\alpha_i}\right) \right]. \tag{4.21}$$

This result is an alternate formulation, in the dual form (in the spirit of (Laurent, 1972), Chapter VII), of the existence of Lagrange-Karush-Kuhn-Tucker multipliers in ($\mathcal{P}$): there exist positive $\alpha_1, \cdots, \alpha_m$ such that

$$v_{opt} = \min_X \left[f_0(x) + \sum_{i=1}^{m} \alpha_i f_i(x) \right].$$

Proof of Corollary 6. By definition,

$$v_{opt} = -(f_0 + i_S)^*(0).$$

Applying Theorem 1, we transform the above into:

$$-v_{opt} = \min_{p \in X^\star} \left[f_0^*(-p) + \sigma_S(p) \right]. \tag{4.22}$$

The optimal p cannot be null, as otherwise we would have $-v_{opt} = f_0^*(0) = -\inf_X f_0$, which is excluded by assumption.

It then remains to apply the result of Corollary 5 to develop $\sigma_S(p)$ in (4.22). ■

Acknowledgments

We would like to thank Professors S. Robinson (University of Wisconsin at Madison) and L. Thibault (Université des sciences et techniques du Languedoc à Montpellier) for their remarks on the first version of this paper.

References

C. Combari, M. Laghdir and L. Thibault, *Sous-différentiel de fonctions convexes composées*, Ann. Sci. Math. Québec 18, 119-148 (1994).

C. Combari, M. Laghdir and L. Thibault, *A note on subdifferentials of convex composite functionals*, Arch. Math. Vol. 67, 239-252 (1996).

J.-B. Hiriart-Urruty and C. Lemaréchal, *Convex analysis and minimization algorithms* (2 volumes), Grundlehren der mathematischen wissenschaften 305 & 306, Springer (1993). New printing in 1996.

S. S. Kutateladze, *Changes of variables in the Young transformation,* Soviet Math. Dokl. 18, 545-548 (1977).

S. S. Kutateladze, *Convex operators*, Russian Math. Surveys 34, 181-214 (1979).

P.-J. Laurent, *Approximation et optimisation*, Hermann (1972).

R. T. Rockafellar and R. J.-B. Wets, *Variational analysis*, Grundlehren der mathematischen wissenschaften 317, Springer (1998).

C. Zalinescu, *Convex analysis in general vector spaces*, World Scientific (2002).

Chapter 5

WHAT IS TO BE A MEAN?

Michel Valadier

Laboratoire ACSIOM, UMR CNRS 5149, Université Montpellier II, Case courier 051, Place Eugène Bataillon, 34095 Montpellier Cedex, France

valadier@darboux.math.univ-montp2.fr

Abstract Bistochastic operators act on functions as bistochastic matrices act on n-vectors. They produce "means". These means are closely connected to inequalities of convexity (Jensen) and, as for the values, to "convex sweeping out" of Potential Theory. We present a synthetic survey, postponing the use of Probability language to the end of the paper.

Keywords: doubly stochastic (or bistochastic) operator, Hardy-Littlewood-Pólya, ordering (or comparison) of measures, balayage, sweeping-out, conditional expectation, Cartier-Fell-Meyer, Ryff, Day, Strassen, Rüschendorf.

1. Introduction

The results we present are essentially Convex Analysis. Convexity is known to be a fundamental notion in Mathematics and Mechanics: see J.J. Moreau (Moreau, 1967). Inequalities relying on convexity were intensively studied in the book of Hardy, Littlewood and Pólya (Hardy et al., 1952).

Comparison of functions and of measures goes back to 1929 when Hardy, Littlewood and Pólya published (Hardy et al., 1929). For two finite sequences of n real numbers $(a_1, \ldots, a_n)$ and $(b_1, \ldots, b_n)$, Hardy, Littlewood and Pólya proved the equivalence of several properties among which:

$$\forall \phi \text{ convex on } \mathbb{R}, \quad \sum_{i=1}^{n} \phi(b_i) \leq \sum_{i=1}^{n} \phi(a_i)$$

and [there exists a bistochastic matrix M such that, $\forall i \in \{1, \ldots, n\}$, $b_i =$

$\sum_{j=1}^{n} M_{ij}\, a_j\,]$. Note that $b_1 = \sum_{j=1}^{n} M_{1j}\, a_j$ is a mean of the a_j with some weights (not necessarily $\frac{1}{n}$). Similarly b_2 is a mean of the a_j, but all the weights are constrained by $\forall j,\ \sum_{i=1}^{n} M_{ij} = 1$ (bistochastic character of M). The arithmetic means are the same: $\sum_{i=1}^{n} \frac{1}{n}\, b_i = \sum_{j=1}^{n} \frac{1}{n}\, a_j$. Using rearrangements, Hardy, Littlewood and Pólya gave another equivalent property which is very popular. Let $(a_i^*)_i$ and $(b_i^*)_i$ denote the increasing rearrangements. This third property is:

$$\forall j \in \{1, \dots, n\}, \quad b_1^* + \cdots + b_j^* \geq a_1^* + \cdots + a_j^* \quad \text{with equality for } j = n. \tag{5.1}$$

For functions, Hardy, Littlewood and Pólya obtained some results, but not as many equivalent properties.

A decisive progress was the famous Ryff theorem (Ryff, 1965) which extends this to real integrable functions on $[0, 1]$. Then P.W. Day (Day, 1973) treated arbitrary spaces, even different spaces for the two functions f_1 and f_2. These works rely on rearrangements, so they need real valued functions. In another direction, Cartier, Fell and Meyer (Cartier et al., 1964) proved an equivalence for comparison of measures on a vector space (*sweeping-out* or in French *balayage*). But compactness of the supports is needed. Knowledge of these only papers is rather disturbing. Fortunately Theorem 8 of Strassen (Strassen, 1965) and an argument of Rüschendorf (Rüschendorf, 1981) permit getting a synthesis. Moreover the integral representation of bistochastic operators is not well known. It is possible and surely useful, gathering several scattered arguments, to get a clearer view: this is our aim.

2. Preliminaries

Let $(\Omega_1, \mathcal{F}_1, \mu_1)$ and $(\Omega_2, \mathcal{F}_2, \mu_2)$ be two probability[1] spaces, U a separable Banach space and $\mathcal{B}(U)$ its Borel tribe.

Definition. A linear continuous operator $T : L^1_{\mathbb{R}}(\Omega_1, \mu_1) \to L^1_{\mathbb{R}}(\Omega_2, \mu_2)$ is *bistochastic* if it is positive (i.e. $f_1 \overset{\mu_1\text{-a.e.}}{\geq} 0$ implies $Tf_1 \overset{\mu_2\text{-a.e.}}{\geq} 0$), if moreover $T\mathbf{1}_{\Omega_1} \overset{\mu_2\text{-a.e.}}{=} \mathbf{1}_{\Omega_2}$ and if $\forall A \in \mathcal{F}_1$, $\|T\mathbf{1}\|_{L^1(\Omega_2)} = \|\mathbf{1}_A\|_{L^1(\Omega_1)}$ (variant for this last property: ${}^tT\mathbf{1}_{\Omega_2} = \mathbf{1}_{\Omega_1}$).

Bistochastic operators generalize bistochastic matrices. They have been studied particularly in (Ryff, 1963, Ryff, 1965, Luxemburg, 1967, Day, 1973, Chong, 1976); see especially (Day, 1973) (page 388).

[1] This is only matter of normalization. One could assume only $\mu_1(\Omega_1) = \mu_2(\Omega_2)$.

We distinguish the "scalar" operator T from its vectorial extension $\overrightarrow{T} : L^1_U(\Omega_1, \mu_1) \to L^1_U(\Omega_2, \mu_2)$ (set $\overrightarrow{T}(x\mathbf{1}_A) = xT(\mathbf{1}_A)$ and extend this to simple functions by linearity).

Let $\mathrm{proj}_i^{\Omega_1 \times \Omega_2}$ denote the projection from $\Omega_1 \times \Omega_2$ onto the i-th factor or proj_i if no confusion may occur. The image measure of μ by a map φ, i.e. the set function $B \mapsto \mu(\varphi^{-1}(B))$, is denoted $\varphi_\sharp(\mu)$.

PROPOSITION 5.1 *Let $T : L^1_{\mathbb{R}}(\Omega_1, \mu_1) \to L^1_{\mathbb{R}}(\Omega_2, \mu_2)$ be a bistochastic operator. The set function Θ defined on the semi-ring $\mathcal{S}$ of all subsets $A \times B$ of $\Omega_1 \times \Omega_2$ where $A \in \mathcal{F}_1$ and $B \in \mathcal{F}_2$ by*

$$\Theta(A \times B) := \int_B T\mathbf{1}_A \, d\mu_2 = \int_{\Omega_2} (T\mathbf{1}_A)\, \mathbf{1}_B \, d\mu_2 \,, \tag{5.2}$$

has an extension into a probability measure on $(\Omega_1 \times \Omega_2, \mathcal{F}_1 \otimes \mathcal{F}_2)$ with marginals μ_1 and μ_2.

PROOF. In (5.2), $T\mathbf{1}_A$ is an equivalence class and $\mathbf{1}_B$ a true function; there is no difficulty. One can prove that Θ is σ-additive on $\mathcal{S}$, hence extends in a probability measure on the σ-algebra $\mathcal{F}_1 \otimes \mathcal{F}_2$. The marginal Θ on Ω_2 is μ_2 since, as $T\mathbf{1}_{\Omega_1} = \mathbf{1}_{\Omega_2}$,

$$\big[(\mathrm{proj}_2)_\sharp(\Theta)\big](B) = \Theta(\Omega_1 \times B) = \int_B T\mathbf{1}_{\Omega_1} \, d\mu_2 = \mu_2(B)\,.$$

Since T is bistochastic, if $A \in \mathcal{F}_1$, $\Theta(A \times \Omega_2) = \int_{\Omega_2} T\mathbf{1}_A \, d\mu_2 = \|T\mathbf{1}_A\|_{L^1}$ $= \|\mathbf{1}_A\|_{L^1} = \mu_1(A)$, which shows that the marginal of Θ on Ω_1 is μ_1. □

3. Comparison relations

Let for $f_i \in L^1_U(\Omega_i, \mathcal{F}_i, \mu_i)$, $\tau_i := (f_i)_\sharp(\mu_i)$ be the image measure (or in the language of Probability Theory, *law* of f_i). The measures τ_i are of *order* 1. We are interested by the relation $\tau_2 \prec \tau_1$ defined by: for any real convex function ϕ on U with linear growth (this implies continuity),

$$\int_U \phi(y)\, d\tau_2(y) \le \int_U \phi(x)\, d\tau_1(x)\,. \tag{5.3}$$

REMARK. This relation still holds if ϕ is an l.s.c. proper convex function. Indeed let $\phi_n(x) := \inf\{\phi(y) + n\|x - y\| \,;\, y \in U\}$. Since ϕ admits a continuous affine minorant, for n large enough, ϕ_n is real valued. And ϕ_n is n-Lipschitz hence has linear growth, and converges increasingly to ϕ as $n \to \infty$. By Fatou lemma the inequality (5.3) holds at the limit.

For emphasizing the role of functions one writes $f_2 \lhd f_1$, which is equivalent to

$$\forall \phi, \quad \int_{\Omega_2} \phi(f_2(\omega_2))\, d\mu_2(\omega_2) \leq \int_{\Omega_1} \phi(f_1(\omega_1))\, d\mu_1(\omega_1)\,.$$

A *kernel* is a measurable family of probability measures.

THEOREM 5.2 *The following are equivalent*

A) $\tau_1 \prec \tau_2$ *(or equivalently* $f_2 \lhd f_1$*),*

B) *there exists a kernel* $(K^y)_{y\in U}$ *satisfying* $\forall y$, $\mathrm{bar}(K^y) = y$ *and*

$$\forall A \in \mathcal{B}(U), \quad \tau_1(A) = \int_U K^y(A)\, d\tau_2(y)\,, \tag{5.4}$$

C) *there exists a bistochastic operator* T *such that* $f_2 = \overrightarrow{T} f_1$,

D) *there exists a probability measure* Θ *on* $\Omega_1 \times \Omega_2$ *with marginals* μ_1 *and* μ_2 *such that*

$$\forall B \in \mathcal{F}_2, \quad \int_B f_2\, d\mu_2 = \int_{\Omega_1 \times B} f_1(\omega_1)\, d\Theta(\omega_1, \omega_2)\,, \tag{5.5}$$

E) *there exists a probability measure* P *on* U^2 *with marginals* τ_1 *and* τ_2 *such that*

$$\forall D \in \mathcal{B}(U), \quad \int_D y\, d\tau_2 = \int_{U\times D} x\, dP(x,y)\,. \tag{5.6}$$

REMARKS. For another equivalent property see (5.9). The equivalences A) $\Leftrightarrow$ B) $\Leftrightarrow$ E) do not concern the function f_1 and f_2 but only the measures τ_1 and τ_2. The direct proof of B) $\Rightarrow$ E) is easy when introducing P by[2] $P(C \times D) = \int_D K^y(C)\, d\tau_2$.

PROOF. A) $\Rightarrow$ B) follows from Theorem 8 of Strassen (Strassen, 1965).

B) $\Rightarrow$ C) is due to Rüschendorf (Rüschendorf, 1981, Theorem 7). We reproduce[3] his arguments. If $h \in L^1_U(U, \mathcal{B}(U), \tau_1)$ we set

$$[\overrightarrow{T_0}(h)](\omega_2) = \int_U h(x)\, dK^{f_2(\omega_2)}(x)\,.$$

[2] More learnedly $P = \int_U (K^y \otimes \delta_y)\, d\tau_2(y)$.

[3] With some more details.

Observe that $\overrightarrow{T_0}$ is a linear contraction from $L^1_U(U, \tau_1)$ to $L^1_U(\Omega_2, \mu_2)$. Consider now for $g \in L^1_U(\Omega_1, \mu_1)$, its conditional expectation[4] $\mathbf{E}(g \mid f_1)$ as an element of $L^1_U(U, \mathcal{B}(U), \tau_1)$ (the map $g \mapsto \mathbf{E}(g \mid f_1)$ is also a linear contraction from $L^1_U(\Omega_1, \mu_1)$ into $L^1_U(U, \tau_1)$). Then, for $g \in L^1_U(\Omega_1, \mu_1)$, we set

$$\overrightarrow{T}(g) = \overrightarrow{T_0}\big(\mathbf{E}(g \mid f_1)\big)\,.$$

Firstly as $\mathbf{E}(f_1 \mid f_1) = \mathrm{id}_U$, $\overrightarrow{T}(f_1)(\omega_2) = \int_U x\, dK^{f_2(\omega_2)}(x) = \mathrm{bar}(K^{f_2(\omega_2)}) = f_2(\omega_2)$. Consider now the operator T acting on $L^1_{\mathbb{R}}(\Omega_1)$ (i.e. consider h and g real valued) and check it is a bistochastic operator. The properties $g \geq 0 \Rightarrow T(g) \geq 0$ and $T(\mathbf{1}_{\Omega_1})(\omega_2) = 1$ are easy. Check $\|T\mathbf{1}_A\|_{L^1(\Omega_2)} = \|\mathbf{1}_A\|_{L^1(\Omega_1)}$. Let $A \in \mathcal{F}_1$ and $h := \mathbf{E}(\mathbf{1}_A \mid f_1)$. One has:

$$\begin{aligned}
\|T\mathbf{1}_A\|_{L^1(\Omega_2)} &= \int_{\Omega_2} \Big[\int_U h(x)\, dK^{f_2(\omega_2)}(x)\Big]\, d\mu_2(\omega_2) \\
&= \int_U \Big[\int_U h(x)\, dK^{y}(x)\Big]\, d\tau_2(y) \\
&= \int_U h\, d\tau_1 \\
&= \int_{\Omega_1} (h \circ f_1)\, d\mu_1 \\
&= \int_{\Omega_1} (\mathbf{E}(\mathbf{1}_A \mid f_1) \circ f_1)\, d\mu_1 \\
&= \mu_1(A)\,.
\end{aligned}$$

C) $\Rightarrow$ D) is easy with the measure Θ obtained from T by Proposition 5.1. Observe that the definition (5.2) of Θ extends into: $\forall h \in L^1_U(\Omega_1)$,

$$\int_{\Omega_2} (\overrightarrow{T}h)\, \mathbf{1}_B\, d\mu_2 = \int_{\Omega_1 \times \Omega_2} h(\omega_1)\, \mathbf{1}_B(\omega_2)\, d\Theta(\omega_1, \omega_2)\,. \tag{5.7}$$

Suppose C) and let $B \in \mathcal{F}_2$. The first term of (5.5) rewrites $\displaystyle\int_{\Omega_2} f_2\, \mathbf{1}_B\, d\mu_2$ which by hypothesis and thanks to (5.7) equals

$$\int_{\Omega_2} (\overrightarrow{T} f_1)\, \mathbf{1}_B\, d\mu_2 = \int_{\Omega_1 \times \Omega_2} f_1(\omega_1)\, \mathbf{1}_B(\omega_2)\, d\Theta(\omega_1, \omega_2)$$

[4] For conditional expectation see Section 4 below. The classical conditional expectation of g with respect to the sub-tribe generated by f_1 in the probability space $(\Omega_1, \mathcal{F}_1, \mu_1)$, denoted by $\mathbf{E}^{\sigma(f_1)}(g)$, is nothing else but $\mathbf{E}(g \mid f_1) \circ f_1$.

hence equals the second term of (5.5).

D) $\Rightarrow$ E) will be proved taking for P the image measure of Θ by $(\omega_1, \omega_2) \mapsto (f_1(\omega_1), f_2(\omega_2))$. Let $D \in \mathcal{B}(U)$. The first term of (5.6) rewrites

$$\begin{aligned}
\int_{\Omega_2} f_2 \, \mathbf{1}_D(f_2) \, d\mu_2 &= \int_{f_2^{-1}(D)} f_2 \, d\mu_2 \\
&= \int_{\Omega_1 \times f_2^{-1}(D)} f_1(\omega_1) \, d\Theta(\omega_1, \omega_2) \quad \text{(thanks to D))} \\
&= \int_{\Omega_1 \times \Omega_2} f_1(\omega_1) \, \mathbf{1}_D(f_2(\omega_2)) \, d\Theta(\omega_1, \omega_2) \\
&= \int_{U^2} x \, \mathbf{1}_D(y) \, dP(x, y)
\end{aligned}$$

which is nothing else but the second term of (5.6).

E) $\Rightarrow$ A) follows from Jensen's inequality for conditional expectation[5]. Firstly (5.6) rewrites $\mathrm{proj}_2 = \mathbf{E}_P\big(\mathrm{proj}_1 \,\big|\, \mathrm{proj}_2\big)$ (we write proj_i instead of $\mathrm{proj}_i^{U^2}$). Jensen's inequality says (Meyer, 1966) ((47.3) page 50) or (Dudley, 1989) (10.2.7 page 274) that for ϕ convex with linear growth

$$\phi\Big(\mathbf{E}_P\big(\mathrm{proj}_1 \,|\, \sigma(\mathrm{proj}_2)\big)\Big) \overset{P\text{-a.e.}}{\leq} \mathbf{E}_P\big(\phi(\mathrm{proj}_1) \,|\, \sigma(\mathrm{proj}_2)\big)$$

hence integrating over U^2

$$\int_{U^2} \phi(\mathrm{proj}_2) \, dP \leq \int_{U^2} \phi(\mathrm{proj}_1) \, dP$$

that is

$$\int_U \phi \, d\tau_2 \leq \int_U \phi \, d\tau_1 \,. \qquad \square$$

REMARKS. A) $\Leftrightarrow$ C) in the vectorial case is Theorem 7 of (Rüschendorf, 1981). Note that in his Theorem 6 he invokes Theorem 3 of (Strassen, 1965) which does not work here: the set he denotes K_x is not closed; but he does refer to Strassen's Theorem 8 in the remark which follows.

[5] For this notion and the meaning of next formula see Section 4 below. In order to use only Analysis arguments, one could prove E) $\Rightarrow$ B) as explained in a remark below and then get A) by the ordinary Jensen inequality.

Among all possible implications some ones could be difficult to be proved directly. Some of them were proved under stronger hypotheses.

A) $\Rightarrow$ B) has been proved by Cartier, Fell and Meyer (Cartier et al., 1964, Théorème 2) (see also (Meyer, 1966, Théorème 36 page 288 attributed to Cartier)) when the τ_i are carried by a convex compact metrizable space (a subset of a locally convex linear topological space).

A) $\Rightarrow$ C) has been proved by Ryff (Ryff, 1965) (Theorem 3) when $U = \mathbb{R}$ and $\Omega_1 = \Omega_2 = [0, 1]$. Then Day (Day, 1973) (Theorem (4.9)) extends[6] the result to arbitrary measured spaces but still under the hypothesis $U = \mathbb{R}$.

E) $\Rightarrow$ B) can be proved taking the horizontal disintegration $(K^y)_y$ of P. As the marginals of P are τ_1 and τ_2, (5.4) is obvious. As for $\mathrm{bar}(K^y) = y$, using (5.6) one easily get $\int_D \mathrm{bar}(K^y)\, d\tau_2(y) = \int_D y\, d\tau_2(y)$. Hence $\mathrm{bar}(K^y) = y$ τ_2-a.e.

D) $\Rightarrow$ C) (begin with $U = \mathbb{R}$) follows from the Radon-Nikodým theorem applied to $B \mapsto \int_{\Omega_1 \times B} g(\omega_1)\, d\Theta(\omega_1, \omega_2)$: this gives Tg.

By Jensen's inequality any of the conditions B) and C') below implies easily A).

4. Probability language

Let $(\Omega, \mathcal{F}, P)$ be a probability space, $\mathcal{G}$ a sub-tribe of $\mathcal{F}$. *Conditional expectation* of a random variable is a classical notion in Probability Theory but is also an elementary notion of Analysis. The conditional expectation $\mathbf{E}^{\mathcal{G}}(X)$, or more precisely $\mathbf{E}^{\mathcal{G}}_P(X)$, of $X \in L^1_{\mathbb{R}}(\Omega, \mathcal{F}, P)$ is characterized by $\mathbf{E}^{\mathcal{G}}(X) \in L^1_{\mathbb{R}}(\Omega, \mathcal{G}, P)$ and $\forall B \in \mathcal{G}$, $\int_B \mathbf{E}^{\mathcal{G}}(X)\, dP = \int_B X\, dP$. If $X \in L^2$, $\mathbf{E}^{\mathcal{G}}(X)$ is the orthogonal projection of X on $L^2_{\mathbb{R}}(\Omega, \mathcal{G}, P)$. Conditional expectation admits a vectorial extension $\overrightarrow{\mathbf{E}}^{\mathcal{G}} : L^1_U(\Omega, \mathcal{F}, P) \to L^1_U(\Omega, \mathcal{G}, P)$. In the sequel the arrow above $\mathbf{E}$ will be forgotten.

Let $f_i \in L^1_U(\Omega_i, \mathcal{F}_i, \mu_i)$ be functions. The following notations will be useful. Whereas f_1 and f_2 are defined on spaces which may be different spaces, the following functions

$$\hat{f}_1 := f_1 \circ \mathrm{proj}_1^{\Omega_1 \times \Omega_2} \quad (\Longleftrightarrow \hat{f}_1(\omega_1, \omega_2) = f_1(\omega_1))$$
$$\check{f}_2 := f_2 \circ \mathrm{proj}_2^{\Omega_1 \times \Omega_2} \quad (\Longleftrightarrow \check{f}_2(\omega_1, \omega_2) = f_2(\omega_2))$$

are both defined on $\Omega_1 \times \Omega_2$. We also write $\check{\mathcal{F}}_2 := \{\emptyset, \Omega_1\} \otimes \mathcal{F}_2$.

Now conditions D) and E) of Theorem 5.2 can be rewritten:

[6] For $[0, +\infty[$ see Theorem 5 of Sakamaki and Takahashi (Sakamaki and Takahashi, 1979).

D) *there exists a probability measure* Θ *on* $\Omega_1 \times \Omega_2$ *with marginals* μ_1 *and* μ_2 *such that*

$$\check{f}_2 = \mathbf{E}_\Theta(\hat{f}_1 \,|\, \check{\mathcal{F}}_2)\,,$$

E) *there exists a probability measure* P *on* U^2 *with marginals* τ_1 *and* τ_2 *such that*

$$\mathrm{proj}_2^{U^2} = \mathbf{E}_P\big(\mathrm{proj}_1^{U^2} \,|\, \mathrm{proj}_2^{U^2}\big)\,.$$

This means that there are two random variables with laws τ_1 and τ_2 such that the second is a conditional expectation of the first. If the probability space is not specified this is a seemingly weaker property which still implies A) (by Jensen), hence is equivalent to the other conditions. For more than two indices (but with the compactness hypothesis) see (Doob, 1968).

5. Integral representation of T

We consider *disintegration* as a slicing of a measure on a product. For example let Q be a probability measure on the product space $(\Xi_1 \times \Xi_2, \mathcal{S}_1 \otimes \mathcal{S}_2)$. If $(\Xi_2, \mathcal{S}_2)$ is abstract Suslin (Dellacherie and Meyer, 1983, III.16 page 72), there exists a kernel $(Q_{\xi_1})_{\xi_1 \in \Xi_1}$ such that for all measurable positive or Q-integrable real function φ,

$$\int_{\Xi_1 \times \Xi_2} \varphi \, dQ = \int_{\Xi_1} \Big[\int_{\Xi_2} \varphi(\xi_1, \xi_2) \, dQ_{\xi_1}(\xi_2) \Big] \, d[(\mathrm{proj}_1)_\sharp(Q)](\xi_1)\,.$$

This is the *vertical disintegration.* If it exists, the *horizontal disintegration* which is characterized similarly will be denoted $(Q^{\xi_2})_{\xi_2 \in \Xi_2}$.

The following is due to Fakhoury (Fakhoury, 1979) (Théorème 1), but the disintegration technique was already in (Arveson, 1974) (Section 1.5 pages 458–466, specially Lemma page 461). See also (Sourour, 1979) (second proof pages 345–346), (Sourour, 1982), and for disintegration on a product (Valadier, 1973).

PROPOSITION 5.3 *Let* $(\theta^{\omega_2})_{\omega_2 \in \Omega_2}$ *be a measurable family of probability measures on* $(\Omega_1, \mathcal{F}_1)$ *such that* $\mu_1 = \int_{\Omega_2} \theta^{\omega_2} \, d\mu_2$. *Set for any* $g \in L^1_{\mathbb{R}}(\Omega_1, \mu_1)$,

$$Tg = \mathrm{Cl}(\omega_2 \mapsto \int_{\Omega_1} g \, d\theta^{\omega_2})\,, \tag{5.8}$$

where Cl *denotes the equivalence class. This defines correctly* T *as a bistochastic operator. Assume that* $(\Omega_1, \mathcal{F}_1)$ *is abstract Suslin. Then any bistochastic operator* $T : L^1_{\mathbb{R}}(\Omega_1, \mu_1) \to L^1_{\mathbb{R}}(\Omega_2, \mu_2)$ *has the foregoing form.*

PROOF. Deriving T from the kernel $(\theta^{\omega_2})_{\omega_2 \in \Omega_2}$ relies on integration with respect to kernels. We turn to the proof of the converse part. Let Θ be defined by Proposition 5.1. Now let $(\theta^{\omega_2})_{\omega_2 \in \Omega_2}$ be an "horizontal disintegration" of Θ. In particular for any $A \in \mathcal{F}_1$ and $B \in \mathcal{F}_2$,

$$\Theta(A \times B) = \int_B \theta^{\omega_2}(A)\, d\mu_2(\omega_2)$$

hence comparing with (5.2), $T\mathbf{1}_A = \mathrm{Cl}(\omega_2 \mapsto \theta^{\omega_2}(A))$. The operator from $L^1(\Omega_1, \mu_1)$ to $L^1(\Omega_2, \mu_2)$ obtained from $(\theta^{\omega_2})_{\omega_2 \in \Omega_2}$ by formula (5.8) coincides with T on the dense subspace of $L^1(\Omega_1, \mu_1)$ consisting of simple functions. So the two operators are the same. □

When $(\Omega_1, \mathcal{F}_1)$ is abstract Suslin another equivalent property[7] can be added to those of Theorem 5.2:

C') *there exists a measurable family of probability measures* $(\theta^{\omega_2})_{\omega_2 \in \Omega_2}$ *on* $(\Omega_1, \mathcal{F}_1)$ *such that* $\mu_1 = \int_{\Omega_2} \theta^{\omega_2}\, d\mu_2(\omega_2)$ *and*

$$f_2(\omega_2) = \int_{\Omega_1} f_1\, d\theta^{\omega_2} \quad \mu_2\text{-a.e.}$$

REMARK. Proving C$'$) $\Rightarrow$ B) is possible with the formula

$$K^y(A) = \int_{\Omega_2} \theta^{\omega_2}(f_1^{-1}(A))\, d\mu_2^y(\omega_2)$$

where $\mu_2^y(\omega_2)$ is the conditional probability on Ω_2 knowing $f_2 = y$ or more precisely the horizontal disintegration of $(\mathrm{id}_{\Omega_2}, f_2)_\sharp(\mu_2)$.

6. A strange property

Denoting f_i^* the increasing rearrangements on $[0,1]$, $f_2 \lhd f_1$ is also equivalent to

$$\forall \xi \in [0,1], \quad \int_0^\xi f_1^*(r)\, dr \le \int_0^\xi f_2^*(r)\, dr, \quad \text{with equality for } \xi = 1. \tag{5.9}$$

This formulation, analogous to (5.1), seems very popular: see for example (Toader, 2000). Here occurs in my opinion a strange coincidence:

[7] For A) $\Rightarrow$ C') when the measures τ_1 and τ_2 are carried by a compact interval of $\mathbb{R}$, see Lemma 4.4 of Toader (Toader, 2000).

the cone of real continuous increasing functions on $[0,1]$ gives an order between probability measures ν_i on $[0,1]$ (this is still balayage), ν_1 is *more on right than* ν_2, which we note $\nu_2 \sqsubset \nu_1$. Suppose further that the f_i are positive and (after normalizing) have integrals 1. If $f_i^*.\mathcal{L}_{\lfloor[0,1]}$ denotes the probability measure having the density f_i^* with respect to Lebesgue measure, (5.9) means $f_2^*.\mathcal{L}_{\lfloor[0,1]} \sqsubset f_1^*.\mathcal{L}_{\lfloor[0,1]}$, so one gets the equivalence $f_2 \lhd f_1 \iff f_2^*.\mathcal{L}_{\lfloor[0,1]} \sqsubset f_1^*.\mathcal{L}_{\lfloor[0,1]}$!

Late remark. The author has just discovered an integral representation theorem of bistochastic operators in the famous paper of V.N. Sudakov *Geometric problems in the theory of infinite-dimensional probability distributions*, Proceed. of Steklov Institute **141** (1979), 1–178 (firstly published in 1976 in Russian). This is Proposition 4.1 page 73.

References

Arveson, W. *Operator algebras and invariant subspaces*, Ann. of Math. **100** (1974), 433–532.

Cartier, P., J.M.G. Fell & P.-A. Meyer, *Comparaison des mesures portées par un ensemble convexe compact*, Bull. Soc. Math. France **92** (1964), 435–445.

Chong, Kong Ming, *Doubly stochastic operators and rearrangement theorems*, J. Math. Anal. Appl. **56** (1976), 309–316.

Day, P.W. *Decreasing rearrangements and doubly stochastic operators*, Trans. Amer. Math. Soc. **178** (1973), 383–392.

Dellacherie, C. & P.-A. Meyer, *Probabilités et potentiel*, Chapitres IX à XI, *Théorie discrète du potentiel*, Hermann, Paris, 1983.

Doob, J.L. *Generalized sweeping-out and probability*, J. Funct. Anal. **2** (1968), 207–225.

Dudley, R.M. *Real Analysis and Probability*, Wadsworth & Brooks/Cole, Pacific Grove, California, 1989.

Fakhoury, H. *Représentations d'opérateurs à valeurs dans $L^1(X,\Sigma,\mu)$*, Math. Ann. **240** (1979), 203–212.

Hardy, G.H., J.E. Littlewood & G. Pólya, *Some simple inequalities satisfied by convex functions*, Messenger of Math. **58** (1929), 145–152.

Hardy, G.H., J.E. Littlewood & G. Pólya, *Inequalities*, Cambridge Univ. Press, 1952 (first edition 1934).

Luxemburg, W.A.J. *Rearrangement-invariant Banach function spaces*, Queen's Papers in Pure and Appl. Math. no10, Queen's University, Kingston, Ont. (1967), 83–144.

Meyer, P.-A. *Probabilités et potentiel*, Hermann, Paris, 1966.

Moreau, J.J. *Fonctionnelles convexes*, Lecture Notes, Séminaire sur les équations aux dérivées partielles, Collège de France, Paris 1967 (108 pages) (second edition: Laboratoire Lagrange, Rome, 2003).

Rüschendorf, L. *Ordering of distributions and rearrangement of functions*, Ann. Probab. **9** (1981), 276–283.

Ryff, J.V. *On the representation of doubly stochastic operators*, Pacific J. Math. **13** (1963), 1379–1386.

Ryff, J.V. *Orbits of L^1-functions under doubly stochastic transformations*, Trans. Amer. Math. Soc. **117** (1965), 92–100.

Sakamaki, K. & W. Takahashi, *Systems of convex inequalities and their applications*, J. Math. Anal. Appl. **70** (1979), 445–459.

Sourour, A.R. *Pseudo-integral operators*, Trans. Amer. Math. Soc. **253** (1979), 339–363.

Sourour, A.R. *Characterization and order properties of pseudo-integral operators*, Pacific J. Math. **99** (1982), 145–158.

Strassen, V. *The existence of probability measures with given marginals*, Ann. Math. Statist. **36** (1965), 423–439.

Toader, A.-M. *Links between Young measures associated to constrained sequences*, ESAIM: Control, Optimisation and Calculus of Variations **5** (2000), 579–590.

Valadier, M. *Désintégration d'une mesure sur un produit*, C. R. Acad. Sci. Paris Sér. A **276** (1973), 33–35.

II

NONSMOOTH MECHANICS

Chapter 6

THERMOELASTIC CONTACT WITH FRICTIONAL HEATING

L.-E. Andersson
Department of Mathematics, University of Linkoping, SE-581 83 Linkoping, Sweden
leand@mai.liu.se

A. Klarbring
Department of Mechanical Engineering, University of Linkoping, SE-581 83 Linkoping, Sweden
andkl@ikp.liu.se

J.R. Barber
Department of Mechanical Engineering, University of Michigan, Ann Arbor, MI 48109-2125, USA
jbarber@engin.umich.edu

M. Ciavarella
CEMEC-PoliBA - Center of Excellence in Computational Mechanics, V.le Japigia 182, Politecnico di Bari, 70125 Bari, Italy
mciava@dimeg.poliba.it

Abstract The paper treats thermoelastic contact problems, where a variable contact heat flow resistance as well as frictional heating are considered. Existence and uniqueness of steady state solutions, for both a one-dimensional and a three-dimensional system, are investigated. Existence is guaranteed if the contact heat flow resistance goes to zero as the pressure goes to infinity or if the frictional heating is sufficiently small. Uniqueness holds in the vicinity of zero frictional heating and thermally insulated contact.

Keywords: thermoelastic contact problems, frictional heating, existence and uniqueness, steady states

1. Introduction

It is well known that contact and friction problems in thermoelasticity may lack solutions or have multiple solutions. Previously, issues related to thermal contact and to frictional heating have been discussed separately. Here they are coupled. We treat both a one-dimensional rod problem and a three-dimensional problem.

For the one-dimensional case we discuss existence and uniqueness of steady state solutions as well as stability of these solutions. The analysis is partly based on (Ciavarella et al. (2003)) and extends, to frictional heating, previous results in (Barber (1978)) and (Barber et al. (1980)). These results, as most of the present ones, rely on the ingenious introduction of an auxiliary variable, which represents contact pressure when there is contact and contact gap when there is no contact and which "removes" the unilateral character of the problem. Such a variable was used also in (Andrews et al. (1992)) and (Andrews et al. (1993)), and a similar idea can be found in (Manners (1998)). Here it is shown how this "trick" relates to fundamentals of non-smooth mechanics, as developed by J.J. Moreau in (Moreau (1962)) and (Moreau (1966)).

For the three-dimensional case we give existence and uniqueness results, the proofs of which are reported in (Andersson et al. (2003)). These results generalize to the case of frictional heating, results of (Duvaut (1979)) and (Duvaut (1981)), which were built on Barber's heat exchange conditions. Two qualitatively different existence results are given. The first one requires that the contact thermal resistance goes to zero at least as fast as the inverse of the contact pressure. The second existence theorem requires no such growth condition, but requires instead that the frictional heating, i.e., the sliding velocity times the friction coefficient, is small enough. Finally, it is shown that a solution is unique if the inverse of the contact thermal resistance is Lipschitz continuous and the Lipschitz constant, as well as the frictional heating, is small enough.

2. The rod model

In this part of the paper we will treat a model of one-dimensional thermoelastic frictional contact. A rod (a one-dimensional thermoelastic body) of length L that may come into contact with a rigid wall is shown in Figure 6.1. Neglecting mechanical inertia but including a time dependency in the thermal part of the problem, the governing equations

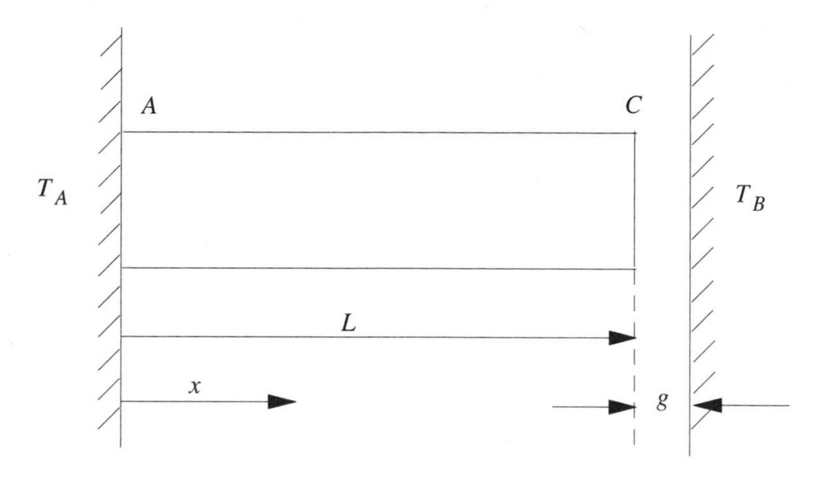

Figure 6.1. The rod model.

are

$$\left.\begin{aligned} &\frac{\partial \sigma}{\partial x} = 0 \\ &\sigma = E\left(\frac{\partial u}{\partial x} - \alpha T\right) \\ &\frac{\partial^2 T}{\partial x^2} = \frac{1}{k}\frac{\partial T}{\partial t} \end{aligned}\right\} \text{ on } (0, L), \tag{6.1}$$

$$u = 0 \text{ and } T = T_A \text{ at } x = 0, \tag{6.2}$$

$$\left.\begin{aligned} &p \geq 0,\ u - g_0 \leq 0,\ p(u - g_0) = 0 \\ &-K\frac{\partial T}{\partial x} = \frac{T_C - T_B}{R} - \mu V p \end{aligned}\right\} \text{ at } x = L. \tag{6.3}$$

Equation $(6.1)^1$ defines equilibrium for the stress field σ. Equation $(6.1)^2$ is the constitutive equation, where E is Young's modulus, α is the heat expansion coefficient, u is the displacement field and T is the temperature field. Equation $(6.1)^3$ is the heat conduction equation, where k is the thermal diffusivity. The right hand side of this equation is the time derivative of the temperature, which is zero in a steady state situation. Equations (6.2) are boundary conditions at the left end of the rod and T_A is a fixed temperature. The inequalities $(6.3)^1$ are Signorini's unilateral contact conditions, where g_0 is an initial contact gap and p is the contact pressure defined by $p = -\sigma(L)$. Equation $(6.3)^2$ is the heat balance of the contact interface. The term on the left hand side contains the thermal conductivity coefficient K and represents the heat flow out of the rod. Friction produces the heat $\mu V p$ where μ is the friction coefficient and V a velocity of the wall, with temperature T_B, that is directed perpendicular to the rod surface. The heat flow term of the rod and the frictionally produced heat leaves into the wall and then passes a thermal resistance, denoted R. The temperature T_C is defined by $T_C = T(L)$.

2.1 Steady states

For the steady state situation the right hand side of $(6.1)^3$ vanishes and the system (6.1) – (6.3) can be integrated. One then finds that the stress field is constant and equal to $-p$, and the temperature is linear going from T_A to T_C. In conclusion, it is easily shown that (6.1) – (6.3) reduce to the problem of finding p and $u = u(L)$ such that

$$-\frac{pL}{E} = u - \Delta, \tag{6.4}$$

$$p \geq 0, \quad u - g_0 \leq 0, \quad p(u - g_0) = 0, \tag{6.5}$$

where

$$\Delta = \alpha L T_A - \frac{\alpha L^2 (T_A - T_B)}{2(KR + L)} + \frac{\alpha L^2 \mu V p R}{2(KR + L)}. \tag{6.6}$$

To proceed further in discussing this problem, we need to say something about how the thermal contact resistance R depends on p and u. Clearly, when the rod is in contact, i.e., when $u = g_0$ and $p > 0$, the resistance to heat flow is smaller then when there is no contact, i.e., when $u < g_0$ and $p = 0$. However, to gain a first understanding of general properties of the problem we, anyway, start by assuming that R is constant. Inserting (6.6) into (6.4) and the result into (6.5) gives

$$0 \leq p \perp \underbrace{g_0 - \alpha L T_A + \frac{\alpha L^2 (T_A - T_B)}{2(KR + L)}}_{q} + \\ + p \underbrace{\left(\frac{L}{E} - \frac{\alpha L^2 \mu V R}{2(KR + L)} \right)}_{M} \geq 0. \tag{6.7}$$

That is, we have a problem with the structure $0 \leq x \perp q + Mx \geq 0$, where $x \perp y$ means $xy = 0$. This is a Linear Complementarity Problem (LCP), or, rather, a scalar version of an LCP. It is easily concluded, either directly or be referring to general results on LCPs, that this problem has a unique solution if M is positive. Thus, without frictional heating, i.e., when μV is zero, the static thermal problem has a unique solution, but for large frictional heating we cannot expect such a property.

To treat a more general R than the constant one, we will make a further rewriting of the problem. Note that Signorini's contact conditions (6.5) can equivalently be written as

$$g_0 - u = \left(g_0 - u - \frac{pL}{E}\right)_+ \quad \text{or} \quad -\frac{pL}{E} = \left(g_0 - u - \frac{pL}{E}\right)_-, \qquad (6.8)$$

where $2(x)_+ = x + |x|$ and $2(x)_- = x - |x|$. That is, the augmented variable

$$r = g_0 - u - \frac{pL}{E},$$

is projected onto two mutually polar cones. This is an idea developed by J. J. Moreau [(Moreau (1962)) and (Moreau (1966))] as a generalization of the classical decomposition of a space into a direct sum of two orthogonal complementary subspaces. Signorini's contact condition now becomes

$$(r)_- = -\frac{pL}{E}, \quad (r)_+ = g_0 - u, \qquad (6.9)$$

so when r is negative it (essentially) represents the negative of the contact pressure and when it is positive it represents the contact gap.

Now, in terms of the augmented variable r, equation (6.4) becomes

$$r = g_0 - \Delta. \qquad (6.10)$$

Therefore, if Δ is a function of r, which follows from (6.6) if R is such a function, then (6.10) represents a scalar equation for r. Solving this equation, we next obtain the solution p and u for the system (6.4) – (6.6) by substituting into (6.9). Thus, remarkably, we have reduced the problem essentially into one single equation in the the augmented variable r and Signorini's contact conditions are simply treated as a post-processing ingredient. This property of the one-dimensional thermoelastic system was first noticed by (Barber (1978)).

2.1.1 Existence and uniqueness of steady states. It will be assumed that the contact resistance is a function of r, i.e., $R = R(r)$, which from the physical point of view is very natural, since it will depend on the the contact pressure when there is contact and on the gap when there is no contact. Then, substituting (6.6) into (6.10) and using (6.9)[1], we find

$$\mathcal{F}(r) \equiv r - (g_0 - \alpha L T_A) + \frac{\hat{V} K R(r)(r)_-}{(KR(r) + L)} - \frac{\alpha L^2 (T_A - T_B)}{2(KR(r) + L)} = 0, \qquad (6.11)$$

where

$$\hat{V} = \frac{\mu V E \alpha L}{2K}$$

is a dimensionless sliding speed. Note that $\mathcal{F}(r)$ is continuous if $R(r)$ is so and this is assumed in the following.

One concludes that

$$\lim_{r \to \infty} \mathcal{F}(r) = \infty.$$

Furthermore, let $\lim_{r \to -\infty} R(r) = R_\infty$, the resistance for large p. Then

$$\lim_{r \to -\infty} \mathcal{F}(r) = -\infty, \tag{6.12}$$

if

$$\hat{V} < \hat{V}_0 \equiv 1 + \frac{L}{KR_\infty}. \tag{6.13}$$

Thus, if (6.13) is satisfied, then $\mathcal{F}(r)$ is a continuous function extending from $+\infty$ to $-\infty$, so it must have an odd number of roots (except in the case of repeated roots) and in particular, at least one root, thus establishing an existence theorem. Note that if R is constant, (6.13) is exactly the condition $M > 0$, which guaranties a unique solution of the LCP (6.7).

For $\hat{V} > \hat{V}_0$, the limit (6.12) is replaced by

$$\lim_{r \to -\infty} \mathcal{F}(r) = \infty, \tag{6.14}$$

and under some conditions, $\mathcal{F}$ may be positive for all r, giving no solutions to $\mathcal{F}(r) = 0$. Under these conditions, (Ciavarella et al. (2003)) showed by a numerical analysis of the transient system (6.1) – (6.3) that the contact pressure would grow without limit due to frictional heating. This state of unbounded increase in pressure or "seizure" is made possible by the interaction between thermal expansion and the zero-displacement boundary condition at the left end of the rod.

In general, we anticipate that $R(r)$ will be a monotonically decreasing function of r and at arbitrarily large pressures it might tend either to a finite limit R_∞, or to zero. In the latter case, (6.12) would hold for all values of μV, thus establishing an unconditional existence theorem in consistency with the results for the three-dimensional case shown in the Section 3.

Turning now to the question of uniqueness, we note that for $1/R(r) = 0$ and $\mu V = 0$, $\mathcal{F}$ degenerates to the linear function $\mathcal{F}(r) = r - (g_0 - \alpha L T_A)$, in which case $\mathcal{F}(r) = 0$ has a unique solution. If we write $1/R(r) = Ak_1(r)$, where A is a constant and $k_1(r)$ a fixed function, the more general expression for $\mathcal{F}$ will be a continuous function of A

and μV and hence there will be a finite neighbourhood of the point $A = 0$, $\mu V = 0$ within which the solution will be unique. This conclusion is also consistent with the general results given below for the three-dimensional case. Note that from the LCP formulation a slightly more precise statement can be given when R is constant.

2.1.2 Stability of steady states. The system (6.1) – (6.3) is piecewise linear. Assuming that the steady state does not occur at the intermediate contact point ($p = 0$ and $u = g_0$), a linear stability analysis can be conducted. In (Ciavarella et al. (2003)) it was concluded that a small temperature perturbation will not grow exponentially in time at a steady state solution point r^* if

$$\frac{d\mathcal{F}(r^*)}{dr} > 0.$$

From this result and the above discussion on existence and uniqueness, we conclude that when a steady state is unique it is also stable. Disregarding situations with repeated roots, we can also conclude that when (6.12) holds, we have an odd number of roots that are alternately stable and unstable and the outermost ones are stable, and when (6.14) holds and there are solutions, they are even in number with the rightmost being stable and the leftmost being unstable.

3. Three-dimensional steady state problem

The system (6.1) – (6.3) can be given a three-dimensional counterpart. In the following we will state such a problem for the case of steady state and give existence and uniqueness results. Proofs of these results are given in (Andersson et al. (2003)).

3.1 Variational formulation

A thermoelastic body occupies a region Ω in R^d, ($d = 2, 3$). The boundary of Ω contains three disjoint parts: Γ_0, Γ_1 and Γ_2. On Γ_0 the displacement vector u and the temperature T are prescribed to be zero. On Γ_1 the traction vector t and the heat flow q are prescribed. In the interior of Ω we have prescribed volume forces f and volume heat sources Q. Moreover, Γ_2 is the potential contact surface.

Now let us introduce function spaces V_1 and V_2 of displacements and temperatures, given by

$$V_1 = \{u \in (H^1(\Omega))^d : u = 0 \text{ on } \Gamma_0\},$$

$$V_2 = \{T \in H^1(\Omega) : T = 0 \text{ on } \Gamma_0\}$$

and the closed, convex subset

$$K_1 = \{u \in V_1 : u_N \leq 0 \quad \text{on} \quad \Gamma_2\}.$$

then, using Green's formula in the usual way, we may reformulate the classical equations for equilibrium of forces and heat production into the following coupled variational problems: Find $(u, T) \in K_1 \times V_2$ such that

$$a_1(u, v - u) - \int_\Omega T s_{ij} \frac{\partial(v_i - u_i)}{\partial x_j} \, dx \geq \langle L_1, v - u \rangle \qquad (6.16)$$

$$a_2(T, \varphi) + \int_{\partial\Omega} [k(p^*)(T - T_0) - \mu V p^*] \, \varphi \, dS = \langle L_2, \varphi \rangle \quad (6.17)$$

for all $(v, \varphi) \in K_1 \times V_2$.

Here $a_1(\cdot,\cdot)$ is the bilinear form of elastic energy, the second term in (6.16) represents the thermoelastic coupling and the linear functional L_1 in the right hand side represents the external volume and traction forces.

In equation (6.17), $a_2(\cdot,\cdot)$ is the bilinear form of heat flow. L_2 in the right hand side represents the external heat sources. The contact pressure is $p = -\sigma_{ij} n_i n_j$, and since this may only be expected to be a positive measure in the function space $H^{-1/2}(\partial\Omega)$ we have introduced a regularization operator

$$H^{-1/2}(\partial\Omega) \ni p \mapsto p^* \in L^2(\partial\Omega) \quad \text{which is linear and bounded} \quad (6.18)$$

with norm C^* so that

$$\|p^*\|_{L^2(\partial\Omega)} \leq C^* \|p\|_{-1/2,\partial\Omega} \qquad (6.19)$$

and that

$$p \geq 0 \Longrightarrow p^* \geq 0. \qquad (6.20)$$

In addition we require that if p is a measure on $\partial\Omega$ then

$$p \geq 0 \Longrightarrow \|p^*\|_{L^\infty(\partial\Omega)} \leq c_0 \|p\|_{\mathcal{M}(\partial\Omega)} \qquad (6.21)$$

where c_0 is independent of p and $\|\cdot\|_{\mathcal{M}(\partial\Omega)}$ denotes the total variation-norm on the space $\mathcal{M}(\partial\Omega)$ of bounded measures on $\partial\Omega$. The conditions (6.18) – (6.21) are certainly satisfied if for example the mapping * is given by a convolution (averaging) with a non-negative, piecewise C^1-function having compact support.

The heat balance of the contact interface is represented by the second term in (6.17). Frictional heating μV passes together with heat flowing from the bulk of the body through a thermal resistance

$$R = 1/k(p^*),$$

where $k(p)$ is non-negative.

3.2 Two Theorems of Existence

In this section we will formulate two theorems of existence for the frictional thermoelastic problem. Proofs can be found in (Andersson et al. (2003)). The first theorem deals with the case when the thermal conductance $k(p)$ is at least linearly increasing as the contact pressure p tends to infinity.

THEOREM 6.1 *If $k(p) = 0$ for $p \leq 0$, if $mp \leq k(p)$ with $m > 0$ and if k is continuous, then the problem defined by (6.16) and (6.17) has at least one solution (u, T). Moreover, for this solution we have*

$$\|T\|_{H^1(\Omega)} \leq C(V\|\mu\|_{1/2,\partial\Omega}, 1/m)$$

with $C(\cdot,\cdot)$ an increasing function of both arguments.

The idea of the growth condition in this theorem came from the fact that if equality is assumed, i.e., $mp = k(p)$, one may define a modified temperature $T_1 = T_0 + \mu V/m \in H^1(\Omega)$, such that the integrand of the integral reads $k(p^*)(T - T_1)$. This means that our problem takes the form already considered by (Duvaut (1979)) for the case of no frictional heating and an existence result follows by analogy.

In the second theorem no growth condition is required for k. Instead there is a restriction on the size of frictional heating. On the other hand the regularity of μ is less restricted, only $\mu \in L^2(\partial\Omega)$ or $\mu \in L^\infty(\partial\Omega)$ is needed.

THEOREM 6.2 *If $k(p) = 0$ for $p \leq 0$, if $k(p) \geq 0$ for $p \geq 0$, if k is continuous and if*

$$V\|\mu\|_{L^2(\partial\Omega)} < c_1c_2/c_0c_4C_1^2\|S\|\|\mathcal{E}\|^2 \tag{6.22}$$

or

$$V\|\mu\|_{L^\infty(\partial\Omega)} < c_1c_2/C^*c_4C_1\|S\|\|\mathcal{E}\| \tag{6.23}$$

then the problem defined by by (6.16) and (6.17) has at least one solution (u, T).

Here the constants c_1, c_2, C_1, C_2 are related to the bilinear forms $a_1(\cdot,\cdot)$ and $a_1(\cdot,\cdot)$, c_4 to a trace theorem and $\|\mathcal{E}\|$ is the norm of an extension operator from $H^{1/2}(\partial\Omega)$ to $H^1(\Omega)$-

3.3 Uniqueness

To prove a theorem on uniqueness of solutions we make the additional assumption that the function k is Lipschitz continuous. The following result can be found in (Andersson et al. (2003)).

THEOREM 6.3 *The solutions proven to exist in Theorems 6.1 and 6.2 are unique if k is Lipschitz continuous with Lipschitz constant L, and if L and the sliding velocity V are small enough*

References

Andersson, L.-E., Klarbring, A., Barber, J.R. & Ciavarella, M. (2003) "On the existence and uniqueness of steady state solutions in thermoelastic contact with frictional heating", submitted.

Andrews, K.T., Mikelić, A, Shi, P., Shillor, M. & Wright, S. (1992) "One-dimensional thermoelastic contact with a stress-dependent radiation condition", *SIAM J. Math. Anal.*, 23, 1393–1416.

Andrews, K.T., Shi, P., Shillor, M. & Wright, S. (1993) "Thermoelastic contact with Barber's heat exchange condition", *Appl. Math. Optim.*, 28, 11–48,

Barber, J.R. (1978) "Contact problems involving a cooled punch", *Journal of Elasticity*, 8(4), 409–423.

Barber, J.R., Dundurs, J. & Comninou, M. (1980) "Stability considerations in thermoelastic contact", *ASME J. Appl. Mech.*,47, 871–874.

Ciavarella, M., Johansson, L., Afferrante, L., Klarbring, A. & Barber, J.R. (2003) "Interaction of thermal contact resistance and frictional heating in thermoelastic instability", *Int. J. Solids Structures*, 40, 5583–5597.

Duvaut, G. (1979) "Free boundary problem connected with thermoelasticity and unilateral contact", In *Free boundary problems*, Vol 11, Pavia.

Duvaut, G. (1981) "Non-Linear boundary value problem in thermoelasticity", In *Proceedings of the IUTAM symposium on Finite Elasticity.* (Eds. D.E. Carlson and R.T. Shild), pp. 151-165. Martinus Nijhoff Publishers.

Manners, W. (1998) "Partial contact between elastic surfaces with periodic profiles", *Proc. R. Soc. London. A*, 454, 3203–3221.

Moreau, J.J. (1962) "Functions convexes duales et points proximaux dans un espace hilbertien", *C. R. Acad. Sci. Paris*, 255, 2857–2899.

Moreau, J.J. (1966) "Quadratic programming in mechanics: dynamics of one-sided constraints", *J. SIAM Control*, 4(1), 153–158.

Chapter 7

A CONDITION FOR STATICAL ADMISSIBILITY IN UNILATERAL STRUCTURAL ANALYSIS

Gianpietro Del Piero

Dipartimento di Ingegneria, Università di Ferrara. Presently on leave at the Centro Interdipartimentale Linceo, Accademia Nazionale dei Lincei, Rome, Italy.

gdpiero@ing.unife.it

Abstract In unilateral structural analysis, the equilibrium problem can be formulated as a minimum problem for a convex functional over a closed convex set. The functional may be either coercive or semi-coercive. In the second case, minimizers do exist only for some particular loads, called *statically admissible*. Sufficient conditions for statical admissibility are available in the literature; here, under some restrictive assumptions peculiar of *linear* structural analysis, I obtain a condition which is both necessary and sufficient, and has a direct mechanical interpretation.

Keywords: Convex analysis, Noncoercive variational problems, Unilateral structural analysis.

1. Introduction

For a linear elastic structure subject to unilateral constraints, the equilibrium problem can be formulated as the minimum problem for a functional of the form

$$F(v) \;=\; \tfrac{1}{2}\, a(v,v) \;+\; f(v)\,, \tag{7.1}$$

defined over a Hilbert space H. Here a is a bilinear form on H, representing the strain energy of the structure, and f is a convex function representing the energy of the constraints and of the applied loads. The effective domain $\mathbb{K}$ of f is determined by the geometry of the constraints.

The problem was formulated independently by (Stampacchia, 1964) and by (Fichera, 1964). They proved the existence of global minimizers for F in the two fundamental cases of a coercive and of $\mathbb{K}$ bounded. These cases cover all situations of interest in structural analysis, except that of constraints insufficient to prevent rigid-body motions. This *semi-coercive* case was studied intensively, and a number of sufficient conditions for the existence of solutions was found.[1] Here I give a condition which, under some supplementary restrictions on f and $\mathbb{K}$, specific of linear structural analysis, is both necessary and sufficient for the existence of minimizers. In Section 2 this condition is shown to be necessary for the existence of minimizers and sufficient for the boundedness from below of F. In Section 3 it is proved that the same condition is sufficient for the existence of minimizers, when a suitable projection of $\mathbb{K}$ is closed and f is piecewise quadratic. In Section 4 these supplementary assumptions are motivated within the context of linear structural analysis, and the proposed condition is compared with other conditions available in the literature.

2. An existence condition

Let H be a Hilbert space, with inner product $u \cdot v$ and norm $\|v\| := (v \cdot v)^{1/2}$, and let a be a bilinear form over H, symmetric, continuous, and with finite-dimensional null-space $\operatorname{nul} a$. I recall that

$$r \in \operatorname{nul} a \quad \Leftrightarrow \quad a(r,v) = 0 \qquad \forall v \in H, \tag{7.2}$$

and that if $\operatorname{nul} a$ is finite-dimensional then every v in H admits a unique decomposition

$$v \;=\; v^{\|} + v^{\perp} \tag{7.3}$$

into the sum of a $v^{\|}$ in $\operatorname{nul} a$ and a $v^{\perp}$ in the orthogonal complement $(\operatorname{nul} a)^{\perp}$. Assume, further, that a is coercive on $(\operatorname{nul} a)^{\perp}$, that is, that there is a positive constant c such that

$$a(v^{\perp}, v^{\perp}) \;\geq\; c\,\|v^{\perp}\|^2 \qquad \forall v^{\perp} \in (\operatorname{nul} a)^{\perp}\,. \tag{7.4}$$

A bilinear form a with a finite-dimensional null-space and coercive in $(\operatorname{nul}\ a)^{\perp}$ is called *semi-coercive.*

Let f be a convex function from H to $\mathbb{R} \cup \{+\infty\}$. Denote by $\mathbb{K}$ the essential domain of f, that is, the set of all v in H such that $f(v) < +\infty$, and assume that $\mathbb{K}$ is closed, convex, non-empty. For any u in $\mathbb{K}$, the set

$$\partial f(u) \;:=\; \{\, w \in H \;\mid\; f(v) - f(u) \geq w \cdot (v-u) \quad \forall v \in H\,\} \tag{7.5}$$

[1](Fichera, 1964; Fichera, 1972; Lions and Stampacchia, 1967; Schatzman, 1973; Baiocchi et al., 1986; Baiocchi et al., 1988; Goeleven 1996).

is the *subdifferential* of f at u.

In this paper I consider the following condition:

> *there is an u_o in $\mathbb{K}$ such that the intersection of $\partial f(u_o)$ and $(\operatorname{nul} a)^\perp$ is not empty.*

In other words, there exist a u_o in $\mathbb{K}$ and w_o in $\partial f(u_o)$ such that

$$w_o \cdot r \;=\; 0 \qquad \forall r \in \operatorname{nul} a\,. \tag{7.6}$$

I begin by showing that this is a necessary condition for the existence of global minimizers for the functional F defined in (7.1).

PROPOSITION 2.1 *Let a, f and $\mathbb{K}$ be as above. If F has a global minimizer, then the condition* (7.6) *is satisfied.*

Proof. Let u_o be a global minimizer for F. Then, for all η in H and for all λ in (0,1),

$$0 \le F(u_o + \lambda\eta) - F(u_o) \;=\; \lambda\, a(u_o,\eta) + \tfrac{1}{2}\lambda^2 a(\eta,\eta) + f(u_o+\lambda\eta) - f(u_o)\,, \tag{7.7}$$

and by the convexity of f,

$$f(u_o+\lambda\eta) - f(u_o) \;\le\; \lambda\,(f(u_o+\eta) - f(u_o))\,. \tag{7.8}$$

The two preceding inequalities imply

$$f(u_o+\eta) - f(u_o) \;\ge\; -a(u_o,\eta) \qquad \forall \eta \in H\,. \tag{7.9}$$

By Lax-Milgram's lemma,[2] if a is continuous and semi-coercive there is a w_o in $(\operatorname{nul} a)^\perp$ such that

$$-a(u_o,\eta) \;=\; w_o \cdot \eta \qquad \forall \eta \in H\,. \tag{7.10}$$

By (7.9), w_o belongs to $\partial f(u_o)$. Then the intersection of $\partial f(u_o)$ and $(\operatorname{nul} a)^\perp$ is not empty. □

That (7.6) is not a general sufficient condition for the existence of global minimizers is proved by the following two-dimensional counterexample. Let $H = \mathbb{K} = \mathbb{R}^2$, and for any $u = (u_1, u_2)$ and $v = (v_1, v_2)$ in H define

$$a(u,v) \;=\; u_1 v_1\,, \qquad f(v) \;=\; \max\{v_1, \exp(v_2)\}\,. \tag{7.11}$$

[2] See e.g. (Brézis, 1983, Corollary V.8.)

Then the vector $w_o = (1, 0)$ belongs to $\partial f(u)$ for every u with $u_1 \geq 1$ and $u_2 = 0$, and satisfies the condition (7.6) because $r_1 = 0$ for every $r = (r_1, r_2)$ in $\operatorname{nul} a$. Nevertheless, the functional

$$F(v) \;=\; \tfrac{1}{2}\, v_1^2 \,+\, \max\{v_1\,,\, \exp(v_2)\} \tag{7.12}$$

has no global minimizer. Indeed, the infimum of F is zero, and this value is not attained.

Next, I show that (7.6) is a sufficient condition for the boundedness from below of F.

PROPOSITION 2.2 *Let a, f and $\mathbb{K}$ be as above. If there exist a u_o in $\mathbb{K}$ and a w_o in $\partial f(u_o)$ which satisfy the condition* (7.6), *then F is bounded from below.*

Proof. For w_o in $\partial f(u_o)$, from (7.5) we have

$$F(v) \;=\; \tfrac{1}{2}\, a(v,v) + f(v) \;\geq\; \tfrac{1}{2}\, a(v,v) + f(u_o) + w_o \cdot (v - u_o)\,. \tag{7.13}$$

Consider the decomposition (7.3) of v. Since $v^{\|} \in \operatorname{nul} a$, we have $a(v,v) = a(v^{\perp}, v^{\perp})$. Moreover, if w_o satisfies the condition (7.6) we have $w_o \cdot v^{\|} = 0$. Then the above inequality reduces to

$$F(v) \;\geq\; \tfrac{1}{2}\, a(v^{\perp}, v^{\perp}) + f(u_o) - w_o \cdot u_o + w_o \cdot v^{\perp}\,, \tag{7.14}$$

and by the semi-coerciveness assumption (7.4) we have

$$\tfrac{1}{2}\, a(v^{\perp}, v^{\perp}) + w_o \cdot v^{\perp} \;\geq\; \tfrac{1}{2}\, c\, \|v^{\perp}\|^2 - \|w_o\|\, \|v^{\perp}\| \;\geq\; -\tfrac{1}{2}\, c^{-1} \|w_o\|^2\,. \tag{7.15}$$

Then substitution into (7.14) shows that F is bounded from below. □

3. Affine constraints

Consider the special case of f *affine*: there are an h in H and a scalar γ such that

$$f(v) \;=\; \begin{cases} h \cdot v + \gamma & \text{if } v \in \mathbb{K}\,, \\ +\infty & \text{if } v \in H \backslash \mathbb{K}\,. \end{cases} \tag{7.16}$$

Let $h = h^{\|} + h^{\perp}$ be the decomposition (7.3) of h, and let $\operatorname{sp} h^{\|}$ be the set of all v in H parallel to $h^{\|}$. Denote by P the orthogonal projection onto $((\operatorname{nul} a)^{\perp} \times \operatorname{sp} h^{\|})$:

$$Pv \;:=\; v^{\perp} \;+\; \frac{h^{\|} \cdot v}{\|h^{\|}\|^2}\; h^{\|}\,. \tag{7.17}$$

I prove below that for f as in (7.16) the condition (7.6) is sufficient for the existence of global minimizers for F. Before this, I prove a property of weak convergence in the projected set $P\mathbb{K}$.

LEMMA 3.1 *Let f be affine, and let the projection $P\mathbb{K}$ of $\mathbb{K}$ be closed. If the condition* (7.6) *is satisfied, then for every sequence $k \mapsto v_k$ in $\mathbb{K}$ such that the numerical sequence $k \mapsto F(v_k)$ is bounded from above, there are a subsequence $k' \mapsto v_{k'}$ and a $v_o \in \mathbb{K}$ such that $k' \mapsto Pv_{k'}$ converges weakly to Pv_o.*

Proof. For f as in (7.16), w_o belongs to $\partial f(u_o)$ if

$$(h - w_o) \cdot (v - u_o) \ \geq \ 0 \qquad \forall v \in \mathbb{K}\,, \tag{7.18}$$

and satisfies (7.6) if $w_o \cdot v = w_o \cdot v^\perp$ for all v in $\mathbb{K}$. If this is the case, by the semi-coerciveness of a we have

$$F(v) \ = \ \tfrac{1}{2}\, a(v^\perp, v^\perp) + h \cdot v + \gamma \ \geq \ \tfrac{1}{2}\, c\ \|v^\perp\|^2 + w_o \cdot v^\perp + \overline{\gamma}\,, \tag{7.19}$$

with $\overline{\gamma} := h \cdot u_o - w_o \cdot u_o^\perp + \gamma$. Thus, in a sequence $k \mapsto v_k$ with $F(v_k)$ bounded from above the sequence $k \mapsto v_k^\perp$ is bounded. Therefore, there is a subsequence $k' \mapsto v_{k'}$ such that $k' \mapsto v_{k'}^\perp$ is weakly convergent.

Its limit element $v_o^\perp$ belongs to $(\mathrm{nul}\, a)^\perp$; indeed, weak convergence implies $(v_{k'}^\perp - v_o^\perp) \cdot r \to 0$ for all r in $\mathrm{nul}\, a$, and because $v_{k'}^\perp \cdot r = 0$ for all k' we have $v_o^\perp \cdot r = 0$ for all r in $\mathrm{nul}\, a$. For each k', consider the energy

$$F(v_{k'}) \ = \ \tfrac{1}{2}\ a(v_{k'}^\perp, v_{k'}^\perp) \ + \ h \cdot v_{k'} \ + \ \gamma\,. \tag{7.20}$$

In it, $k' \mapsto F(v_{k'})$ is bounded from above by assumption, and is bounded from below as a consequence of (7.6), as proved in Proposition 2.2. Moreover, $k' \mapsto a(v_{k'}^\perp, v_{k'}^\perp)$ is bounded because $k' \mapsto v_{k'}^\perp$ is bounded and a is continuous. Then we may extract a further subsequence, again denoted by $k' \mapsto v_{k'}$, such that all terms in (7.20) converge, except $h \cdot v_{k'}$. Then this term must converge as well.

Denote the limit by δ. Then the numerical sequence $h^\parallel \cdot v_{k'} = h \cdot v_{k'}^\parallel = h \cdot v_{k'} - h \cdot v_{k'}^\perp$ converges to $\delta - h \cdot v_o^\perp$, and from (7.17) we have

$$Pv_{k'} \ \rightharpoonup \ v_o^\perp \ + \frac{\delta - h \cdot v_o^\perp}{\|h^\parallel\|^2}\, h^\parallel \ \ =: \ v_1\,, \tag{7.21}$$

where the symbol $\rightharpoonup$ denotes weak convergence. But each $Pv_{k'}$ belongs to $P\mathbb{K}$, $P\mathbb{K}$ is closed by assumption, and $P\mathbb{K}$ closed and $\mathbb{K}$ convex imply $P\mathbb{K}$ weakly closed.[3] Then $v_1 \in P\mathbb{K}$, that is, there is a v_o in $\mathbb{K}$ such that

[3] See e.g. (Brézis, 1983, Theorem III.7).

$Pv_o = v_1$. □

Now I prove that condition (7.6) is necessary and sufficient for the existence of global minimizers for F when f is affine.

PROPOSITION 3.2 *Let a be a bilinear form on H, symmetric, continuous, and semi-coercive. Let f be affine, with essential domain $\mathbb{K}$ closed, convex, non-empty, and let the projection $P\mathbb{K}$ be closed. Then there are global minimizers for F if and only if the condition* (7.6) *is satisfied.*

Proof. The *only if* part has been proved in Proposition 2.1. Now assume that (7.6) holds. Since F is bounded from below by Proposition 2.2, there is a minimizing sequence $k \mapsto v_k$, and by Lemma 3.1 we may extract a subsequence $k' \mapsto v_{k'}$ such that $Pv_{k'} \rightharpoonup Pv_o$ for some v_o in $\mathbb{K}$. Moreover, the equality $h \cdot Pv = h \cdot v$ which follows from (7.17) tells us that $k' \mapsto h \cdot v_{k'}$ converges to $h \cdot v_o$. Thus, from (7.20) in the limit for $k' \to +\infty$ we get

$$\inf \{ F(v), \ v \in H \} \ = \ \tfrac{1}{2} \lim_{k' \to +\infty} a(v_{k'}^{\perp}, v_{k'}^{\perp}) \ + \ h \cdot v_o \ + \ \gamma \, . \tag{7.22}$$

After recalling that the restriction of a to $(\operatorname{nul} a)^{\perp}$ is coercive, so that $\|v\|$ and $(a(v,v))^{1/2}$ are equivalent norms in $(\operatorname{nul} a)^{\perp}$, and that by (7.17) $Pv_{k'} \rightharpoonup Pv_o$ implies $v_{k'}^{\perp} \rightharpoonup v_o^{\perp}$, we have[4]

$$\lim_{k' \to +\infty} a(v_{k'}^{\perp}, v_{k'}^{\perp}) \ \geq \ a(v_o^{\perp}, v_o^{\perp}) \, . \tag{7.23}$$

Then, from (7.22),

$$\inf \{ F(v), \ v \in H \} \ \geq \ \tfrac{1}{2} a(v_o^{\perp}, v_o^{\perp}) \ + \ h \cdot v_o \ + \ \gamma \ = \ F(v_o) \, , \tag{7.24}$$

that is, v_o is a global minimizer for F. □

That the assumption of $P\mathbb{K}$ closed is indeed necessary, is shown by the following counterexample.[5] Take $H = \mathbb{R}^2$ and

$$\begin{gathered} a(u,v) \ = \ u_1 v_1 \, , \qquad f(v) = \begin{cases} 0 & \text{if } v \in \mathbb{K} \, , \\ +\infty & \text{if } v \in \mathbb{R}^2 \backslash \mathbb{K} \, , \end{cases} \\ \mathbb{K} \ = \ \{ (v_1, v_2) \in \mathbb{R}^2 \ | \ v_1 > 0 \, , \ v_1 v_2 \geq 1 \} \, . \end{gathered} \tag{7.25}$$

Then $P\mathbb{K}$ is the (open) half-line $(0, +\infty) \times \{0\}$, and the condition (7.6) is satisfied by $w_o = 0$, which belongs to $\partial f(x_o)$ for all u_o in $\mathbb{K}$. However, the infimum in $\mathbb{K}$ of $F(v) = v_1^2/2$ is zero, and is not attained.

---

[4] See e.g. (Dacorogna, 1989, Sect. 2.1.1.)
[5] From (Fichera 1972), modified.

The sufficient condition proved above can be extended to *convex and piecewise quadratic* functions. These are functions f whose essential domain $\mathbb{K}$ is a finite union of n pairwise disjoint regions $\mathbb{K}_i$, on each of which f has the form

$$f(v) \;=\; f_i(v) \;:=\; \frac{1}{2}\, b_i(v,v) + h_i \cdot v + \gamma_i\,, \tag{7.26}$$

where γ_i are scalars, h_i are elements of H, and b_i are bilinear forms over H, symmetric and non-negative.

PROPOSITION 3.3 *Let a and $\mathbb{K}$ be as in Proposition 3.2, and let f be convex and piecewise quadratic. If the condition* (7.6) *is satisfied, then there are global minimizers for F.*

Proof. Consider first the case $n = 1$. For it, we have

$$F(v) \;=\; \tfrac{1}{2}\,(a+b)(v,v) + h\cdot v + \gamma \qquad \forall v \in \mathbb{K}\,, \tag{7.27}$$

that is, F has the same form as in (7.1), with a replaced by $(a + b)$. Because the semi-coerciveness of a implies that of $(a+b)$,[6] from Proposition 3.2 it follows that F has global minimizers if there is an u_o in $\mathbb{K}$ such that the intersection of $(\operatorname{nul}(a+b))^{\perp}$ and $\partial f(u_o)$ is not empty.

But $\operatorname{nul}(a + b) = \operatorname{nul} a \cap \operatorname{nul} b$ is included in $\operatorname{nul} a$, and therefore $(\operatorname{nul} a)^{\perp}$ is included in $(\operatorname{nul}(a + b))^{\perp}$. Then F has global minimizers if the intersection of $(\operatorname{nul} a)^{\perp}$ and $\partial f(u_o)$ is not empty.

Let now $n > 1$. Take a minimizing sequence $k \mapsto v_k$. Because n is finite, at least one of the $\mathbb{K}_i$ contains a subsequence. In the subsequence, f may be replaced by the corresponding f_i all over $\mathbb{K}$, and so we are back to the case $n = 1$. □

4. The case of structural analysis

In linear structural analysis the energy of the applied loads is linear, the energy of the constraints is convex and piecewise quadratic, and its effective domain is a *polyhedral convex set.*[7] Accordingly, one may decompose f into the sum of a linear part and of a piecewise quadratic part

$$f(v) \;=\; -p\cdot v + g(v)\,, \tag{7.28}$$

[6] See e.g. (Del Piero and Smaoui, in preparation).

[7] A polyhedral convex set is a finite union of closed half-spaces (Rockafellar, 1970, Sect. 2). The restriction on the energy of the constraints is dictated by consistency with the strain energy a of the structure, which is assumed to be quadratic. For examples see (Del Piero and Smaoui, in preparation).

representing the energy of the applied loads and the energy of the constraints, respectively.

A load p is said to be *statically admissible* for given a and g if the total energy F has global minimizers. As already said in the Introduction, if a is coercive or if $\mathbb{K}$ is bounded then all p in H are statically admissible. For a semi-coercive and $\mathbb{K}$ unbounded, we may take advantage of Propositions 2.1 and 3.3 and of the fact that any orthogonal projection of a closed polyhedral set is closed,[8] to assert that the condition (7.6) is necessary and sufficient for statical admissibility. With f as in (7.28), the relation $w_o \in \partial f(u_o)$ becomes:

$$-p \cdot (v - u_o) + g(v) - g(u_o) \;\geq\; w_o \cdot (v - u_o) \qquad \forall v \in H\,, \tag{7.29}$$

and after setting $\overline{w}_o := w_o + p$ the condition (7.6) can be re-stated as follows: there are a u_o in $\mathbb{K}$ and a $\overline{w}_o$ in $\partial g(u_o)$ such that

$$\overline{w}_o \cdot r \;=\; p \cdot r \qquad \forall r \in \operatorname{nul} a\,, \tag{7.30}$$

or, equivalently,

$$\overline{w}_o^{\|} \;=\; p^{\|}\,. \tag{7.31}$$

This is a characterization of the set of all admissible loads in linear structural analysis. The condition (7.30) can be interpreted as a virtual work equation, in which the two sides are the virtual work done by the reactions $\overline{w}_o$ of the constraints in a configuration u_o of the body and by the applied loads p, respectively. That equality holds for all rigid displacements r means that p and $\overline{w}_o$ form an equilibrated system of forces. Thus, equation (7.30) tells us that the equilibrium problem has solutions if there are a configuration u_o of the body and reactions $\overline{w}_o$ permitted in that specific configuration, such that the reactions $\overline{w}_o$ and the applied load p form an equilibrated system of forces.[9]

Of particular interest is the case of *perfect constraints*, in which g is the characteristic function of $\mathbb{K}$

$$j_{\mathbb{K}}(v) \;:=\; \begin{cases} 0 & \text{if } v \in \mathbb{K}\,, \\ +\infty & \text{if } v \notin \mathbb{K}\,, \end{cases} \tag{7.32}$$

and the subdifferential $\partial j_{\mathbb{K}}(u_o)$ is the *outward normal cone* to $\mathbb{K}$ at u_o

$$\partial j_{\mathbb{K}}(u_o) \;:=\; \{ n \in H \;\mid\; n \cdot (v - u_o) \leq 0 \;\; \forall v \in \mathbb{K} \}\,. \tag{7.33}$$

This case has been studied by (Fichera, 1964) and by (Baiocchi et al., 1986). In our notation, their results can be stated as follows.

[8](Rockafellar, 1970, Theorem 19.3.)

[9]Note that u_o need not be an equilibrium configuration under the prescribed load.

THEOREM 4.1 (Fichera, 1972). *Let H be a Hilbert space, let $\mathbb{K}$ be a closed convex subset of H, let a be a bilinear form on H, symmetric, continuous and semicoercive, and let g be equal to $j_{\mathbb{K}}$. If a load p is statically admissible, then*

$$p \cdot r \leq 0 \quad \forall r \in \operatorname{nul} a \cap \operatorname{rc} \mathbb{K} . \tag{7.34}$$

Moreover, if $P\mathbb{K}$ is closed and if the above inequality holds, then p is statically admissible.

Here P is the projection defined in (7.17) and $\operatorname{rc}\mathbb{K}$ is the *recession cone* of $\mathbb{K}$:

$$\operatorname{rc}\mathbb{K} := \{ \eta \in H \mid u + \lambda\eta \in \mathbb{K} \quad \forall u \in \mathbb{K}, \forall \lambda \geq 0 \} . \tag{7.35}$$

In the next theorem, $\operatorname{nul} p$ denotes the set of all elements of H which are orthogonal to p.

THEOREM 4.2 (Baiocchi et al., 1986). *Let H, $\mathbb{K}$, a and g be as above. If a load p is statically admissible, then* (7.34) *holds. Moreover, p is statically admissible if* (7.34) *holds and the set*

$$\operatorname{nul} a \cap \operatorname{rc}\mathbb{K} \cap \operatorname{nul} p$$

is a subspace.

When compared with these two theorems, Proposition 3.3 shows some distinctive properties: (i) there is no gap between necessary and sufficient condition for statical admissibility,[10] (ii) no role is plaid by the recession cone, (iii) constraints more general than perfect constraints are considered, (iv) the characterization (7.30), (7.31) of the set of all admissible loads has a direct mechanical interpretation.

Acknowledgments

This research was supported by the Programma Cofinanziato 2002 *Modelli Matematici per la Scienza dei Materiali* of the Italian Ministry for University and Scientific Research.

[10] In the case of structural analysis, in which $\mathbb{K}$ is a closed polyhedral set, this is also true for Theorem 4.1.

References

Baiocchi C., Gastaldi F., Tomarelli F., Some existence results on noncoercive variational inequalities, Annali Scuola Normale Superiore, Pisa, Serie IV, vol. XIII, pp. 617-659 (1986).

Baiocchi C., Buttazzo G., Gastaldi F., Tomarelli F., General existence theorems for unilateral problems in continuum mechanics, Arch. Rational Mech. Analysis, vol. 100, pp. 149-189 (1988).

Brézis H., Analyse Fonctionnelle. Théorie et Applications. Masson, Paris, 1983.

Dacorogna B., Direct Methods in the Calculus of Variations, Springer 1989.

Del Piero G., Smaoui H., Unilateral Problems in Structural Analysis. Theory and Applications. Springer-Verlag Heidelberg, in preparation.

Fichera G., Problemi elastostatici con vincoli unilaterali: il problema di Signorini con ambigue condizioni al contorno, Atti Accademia Naz. dei Lincei, sez. I, vol. 7, pp. 71-140 (1964).

Fichera G., Boundary value problems in elasticity with unilateral constraints, in: Handbuch der Physik, vol VIa/2, Springer-Verlag, Berlin 1972.

Goeleven D., Noncoercive Variational Problems and Related Results, Longman, Harlow 1996.

Lions J.-L., Stampacchia G., Variational Inequalities, Comm. Pure Appl. Mathematics, vol. XX, pp. 493-519 (1967).

Rockafellar R.T., Convex Analysis, Princeton University Press, Princeton, N.J. 1970.

Schatzman M., Problèmes aux limites non linéaires, non coercifs, Annali Scuola Normale Superiore, Pisa, Serie III, Vol. XXVII, pp. 641-686 (1973).

Stampacchia G., Formes bilinéaires coercitives sur les ensembles convexes, C. R. Acad. Sci. Paris vol. 258 pp. 4413-4416 (1964).

Chapter 8

MIN-MAX DUALITY AND SHAKEDOWN THEOREMS IN PLASTICITY

Quoc-Son Nguyen

Laboratoire de Mécanique des Solides, CNRS-UMR7649, Ecole Polytechnique, Palaiseau, France

son@lms.polytechnique.fr

Abstract This paper gives an overall presentation of shakedown theorems in perfect and in hardening plasticity. General results on shakedown theorems are discussed in the framework of generalized standard materials. The starting point is a static shakedown theorem available for perfect and for hardening plasticity. It leads by min-max duality to the dual static and kinematic approaches to compute the safety coefficient with respect to shakedown. These approaches are discussed for common models of isotropic and of kinematic hardening. In particular, the kinematic approach leads to some new results on the expressions of the safety coefficient.

Keywords: Plasticity; shakedown analysis; min-max duality; kinematic/isotropic hardening; static and kinematic approaches; safety coefficient.

Introduction

The shakedown phenomenon is related to the long-term behavior of an elastic-plastic solid under variable loads and expresses the fact that the mechanical response of the solid becomes purely elastic if the load amplitude is small enough or if the hardening effect is strong enough, immaterial of the initial state of the evolution. Shakedown conditions have been discussed in a large number of papers of the literature. Classical shakedown theorems in a quasi-static deformation takes its definitive form from the pioneering works of Bleich (1932), Melan (1936) and Koiter (1960). Its generalization to dynamics has been discussed, cf. (Corradi and Maier, 1974). Further extensions to hardening plasticity, to visco-plasticity or to damage mechanics and poro-plasticity can be found in a large number of references, e.g. (Mandel, 1976, Konig, 1987, Polizzotto et al., 1991, Debordes, 1976, Maier, 2001).

The objective of this paper is to give an overall presentation of shakedown theorems in elasto-plasticity, available for common models of strain hardening in the framework of generalized standard models of plasticity. This framework is a straightforward extension of perfect plasticity, with the same ingredients of convexity and normality and has been shown to be large enough to cover most common models of hardening plasticity, cf. (Halphen and Nguyen, 1975, Nguyen, 2000). As in perfect plasticity, e.g. (Koiter, 1960, Debordes, 1976), the method of min-max duality can be applied to obtain shakedown theorems. The starting point is an extended version of Melan's static shakedown theorem, available in perfect plasticity and in hardening plasticity. This theorem leads to the definition of the safety coefficient with respect to shakedown and, by a min-max duality, to dual safety coefficients obtained respectively from static and kinematic approaches. This method is then discussed in the particular cases of strain hardening materials for which hardening parameters are the plastic strain or equivalent plastic strain.

1. Perfect and hardening plasticity

This discussion is limited to the case of generalized standard models of plasticity. A generalized standard material is defined by the following conditions:

- State variables are (ϵ, α), which represents respectively the strain tensor and a set of internal parameters. Internal parameters α include the plastic strain and other hardening variables, $\alpha = (\epsilon^p, \beta)$. There exists an energy potential $W(\epsilon, \alpha)$ which leads to associated forces

$$\sigma = W_{,\epsilon}\,, \quad A = -W_{,\alpha} \tag{8.1}$$

and to a dissipation per unit volume d_{in}

$$d_{in} = A \cdot \dot{\alpha}. \tag{8.2}$$

- The force A must be plastically admissible, this means that physically admissible forces A must remain inside a convex domain C, called the elastic domain.
- Normality law is satisfied for α:

$$\dot{\alpha} \;\in\; \partial\Psi_C(A). \tag{8.3}$$

Thus, the following maximum dissipation principle is satisfied:

$$A \cdot \dot{\alpha} = D(\dot{\alpha}) = \max_{A^* \in C} \quad A^* \cdot \dot{\alpha} \tag{8.4}$$

in the spirit of Hill's maximum principle in perfect plasticity. Finally, as in perfect plasticity, the notions of convexity, normality, energy and generalized forces are four principal ingredients of the generalized standard models, cf. (Moreau, 1970, Halphen and Nguyen, 1975). The dissipation potential $D(\dot{\alpha})$ is convex, positively homogeneous of degree 1. This function is state-independent, i.e. independent of the present value of state variables (ϵ, α), if the plastic criterion is assumed to be state-independent.

In particular, the case of generalized standard models satisfying the *assumption of linear stress/elastic-strain relationship* is considered here. For such a model, the energy is

$$\begin{cases} W = W_E(\epsilon - \epsilon^p) + W_H(\epsilon^p, \beta), \\ \text{where} \quad W_E = \frac{1}{2}(\epsilon - \epsilon^p) : L : (\epsilon - \epsilon^p) \end{cases} \tag{8.5}$$

is a quadratic, positive function representing the stored energy by elastic deformation, L denotes the elastic tensors. The second term $W_H(\epsilon^p, \beta)$ is an arbitrary convex, differentiable function of the internal parameters and represents the energy stored by hardening effects. With the notation $B = -W_{H,\beta}$, $X = W_{H,\epsilon^p}$, $A_H = -W_{H,\alpha}$, the thermodynamic force A is: $A = (\sigma - X, B) = (\sigma, 0) + A_H$.

2. Static shakedown theorem

The problem of elastic-plastic response of a solid of volume V in small quasi-static deformation, starting from any given initial conditions of displacement and of the internal state under the action of a loading history is considered. It is assumed as usual in shakedown analysis that an admissible loading history consists of a linear combination $P_i(g_i, F_i, u_i^d)$ of n load systems (of volume force g_i and surface forces F_i on S_F and of given displacements u_i^d on S_u) where $P_i(t)$, $i = 1, n$ are n load parameters. These parameters may be considered as the components of a load point P defined in the load parameter space R^n and the loading history is described by a given curve $P(t)$, $0 \leq t \leq +\infty$. It is assumed that $P(t)$ takes its value arbitrarily in an allowable polyhedral and convex domain S of corners S^j, $j = 1, J$.

Let $(u(t), \sigma(t), \epsilon^p(t), \beta(t))$ be the elastic-plastic response of the solid starting from a given initial state on the interval $t \geq 0$ under the action of an admissible loading history.

By definition, this response shakes down if the existence of $\lim_{t\to\infty} \epsilon^p(t)$ *is ensured. This property also ensures that* $\lim_{t\to\infty} \; u(t) - u_{el}(t)$ *and* $\lim_{t\to\infty} \; \sigma(t) - \sigma_{el}(t)$ *exist*

where $\sigma_{el}(t)$, $u_{el}(t))$ denote the fictitious response to the same loading of the same solid, assumed to be purely elastic and admitting the same elastic coefficients L. The proof of this statement is clear for discrete systems while for continua, some difficulties remain concerning the choice of a relevent functional space in the case of perfect plasticity, cf. (Debordes, 1976, Nayroles, 1977, Suquet, 1981).

The following extension of Melan's theorem is an enhanced version of some previous results given in the literature, cf. for example (Nguyen, 1976, Polizzotto et al., 1991):

THEOREM 8.1 **(static shakedown theorem in perfect and in hardening plasticity)**
It is assumed that the plastic criterion is state-independent and that the assumption of linear stress/elastic-strain relationship is fulfilled. Then there is shakedown whatever be the initial state and the admissible loading history if there exists a self-stress field $s*$*, a constant internal parameter field* $\alpha^* = (\epsilon^{p*}, \beta^*)$ *and a coefficient* $m > 1$ *such that the force field* mA^{*j} *is plastically admissible for all* $j = 1, J$*, where* $A^{*j} = (s * + \sigma^j_{el} - X(\alpha^*), B(\alpha^*))$*, and* σ^j_{el} *denotes the elastic response in stress of the solid, associated with the values of load parameters at corners* S^j.

The proof of this theorem can be obtained in two steps. In the first step, it is shown that under the introduced assumptions, the dissipated energy W^d is necessarily bounded. In the second step, this property ensures the existence of $\lim_{t\to\infty} \; \epsilon^p(t)$.

- first step:

Under the assumptions of the theorem and from the convexity of the elastic domain, it is clear that mA^* is plastically admissible for all $t \geq \tau$, where $A^*(t)$ denotes

$$A^*(t) = (s^* + \sigma_{el}(t) - X(\alpha^*), B(\alpha^*)) \tag{8.6}$$

the following inequality holds

$$\int_V (A - A^*) \cdot \dot{\alpha} \, dV \; \geq \; \frac{m-1}{m} \int_V A \cdot \dot{\alpha} \, dV.$$

More explicitly,

$$(A - A^*) \cdot \dot{\alpha} = (\sigma - \sigma^*) : \dot{\epsilon}^p - (X - X^*) : \dot{\epsilon}^p + (B - B^*) : \dot{\beta},$$

where $\sigma^* = s^* + \sigma_{el}$. Since $\sigma - \sigma^*$ is a self-stress field and since $\dot{u} - \dot{u}_{el} = 0$ on S_u, one obtains

$$0 = \int_V (\sigma - \sigma^*) : (\dot{\epsilon} - \dot{\epsilon}_{el}) \, dV \quad \text{thus}$$

$$\int_V (\sigma - \sigma^*) : \dot{\epsilon}^p + \int_V (\sigma - \sigma^*) : L^{-1} : (\dot{\sigma} - \dot{\sigma}*) \, dV = 0.$$

It follows that

$$\left\{ \begin{array}{l} -\frac{d}{dt} \int_V \{\frac{1}{2}(\sigma(t) - \sigma^*(t)) : L^{-1} : (\sigma(t) - \sigma^*(t)) + \\ W_H(\alpha) - W_H(\alpha^*) - W_{H,\alpha}(\alpha^*) \cdot (\alpha - \alpha^*) \} \, dV \quad \geq \quad \frac{m-1}{m} \int_V d_{in} \, dV. \end{array} \right.$$

or that

$$\left\{ \begin{array}{l} I(0) - I(t) \geq \frac{m-1}{m} W^d \quad \text{with} \quad W^d(t) = \int_0^t \int_V A \cdot \dot{\alpha} \, dV d\tau \\ I(t) = \int_V \{\frac{1}{2}(\sigma(t) - \sigma^*(t)) : L : (\sigma(t) - \sigma^*(t)) + \\ W_H(\alpha(t)) - W_H(\alpha^*) - W_{H,\alpha}(\alpha^*) \cdot (\alpha(t) - \alpha^*) \} \, dV. \end{array} \right. \tag{8.7}$$

Since $I(t)$ is non-negative as a sum of two non-negative terms, it is concluded that the dissipated energy $W^d(t)$ is bounded by $I(0)$ for all t.

- second step:

The second step consists of proving that $\alpha(t) = (\epsilon^p(t), \beta(t))$ tends to a limit. The fact that the dissipated energy remains bounded already ensures the existence of a limit of $\alpha(t)$ for any appropriate functional space which is complete with respect to the norm associated with the dissipation, since there exists a constant $c > 0$ such that $A \cdot \dot{\alpha} \geq c\|\dot{\alpha}\|$. This inequality follows from the fact that the origin of forces is strictly inside the elastic domain.

Theorem 8.1 gives a sufficient condition of shakedown. Melan's theorem in perfect plasticity is recovered when $W_H = 0$. In this case, it is well known that, since the stress solution $\sigma(t)$ belongs to the convex set of plastically admissible fields PA and to the set of statically admissible fields $SA(t)$ for all t, the self-stress $s(t) = \sigma(t) - \sigma_{el}(t)$ satisfies

$$\int_V (\tilde{s} - s) : L^{-1} : \dot{s} dV \geq 0, \quad \forall \ \tilde{s} \in K(t) = \{\sigma_{el}(t)\} + PA \cap SA(t).$$

Thus, $s(t)$ is a solution of the equation $-\dot{s} \in \partial \Psi_{K(t)}(s)$, a sweeping process by the variable convex $K(t)$, cf. (Moreau, 1971, Moreau, 1974) in the stress space with energy norm.

3. Safety coefficient and min-max duality

Min-max duality is a well known method of Convex Analysis, cf. for example (Rockafellar, 1970). It has been discussed in perfect plasticity by (Debordes and Nayroles, 1976) for shakedown analysis. This method is applied here following the same ideas. For this, the definition of a static safety coefficient with respect to shakedown is first considered and dual static and kinematic approaches are introduced.

3.1 Static safety coefficient

A safety coefficient with respect to shakedown can be introduced from the previous theorem.

THEOREM 8.2 **(static safety coefficient)**
Let m_s be the safety coefficient defined by the maximum

$$m_s = \max_{s^*, \alpha^*, m} \quad m \tag{8.8}$$

among constant coefficient m, internal parameters α^* and self-stress fields s^* such that

$$\left\{ \begin{array}{l} \forall \ j, \quad mA^{*j} \ \text{is plastically admissible}, \\ \text{with} \ \ A^{*j} = (s^* + \sigma^j_{el} - X(\alpha^*), B(\alpha^*)). \end{array} \right. \tag{8.9}$$

Then there is shakedown if $m_s > 1$.

Indeed, if $m_s > 1$ then the assumptions of Theorem 8.1 are fulfilled and the conclusion holds.

3.2 Min-max duality

The minimization (8.8) is a convex optimization problem. The dual approach consists of considering dual problems obtained by relaxing some constraints. For this, the initial problem (8.8) is first written as the search of maximum of m on the set of self-stress fields s^*, internal parameter fields α^*, coefficients m and of force fields $\tilde{A}^j$ such that

$$\left\{ \begin{array}{l} \forall \ j, \ \ \tilde{A}^j \ \ \text{is plastically admissible}, \\ \tilde{A}^j = m \ (s^* + \sigma^j_{el} - X(\alpha^*), B(\alpha^*)). \end{array} \right. \tag{8.10}$$

The last constraint is relaxed by the introduction of Lagrange multipliers $a^j = (d^{pj}, b^j)$ associated with the Lagrangian

$$\left\{ \begin{array}{l} \mathbf{\Lambda} = m + \int_V (\tilde{A}^j - m \ (s^* + \sigma^j_{el} - X(\alpha^*), B(\alpha^*))) \cdot a^j \ dV = \\ = m(1 - \int_V \sigma^j_{el} : d^{pj} dV) - m \int_V s^* : (\sum_j d^{pj}) dV + \\ \int_V \tilde{A}^j \cdot a^j dV + m \int_V (X^* : (\sum_j d^{pj}) - B^* \cdot (\sum_j \ b^j)) dV. \end{array} \right. \tag{8.11}$$

The max-min problem

$$\max_{s^*, \alpha^*, \tilde{A}, m} \quad \min_{a} \quad \mathbf{\Lambda}(a, m, s^*, \alpha^*, \tilde{A}^j) \tag{8.12}$$

in the set of arbitrary rate fields a^j and of self-stress fields s^*, int parameter fields α^*, coefficients m and of plastically admissible fields leads to the initial problem. Indeed, the result of the minimization wit respect to a gives $\min_a \mathbf{\Lambda} = m$ if $\tilde{A}^j = m\, A^{*j}$ for all j and $\min_a \mathbf{\Lambda} = -\infty$ otherwise.

3.3 Kinematic safety coefficient

The dual problem

$$m_k = \min_{a} \max_{s^*,\alpha^*,\ \tilde{A},\ m} \mathbf{\Lambda} \tag{8.13}$$

defines the dual approach to compute the safety coefficient. It is clear that $m_s \leq m_k$ and that it is not possible to find a self-stress satisfying the condition of the previous propositions if $m_k < 1$.

For a given rate $a^j = (d^{pj}, b^j),\quad j = 1, J$, the first operation consists of searching for the maximum of Λ among coefficients m, self-stress field s^*, internal parameters α^* and plastically admissible fields $\tilde{A}^j$. The following notation for the cycle residues is introduced

$$\Delta\alpha = \sum_j a_j = (\Delta\epsilon^p, \Delta\beta), \quad \Delta\epsilon^p = \sum_j d^{pj}, \quad \Delta\beta = \sum_j b^j, \tag{8.14}$$

and let $Q(\Delta\alpha)$ denote the quantity

$$Q(\Delta\alpha) = \max_{\alpha^*} -A^{H*} \cdot \Delta\alpha = \max_{\epsilon^{p*},\beta^*} X^* : \Delta\epsilon^p - B^* \cdot \Delta\beta. \tag{8.15}$$

The result of the maximization is:

$$\max_{m,s^*,\alpha^*,\tilde{A}^j} \mathbf{\Lambda} = \sum_j \int_V D(a^j) dV$$

where $D(a)$ denotes the dissipation function, under the conditions:

$$\begin{cases} \Delta\epsilon^p \quad \text{must be a compatible strain field,} \\ \Delta\epsilon^p \ \textit{and}\ \Delta\beta \quad \text{must ensure} \quad Q(\Delta\alpha) < +\infty, \\ \int_V (\sigma^j_{el} : d^{pj} - Q(\Delta\alpha)) = 1 \end{cases} \tag{8.16}$$

A compatible field means that there exists a displacement field u^p with $u^p = 0$ on S_u such that $\Delta\epsilon^p = (\nabla u^p)$. The fact that $(\Delta\epsilon^p,\ \Delta\beta)$ must ensure a finite value of Q leads to additional conditions as it will be shown later in some examples. Finally, the following theorem is obtained

THEOREM 8.3 **(kinematic safety coefficient)**
Let m_k be the kinematic safety coefficient, defined by the minimization

em

$$m_k = \min_{a^j} \sum_j \int_V D(a^j) dV \tag{8.17}$$

among the fields $a^j = (d^{pj}, b^j)$, $j = 1, J$ such that the conditions (8.16) are fulfilled. It is concluded that $m_k \geq m_s$ and the assumptions of Theorem 8.1 is impossible if $m_k < 1$.

As usually established in min-max duality, equality $m_s = m_k$ is generically satisfied. In particular, this equality holds if the plastic domain is bounded in the stress space as it has been shown by (Debordes and Nayroles, 1976) in perfect plasticity. It is expected that this equality always holds although a general proof available for any standard model is lacking for continuum solids. Under this equality, it is clear that $m_k > 1$ is then a sufficient condition of shakedown.

4. Kinematic approach

In the following sections, the kinematic approach is considered for common models of plasticity.

4.1 Perfect plasticity

The elastic perfectly plastic model admits as internal parameter the plastic strain ϵ^p and the hardening energy is $W_H = 0$. Theorem 8.1 is reduced to Melan's theorem and the kinematic approach leads to the following result, cf. (Koiter, 1960, Debordes, 1976):

STATEMENT 1 *In perfect plasticity, the dual kinematic approach leads to a coefficient $m_k \geq m_s$, defined as the minimum*

$$m_k = \min_{d^p} \sum_j \int_V D(d^{pj}) \; dV \tag{8.18}$$

among the plastic rates d^p satisfying

$$\Delta\epsilon^p = \sum_j d^{pj} \; \textit{compatible}, \qquad \int_V \sigma^j_{el} : d^{pj} \; dV = 1. \tag{8.19}$$

4.2 Isotropic hardening

For the standard model with Mises criterion, the internal parameter $\alpha = (\epsilon^p, \beta)$ represents respectively the plastic strain and the equivalent plastic deformation β. By definition, $\dot{\beta} = \|\dot{\epsilon}^p\|$. It is assumed that the derivative of the hardening energy $R = -B = W'_H(\beta)$ is a positive, increasing function admitting a finite limit R_{max}. The plastic cri[illegible] is $\|\sigma'\| + B - k \leq 0$. The normality law gives $\dot{\epsilon}^p = \lambda \frac{\sigma'}{\|\sigma'\|}$, $\dot{\beta} = \lambda$. [illegible] dissipation is $D = \sigma' : \dot{\epsilon}^p + B\dot{\beta} = k\ \|\dot{\epsilon}^p\|$. Since

$$Q = \max_{\beta^*} -B^* \Delta\beta = R_{max}\Delta\beta = R_{max} \sum_j \|d^{pj}\| \tag{8.20}$$

the min-max duality leads to a well known result

STATEMENT 2 *In isotropic hardening, the dual kinematic approach leads to a coefficient $m_k \geq m_s$, defined as the minimum*

$$m_k = \min_{d^p} \quad \int_V k \sum_j \|d^{pj}\|\ dV \tag{8.21}$$

among the plastic rates d^p satisfying

$$\Delta\epsilon^p = \sum_j d^{pj}\ compatible, \qquad \int_V (\sigma^j_{el} : d^{pj} - R_{max} \sum_j \|d^{pj}\|)\ dV = 1, \tag{8.22}$$

i.e. the shakedown behaviour of the solid is obtained as in perfect plasticity with the yield value $R_{max} + k$.

4.3 Kinematic hardening

For this model, the internal parameter is ϵ^p and the hardening energy $W_H = V(y)$, with $y = \|\epsilon^p\| = \sqrt{\epsilon^p : \epsilon^p}$, is such that $R(y) = V'(y)$ is a monotone increasing function varying in $[0, R_{max}]$ when $y \in R^+$. If ϵ^p is assumed to be incompressible ($\epsilon^p_{kk} = 0$), the generalized force is $A = -W_{,\epsilon^p} = \sigma' - X$ with $X = R(\|\epsilon^p\|)\epsilon^p/\|\epsilon^p\|$. With Mises criterion $\|A\| - k \leq 0$, the elastic domain is a sphere of radius k and with center X in the force space. This center X remains near the origin since $\|X\| \leq R_{max}$. Since

$$Q = \max_{\epsilon^{p*j}} \ R(\|\epsilon^{p*}\|) : (\sum_j d^{pj}) = R_{max}\|\Delta\epsilon^p\|,$$

It follows that:

STATEMENT 3 *For this limited kinematic hardening model, the dual kinematic approach leads to a coefficient $m_k \geq m_s$, defined as the minimum*

$$m_k = \min_{d^{pj}} \quad \sum_j \int_V k\|d^{pj}\|\ dV \tag{8.23}$$

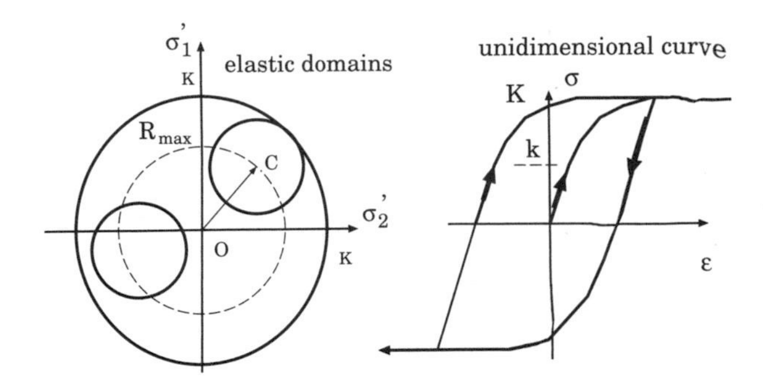

Figure 8.1. A model of limited kinematic hardening

among plastic rates d^{pj} satisfying

$$\begin{cases} d^{pj}_{kk} = 0, \quad and \quad \Delta\epsilon^p \quad compatible, \\ \int_V (\sigma^j_{el} : d^{pj} - R_{max}\|\Delta\epsilon^p\|)\, dV = 1. \end{cases} \tag{8.24}$$

If Zieger-Prager's model of linear kinematic hardening is considered, then $V = 1/2h\|\epsilon^p\|^2$ and the kinematic hardening is unlimited. Since

$$Q = \max_{\epsilon^{p*j}} \; h\epsilon^{p*} : (\sum_j d^{pj}) = +\infty \;\; if \;\; \Delta\epsilon^p \neq 0,$$

It is concluded that

STATEMENT 4 *In linear kinematic hardening with Mises criterion, the dual kinematic approach leads to a coefficient $m_k \geq m_s$, defined as the minimum*

$$m_k = \min_{d^{pj}} \quad \sum_j \int_V k\|d^{pj}\|dV. \tag{8.25}$$

among the plastic rates d^{pj} satisfying

$$d^{pj}_{kk} = 0, \quad \Delta\epsilon^p = 0, \quad \int_V \sigma^j_{el} : d^{pj}\; dV = 1. \tag{8.26}$$

Thus, closed cycles of plastic rates must be considered instead of compatible plastic cycles as in perfect plasticity or in isotropic hardening.

In particular, if m^0_k, m^{ukh}_k and m^{lkh}_k denote respectively the kinematic safety coefficients in perfect plasticity, in linear kinematic hardening and in limited kinematic hardening with the same yield stress k, it follows from their definition that

$$m^0_k \leq m^{lkh}_k \leq m^{ukh}_k. \tag{8.27}$$

The case of limited hardening appears as the penalization of the unlimited case, R_{max} is the penalty parameter associated with the constraint $||\Delta\epsilon^p|| = 0$.

5. Conclusion

In this paper, an overall presentation of shakedown theorems is given by min-max duality. In particular, new results concerning the expressions of kinematic safety coefficients, available for common models of isotropic/kinematic hardening, are presented.

References

Corradi, L., Maier, G., 1974. Dynamic non-shakedown theorem for elastic perfectly-plastic continua. J. Mech. Phys. Solids 22, 401–413.

Debordes, O., 1976. Dualité des théorèmes statique et cinématique sur la théorie de l'adaptation des milieux continus élasto-plastiques. C.R. Acad. Sc. 282, 535–537.

Debordes, O., Nayroles, B., 1976. Sur la théorie et le calcul à l'adaptation des structures élasto-plastiques. J. Mécanique 20, 1–54.

Halphen, B., Nguyen, Q.-S., 1975. Sur les matériaux standard généralisés. J. Mecanique 14, 1–37.

Koiter, W., 1960. General problems for elastic-plastic solids. In: Sneddon, J., Hill, R. (Eds.), Progress in Solid Mechanics. Vol. 4. North Holland, pp. 165–221.

Konig, J., 1987. Shakedown of elastic-plastic structures. Elsevier, Amsterdam.

Maier, G., 2001. On some issues in shakedown analysis. J. Appl. Mech. 68, 799–808.

Mandel, J., 1976. Adaptation d'une structure plastique écrouissable. Mech. Res. Com 3, 251–256.

Moreau, J., 1970. Sur les lois de frottement, de plasticité et de viscosité. C. R. Acad. Sciences 271, 608–611.

Moreau, J., 1971. Rafle par un convexe variable. Séminaire d'Analyse Convexe.

Moreau, J., 1974. On unilateral constraints, friction and plasticity. In: New variational techniques in Mathematical Physics. CIME Course, Springer-Verlag, pp. 173–322.

Nayroles, B., 1977. Tendances récentes et perspectives à moyen terme en élastoplasticité asymptotique des constructions. In: Proceedings, Congrès Français de Mécanique, Grenoble, France. Grenoble.

Nguyen, Q.-S., 1976. Extension des théorèmes d'adaptation et d'unicité en écrouissage non linéaire. C.R. Acad. Sc. 282, 755–758.

Nguyen, Q.-S., 2000. Stability and Nonlinear Solid Mechanics. Wiley, Chichester.

Polizzotto, C., Borino, G., Cademi, S., Fuschi, P., 1991. Shakedown problems for mechanical models with internal variables. Eur. J. Mech. A/Solids 10, 621–639.

Rockafellar, R., 1970. Convex analysis. Princeton University Press, Princeton.

Suquet, P., 1981. Sur les équations de la plasticité: existence et régularité des solutions. J. Mécanique 20, 3–40.

Chapter 9

FRICTION AND ADHESION

Michel Raous

Laboratoire de Mécanique et d'Acoustique, CNRS, 31, chemin Joseph Aiguier, 13402 Marseille Cedex 20, France

raous@lma.cnrs-mrs.fr

Abstract The studies carried out on adhesion by the group "Modeling in Contact Mechanics" at the LMA are reviewed in this paper and recent applications are presented. Based on the introduction of the adhesion intensity variable developed by M. Frémond, different forms of a model coupling adhesion to unilateral contact and friction have been developed. The formulations are given either under the form of implicit variational inequalities or the one of complementarity problems. Both quasi-static and dynamic formulations are considered.

The model is non smooth because we do not use any regularization for the unilateral conditions and for the friction, i.e. Signorini conditions and strict Coulomb law are written. In the thermodynamics analysis, the state and the complementarity laws are then written using subdifferentials and differential inclusions because of the non convexity and non differentiability of the potentials. For the dynamics, the formulation is given in term of differential measures in order to deal with the non continuity of the velocities that may occur in the solutions.

This work therefore owes much to the theories and the numerical scheme developed by J. J. Moreau and M. Jean.

Keywords: Unilateral contact, friction, adhesion, non smooth dynamics

Introduction

In order to describe the smooth transition from a completely adhesive contact to a usual unilateral contact (Signorini conditions) with Coulomb friction, a model based on interface damage has been first developed for quasi-static problems in (Raous et al, 1997, Cangémi, 1997, Raous et al, 1999, Raous, 1999). Using a dynamic formulation, the model was then extended to account for the brittle behaviour occurring when a crack

...cts with fiber-matrix interfaces in composite materials in (Raous-...erie, 2002, Monerie, 2000). More recently, the model has been used ...study metal/concrete interfaces in reinforced concrete in civil engi...neering (Karray et al, submitted), delamination of coated bodies (Raous et al, 2002), delamination of glued assembling in civil engineering (Raous et al, 2004), cohesive masonry (Jean et al, 2001, Acary, 2001) and production of wear particles in bio-engineering (Baudriller, 2003).

The quasi-static formulation was extended to deal with hyperelasticity in (Bretelle et al, 2001). Mathematical results about the existence of the solutions were given in (Cocou-Rocca, 2000) without using any regularization on the contact conditions.

1. The model

The RCC model (Raous-Cangémi-Cocou) has been first given in (Raous et al, 1997, Cangémi, 1997) and then extensively presented in (Raous et al, 1999). It has been extended to the present form including progressive friction with the term $(1-\beta)$ in (Monerie, 2000, Raous-Monerie, 2002). Adhesion is characterized in this model by the internal variable β, introduced by Frémond (Frémond, 1987, Frémond, 1988), which denotes the intensity of adhesion. The introduction of a damageable stiffness of the interface ensures a good continuity between the two models during the competition between friction and adhesion. The behaviour of the interface is described by the following relations, where (9.1) gives the unilateral contact with adhesion, (9.2) gives the Coulomb friction with adhesion and (9.3) gives the evolution of the adhesion intensity β. Initially, when the adhesion is complete, the interface is elastic as long as the energy threshold w is not reached. After that, damage of the interface occurs and consequently, on the one hand, the adhesion intensity β and the apparent stiffness $\beta^2 C_{\mathrm{N}}$ and $\beta^2 C_{\mathrm{T}}$ decrease, and on the other hand, friction begins to operate. When the adhesion is completely broken ($\beta = 0$), we get the classical Signorini problem with Coulomb friction.

$$-R_{\mathrm{N}}^r + \beta^2 C_{\mathrm{N}}\, u_{\mathrm{N}} \geq 0\,, \quad u_{\mathrm{N}} \geq 0\,, \quad \left(-R_{\mathrm{N}}^r + \beta^2 C_{\mathrm{N}}\, u_{\mathrm{N}}\right).\, u_{\mathrm{N}} = 0, \tag{9.1}$$

$$\begin{aligned} R_{\mathrm{T}}^r = \beta^2 C_{\mathrm{T}}\, u_{\mathrm{T}}\,, \quad R_{\mathrm{N}}^r = R_{\mathrm{N}}\,, \\ \|R_{\mathrm{T}} - R_{\mathrm{T}}^r\| \leq \mu(1-\beta)\left|R_{\mathrm{N}} - \beta^2 C_{\mathrm{N}}\, u_{\mathrm{N}}\right|, \end{aligned} \tag{9.2}$$

with

$$\begin{aligned} &\text{if } \|R_{\mathrm{T}} - R_{\mathrm{T}}^r\| < \mu(1-\beta)\left|R_{\mathrm{N}} - \beta^2 C_{\mathrm{N}}\, u_{\mathrm{N}}\right| \Rightarrow \dot{u}_{\mathrm{T}} = 0, \\ &\text{if } \|R_{\mathrm{T}} - R_{\mathrm{T}}^r\| = \mu(1-\beta)\left|R_{\mathrm{N}} - \beta^2 C_{\mathrm{N}}\, u_{\mathrm{N}}\right| \Rightarrow \exists \lambda \geq 0,\ \dot{u}_{\mathrm{T}} = \lambda(R_{\mathrm{T}} - R_{\mathrm{T}}^r), \end{aligned}$$

$$\dot{\beta} = -\left[\left(w - \beta\,(C_{\rm N}\,u_{\rm N}^2 + C_{\rm T}\,\|u_{\rm T}\|^2) - k\Delta\beta\right)^{-}/b\right]^{1/p} \text{ if } \beta \in [0,1[,$$
$$\dot{\beta} \leq -\left[\left(w - \beta\,(C_{\rm N}\,u_{\rm N}^2 + C_{\rm T}\,\|u_{\rm T}\|^2) - k\Delta\beta\right)^{-}/b\right]^{1/p} \text{ if } \beta = 1\,.$$

$R_{\rm N}$ and $u_{\rm N}$ are the algebraic values of the normal components of the contact force and those of the relative displacement between the two bodies (occupying domains Ω^1 and Ω^2) defined on the contact boundary Γ_c, and $R_{\rm T}$ and $u_{\rm T}$ are the tangential components of this contact force and those of this relative displacement. Subscript r denotes the reversible parts. The constitutive parameters of this interface law are as follows: $C_{\rm N}$ and $C_{\rm T}$ are the initial stiffnesses of the interface (full adhesion), μ is the friction coefficient, w is the decohesion energy, p is a power coefficient ($p = 1$ in what follows) and $k = 0$ in what follows.

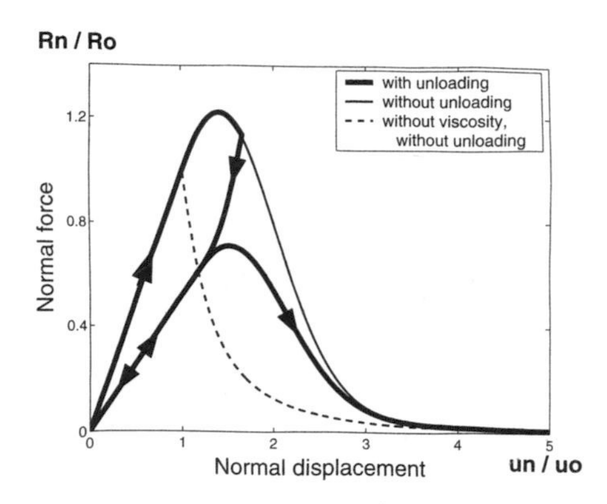

Figure 9.1. Normal behaviour of the interface

Fig. 9.1 and Fig. 9.2 give the normal and tangential behaviour of the interface during loading and unloading ($C_{\rm N} = C_{\rm T} = C$, $u^0 = \sqrt{\omega/C}$ and $R^0 = \sqrt{\omega C}$) . It should be noted that the Signorini conditions are strictly imposed when compression occurs. References on other models can be found in (Raous, 1999, Raous et al, 1999) and a comparison between some of them is made in (Monerie et al, 1998) (models developed by Tvergaard-Needleman, Girard-Feyel-Chaboche, Michel-Suquet, Allix-Ladevèze). Using penalization and augmented Lagrangian on a similar model to the RCC one, Talon-Curnier have solved the quasi-static problem using generalized Newton method (Talon-Curnier, 2003).

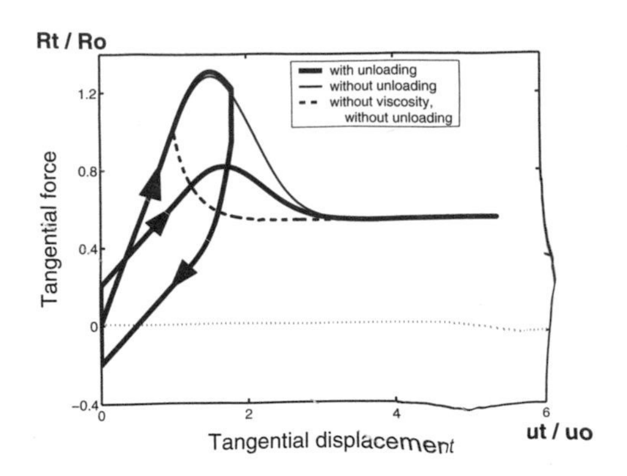

Figure 9.2. Tangential behaviour of the interface

2. The thermodynamics

In the framework of continuum thermodynamics, the contact zone is assumed to be a material surface and the local constitutive laws are obtained by choosing two specific forms of the free energy and the dissipation potential associated to the surface. The following thermodynamic variables are introduced: the relative displacements $(u_{\text{N}}\, n, u_{\text{T}})$ and the adhesion intensity β are chosen as the state variables, and the contact force R and a decohesion force G, as the associated thermodynamic forces. The thermodynamic analysis given for the RCC model in (Raous et al, 1999) has been extended to the present model in (Monerie, 2000, Raous-Monerie, 2002) in order to obtain relations (9.1) to (9.3), where friction is progressively introduced in the form of the term $(1-\beta)\mu$ into (9.2).

Expressions (9.4) and (9.5) are adopted for the free energy $\Psi(u_{\text{N}}, u_{\text{T}}, \beta)$ and the potential of dissipation $\Phi\left(\dot{u}_{\text{T}}, \dot{\beta}\right)$. In (9.4), the indicator function $I_{\widetilde{K}}$ (where $\widetilde{K} = \{v \;/\; v \geq 0\}$) imposes the unilateral condition $u_{\text{N}} \geq 0$ and the indicator function I_P (where $P = \{\gamma \;/\; 0 \leq \gamma \leq 1\}$) imposes the condition $\beta \in [0,1]$. In (9.5), the indicator function $I_{C^-}(\dot{\beta})$ (where $C^- = \{\gamma / \gamma \leq 0\}$) imposes that $\dot{\beta} \leq 0$: the adhesion can only decrease and cannot be regenerated (it is irreversible) in the present model.

$$\Psi(u_{\text{N}}, u_{\text{T}}, \beta) = \frac{1}{2}\beta^2\, C_{\text{N}}\, u_{\text{N}}^2 + \frac{1}{2}\beta^2\, C_{\text{T}}\, \|u_{\text{T}}\|^2 - w\,\beta + I_{\widetilde{K}}(u_{\text{N}}) + I_P(\beta) \quad (9.4)$$

$$\Phi\left(\dot{u}_{\text{T}}, \dot{\beta}\right) = \mu(1-\beta)\left|R_{\text{N}} - \beta^2\, C_{\text{N}}\, u_{\text{N}}\right|\, \|\dot{u}_{\text{T}}\| + \frac{b}{p+1}\left|\dot{\beta}\right|^{p+1} + I_{C^-}(\dot{\beta}) \quad (9.5)$$

Ψ has a part which is convex but not differentiable and another pa which is differentiable but not convex with respect to the pair (u, β). Φ is convex but has a part which is not differentiable. The state laws and the complementarity laws are then written as follows in order to obtain the contact behaviour laws given in section 1 (Raous et al, 1999).

$$\begin{cases} R_T^r = \frac{\partial \Psi^d}{\partial [u_T]} & R_N^r - \frac{\partial \Psi^d}{\partial [u_N]} \in \partial I_{\widetilde{K}}([u_N]) \\ -G_\beta - \frac{\partial \Psi^d}{\partial \beta} \in \partial I_{[0,1]}(\beta) & \end{cases} \tag{9.6}$$

$$R_N = R_N^r \qquad (R_T^{ir}, G_\beta) \in \partial \Phi([\dot{u_T}], \dot{\beta}) \tag{9.7}$$

3. The quasi-static formulation

The formulation and the approximation of quasi-static frictional problems given in (Cocou et al, 1996) have been extended to adhesion problems in (Raous et al, 1997, Raous et al, 1999, Raous, 1999). The problem can be here set as the coupling between two variational inequalities (one of which is implicit) and a differential equation.

Problem (P_1): Find $(\widetilde{u}, \beta) \in W^{1,2}(0,T;V) \times W^{1,2}(0,T;H)$ such that $\widetilde{u}(0) = \widetilde{u}_0 \in K$, $\beta(0) = \beta_0 \in H \bigcap [0,1[$ and for $\forall t \in [0,T]$, $\widetilde{u}(t) \in K$, and

$$\forall v \in V \qquad a(\widetilde{u}, v - \dot{\widetilde{u}}) + j(\beta, u_{\mathrm{N}}, v_{\mathrm{T}}) - j(\beta, u_{\mathrm{N}}, \dot{u}_{\mathrm{T}}) +$$

$$\int_{\Gamma_C} \beta^2 C_{\mathrm{T}} u_{\mathrm{T}} \,.\, (v_{\mathrm{T}} - \dot{u}_{\mathrm{T}}) ds \geq (\widetilde{F}, v - \dot{\widetilde{u}}) - \langle R_{\mathrm{N}}, v_{\mathrm{N}} - \dot{u}_{\mathrm{N}} \rangle, \tag{9.8}$$

$$-\langle R_{\mathrm{N}}, z - u_{\mathrm{N}} \rangle + \int_{\Gamma_C} \beta^2 C_{\mathrm{N}} u_{\mathrm{N}}.(z - u_{\mathrm{N}}) ds \geq 0 \qquad \forall z \in K, \tag{9.9}$$

$$\dot{\beta} = -1/b \left[w - (C_{\mathrm{N}}\, u_{\mathrm{N}}^2 + C_{\mathrm{T}} \, \|u_{\mathrm{T}}\|^2) \beta \right]^- \quad a.e. \ on \ \Gamma_C, \tag{9.10}$$

where:
- $\widetilde{u} = (u^1, u^2)$ where u^1 and u^2 define the displacements in Ω^1 and Ω^2,
- $V = (V^1, V^2)$, $V^\alpha = \left\{ v^\alpha \in \left[H^1(\Omega^\alpha)\right]^3 ; v^\alpha = 0 \text{ a.e. on } \Gamma_U^\alpha \right\}$, $\alpha = 1, 2$,
- $H = L^\infty(\Gamma_c)$,
- $K = \left\{ v = (v^1, v^2) \in V^1 \times V^2 ; v_{\mathrm{N}} \geq 0 \text{ a.e. on } \Gamma_c \right\}$, where Γ_c is the contact boundary between the two solids Ω_1 and Ω_2,
- $a(.,.)$ is the bilinear form associated to the elasticity mapping,
- $j(\beta, u_{\mathrm{N}}, v_{\mathrm{T}}) = \int_{\Gamma_C} \mu(1-\beta) \left| R_{\mathrm{N}} - \beta^2 C_{\mathrm{N}} u_{\mathrm{N}} \right| \|v_{\mathrm{T}}\| ds$,
- $\widetilde{F} = (F^1, F^2)$ are the given force densities applied to solid 1 and to solid 2 respectively.

By using an incremental approximation, it has been established in Raous et al, 1997, Raous et al, 1999) that the numerical solutions can be obtained by adapting the methods that we have developed for dealing with classical frictional unilateral contacts (Raous, 1999). The solutions have mainly been obtained as follows:

- either by taking a fixed point on the sliding threshold and solving a sequence of minimization problems with the choice between a projected Gauss-Seidel method (accelerated by relaxation or Aitken processes) and a projected conjugate gradient method (Raous-Barbarin, 1992),
- or by taking a complementarity formulation, using a mathematical programming method (Lemke's method).

With the adhesion model, these solvers are coupled with the numerical integration of the differential equation on β by using θ-methods ($\theta = 1$ is often chosen). Implementation of the algorithms has been conducted in our finite element code GYPTIS90 (Latil-Raous, 1991).

4. The dynamics

Depending on the characteristics of the interface, especially when b tends towards zero (no viscosity for the evolution of the intensity of adhesion), brittle behaviour can be obtained and the inertia effects have to be taken into account. A 3D dynamic formulation has been developed (Raous-Monerie, 2002, Monerie-Raous, 1999, Monerie-Acary, 2001). Because of the non smooth character of these interface laws, the dynamics is written in terms of differential measures as follows (where q denotes the displacement and r the contact force):

$$M(q,t)d\dot{q} = F(q,\dot{q},t)dt - rd\nu\,, \tag{9.11}$$

where $d\dot{q}$ is a differential measure associated with $\dot{q}(t)$:

$$\int_{t_1}^{t_2} d\dot{q} = q(t_2^+) - q(t_1^-) \;\; \forall t_2 > t_1\,, \tag{9.12}$$

and hence:

$$\int_{t_1}^{t_2} M(q,t)d\dot{q} = \int_{t_1}^{t_2} F(q,\dot{q},t)dt - \int_{]t_1,\,t_2]} rd\nu, \tag{9.13}$$

$$q(t_2) = q(t_1) + \int_{t_1}^{t_2} \dot{q}(\tau)d\tau. \tag{9.14}$$

The Non Smooth Contact Dynamics method developed by (Moreau, 1998, Moreau, 1994, Moreau, 1999, Jean, 1999) has been extended to the treatment of the RCC model. A solver for dealing with adhesion has been

implemented in the finite element code LMGC (Jean, 1999, ...
Acary, 2001). Another solver based on complementarity form...
(Lemke's method) dealing with non smooth dynamics problems has ...
implemented in the finite element code SIMEM3 at the LMA (Vola ...
al, 1998).

5. Applications

- **Delamination benchmarks**
  Various benchmarks for simulating delamination have been developed in the framework of a joint project with the Laboratoire Central des Ponts et Chaussées (Raous et al, 2004) focusing on adhesion and gluing in civil engineering.

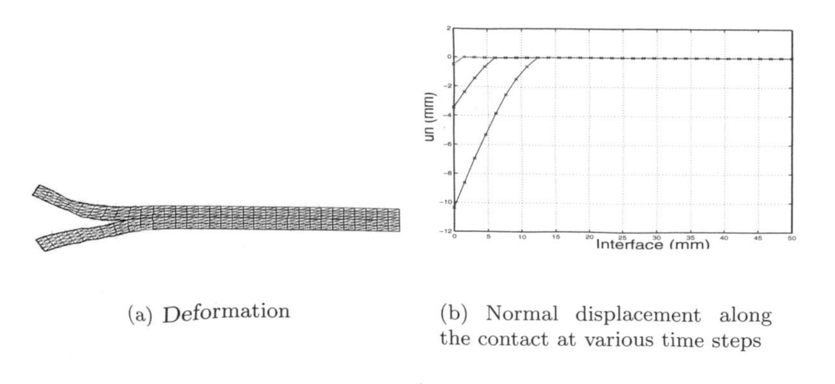

(a) Deformation

(b) Normal displacement along the contact at various time steps

Figure 9.3. Delamination of two layers submitted to vertical traction

- **Micro-indentation of a single fiber in a composite material**
  This model was first developed in order to describe the behaviour of the fiber-matrix interface of composite materials (Cangémi, 1997, Raous et al, 1999). The parameters of the model were identified on micro-indentation experiments performed at the ONERA on single fibers. Physical and mechanical considerations are taken into account in the identification procedure: for example, the values of the initial interface stiffness C_N, C_T are taken to be in the range corresponding to the elastic properties of the oxides located at the interfaces. The viscosity parameter b is particularly difficult to identify: experiments with different loading velocities are required for this purpose. The validity of the model was then confirmed by taking various kinds of loadings (especially cyclic loadings) and comparing experimental results with those obtained in the numer-

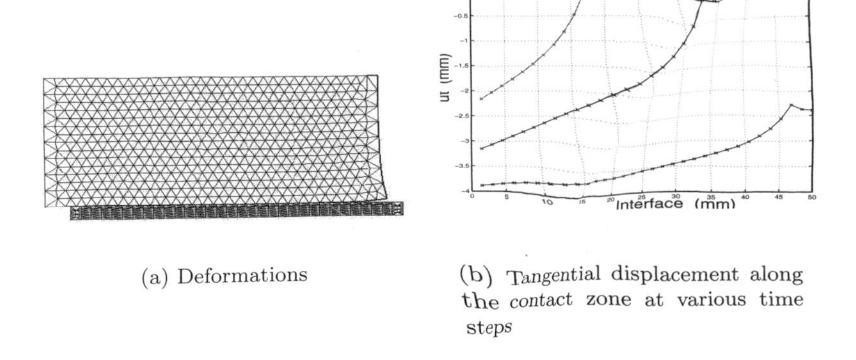

(a) Deformations

(b) Tangential displacement along the contact zone at various time steps

Figure 9.4. Shear delamination of a block submitted to horizontal loading

ical simulation whith the same interface parameters. Details of these studies can be found in (Cangémi, 1997, Raous et al, 1999) and later in (Monerie, 2000).

- **Interaction between cracks and fiber/matrix interfaces in composite materials**
  Again in collaboration with ONERA, during the thesis (Monerie, 2000), the RCC model has been used to investigate the different ways in which cracks propagate through a composite material and how they depend on the interface properties. The adhesion model has been used to account both the crack propagation (decohesive crack with no viscosity, i.e. $b = 0$) and the interface behaviour. Crack bridging, crack trapping and fiber breaking can be observed depending on the interface characteristics (Raous-Monerie, 2002, Monerie, 2000). On Fig. 9.5 adhesion is broken in the black zones.

- **Metal/concrete interfaces in reinforced concrete**
  In civil engineering, we are now testing the RCC model for the adhesive contact between steel and concrete in reinforced concrete. Pull-out tests of a steel shaft are being simulated in the framework of a joint project between LMA and ENIT in Tunisia (Karray et al, submitted). In that case, a variable friction coefficient (depending on the sliding displacement) has been used in order to take into account the wear of the surface which occurs during the sliding which seems to be quite significant with concrete. Another version of the way used to introduce friction has been used: $(1-\beta^2)$ instead

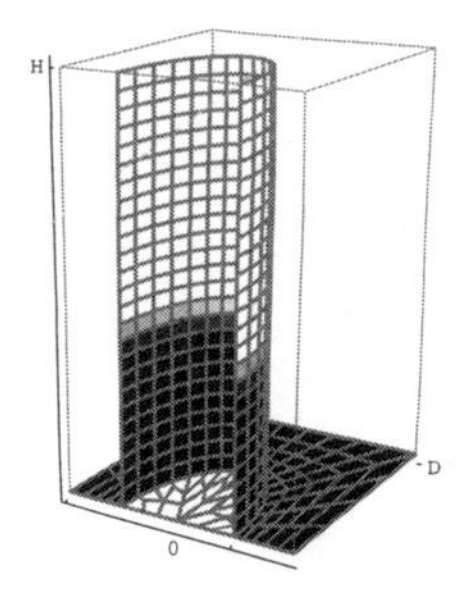

Figure 9.5. Interaction of a crack and a fiber/matrix interface

of $(1 - \beta)$. Results with $C_T = C_N$=16N/mm^3 , μ varying from 0.45 to 0.3 and f(β) = (1 - β^2) are given in Fig. 9.6.

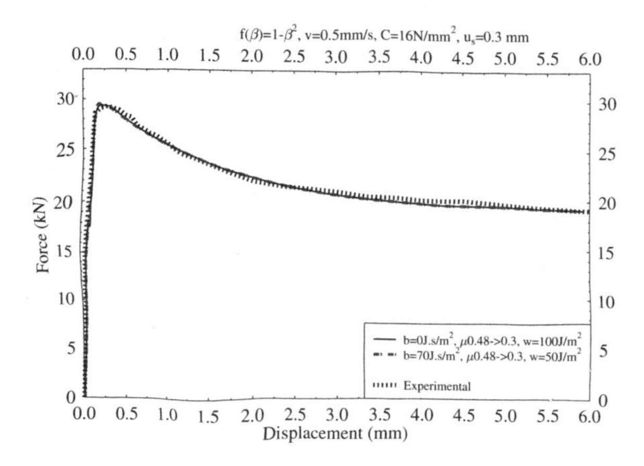

Figure 9.6. Simulation of a pull-out test on reinforced concrete

- **Delamination of a coated body**
  A simplified approach to the delamination of a coated body has been conducted to simulate the indentation of a thin Chrome layer adhering to a metal body (Raous et al, 2002).

- **Cohesive masonry**
  The RCC model has been used in (Jean et al, 2001, Acary, 2001) to simulate the behaviour of a cohesive dome. In Fig. 9.8, the deformation of a dome let on pillars and submitted to gravity is shown.

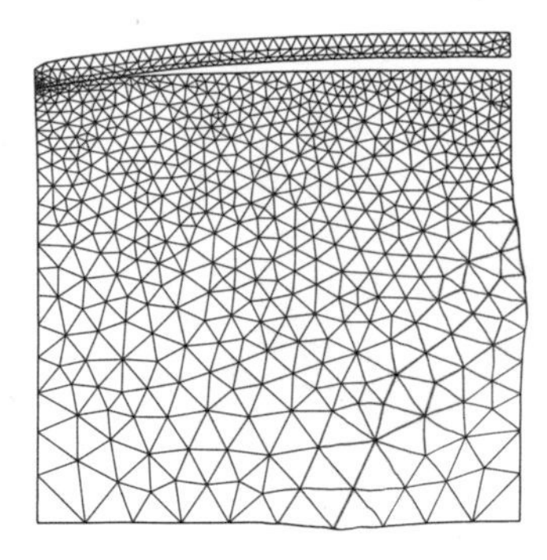

Figure 9.7. Indentation of a coated surface

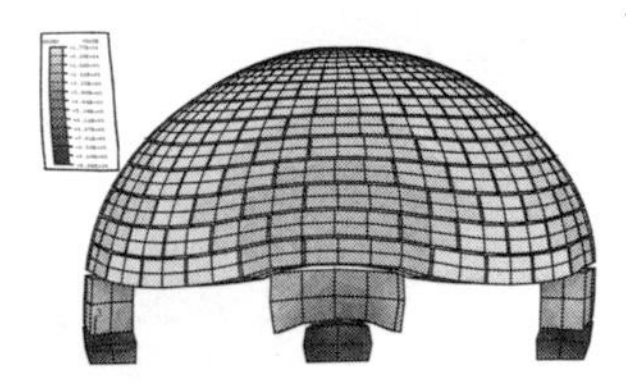

Figure 9.8. Behaviour of a cohesive dome

- **Wear in biomechanics**
  In order to simulate the wear occurring in biomaterials and especially the production of wear particles, H. Baudriller (Baudriller, 2003) has used the RCC model in a volumic sense. The adhesive model is used on all the interfaces between the finite elements. Fig 9.9 gives an example of production of fragments.

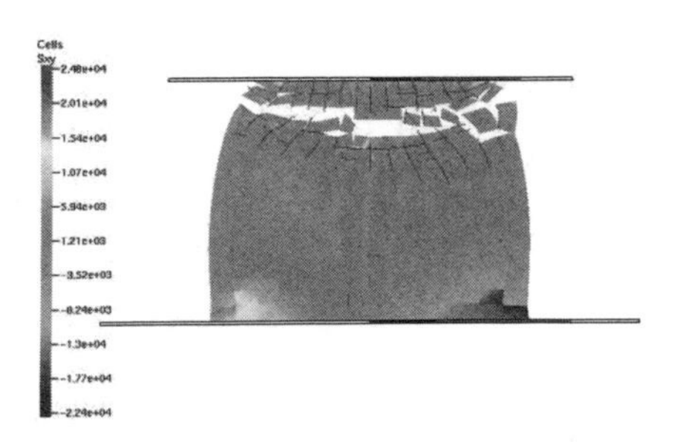

Figure 9.9. Fragmentation of a block

Acknowledgments

M. Cocou, J.-C. Latil, I. Rosu, in the group MMC at the LMA, L. Cangémi, Y. Monerie, during the preparation of their thesis, and M.A Karray (ENIT) have contributed to the works presented here. Thanks to V. Acary and H. Baudrier for providing figures 9.8 and 9.9.

References

Acary, V. (2001). Contribution la modélisation mécanique et numérique des édifices maçonnés . *Thesis*, Jean, M., supervisor, Université de la Méditerranée, Marseille.

Baudriller, H. (2003). Approche micro-mécanique et modélisation de l'usure : application aux biomatériaux des prothéses totales de hanche. *Thesis*, Chabrand, P., supervisor, Univ. de la Méditerranée, Marseille.

Alart, P., and Curnier, A. (1991). A generalised Newton method for contact problems with friction. *J. Mech. Théor. Appliq.*, (7): 67–82.

Bretelle, A.-S., Cocou, M., and Monerie, Y. (2001). Unilateral contact with adhesion and friction between two hyperelastic bodies. *Int. J. Eng. Sci.*, (39): 2015–2032.

Cangémi, L. (1997). Frottement et adhérence : modèle, traitement numérique et application à l'interface fibre/matrice, *Thesis*, Raous, M., supervisor, Université de la Méditerranée, Marseille.

Cocou, M., Pratt, E., and Raous, M. (1996). Formulation and approximation of quasistatic frictional contact. *Int. J. Eng. Sci.*, (34)7: 783–798.

Cocou, M., Cangémi, L., and Raous, M. (1997). Approximation results for a class of quasistatic contact problem including adhesion and friction. In Argoul, A., Frémond, M. and Nguyen, Q.S., editors, *Variation of domains and free-boundary problems in solids mechanics*, Solid Mechanics and its Applications, Number 66, Kluwer, 211–218.

Cocou, M., and Rocca, R. (2000). Existence results for unilateral quasistatic contact problems with friction and adhesion. *Math. Model. Num. Anal.*, (34)5: 981–1001.

Frémond, M. (1987). Adhérence des solides. *J. Méc. Thé. Appl.* (6)3, 383–407.

Frémond, M. (1988) Contact with adhesion. In Moreau, J. J. and Panagiotopoulos, P.D., editors, *Non Smooth Mechanics and Applications*, CISM Courses and lectures, Number 302, Springer, 177–221.

Chaboche, J.-L., Feyel, F., and Monerie, Y. (2001). Interface debonding model: a viscous regularization with a limited rate dependency. *Int. J. Solids Structures*, (38)18: 3127–3160.

, M. (1999). The Non Smooth Contact Dynamics method. *Computer Meth. Appl. Mech. and Engrg.*, (177)3-4: 235–257.

ean, M., Acary, V. and Monerie, Y. (2001). Non-smooth contact dynamics approach of cohesive materials. *Phil. Trans. R. Soc. Lond. A*, (359): 2497–2518.

Karray, M. A., Barbarin, S. and Raous, M. (submitted). Liaison béton-acier par un modèle d'interface couplant adhésion et frottement.

Latil, J.-C. and Raous, M. (1991). Module GYPTIS 1.0 - Contact unilatéral avec frottement en mécanique des structures. *Note interne LMA*, Number 132, CNRS.

Monerie, Y., Raous, M., Leroy, F.-H., Sudre, O., Feyel, F., and Chaboche, J.-L. (1998). Comparaison de lois d'interface fibre/matrice sur la base d'un modèle uni-axial d'expérience de micro-indentation. In Lamon, J. and Baptiste D., *XI J. Nat. Composites*, AMAC publisher, 565–574.

Monerie, Y., and Raous, M. (2000). A model coupling adhesion to friction for the interaction between a crack and a fibre-matrix interface. *Zeitschrift für Angewandte Mathematik und Mechanik*, (80): 205–209.

Monerie, Y. (2000) Fissuration des matériaux composites : rôle de l'interface fibre-matrice, *Thesis*, Raous, M., supervisor, Université de la Méditerranée, Marseille.

Monerie, Y., and Acary, V. (2001). Formulation dynamique d'un modèle de zone cohésive tridimensionnel couplant endommagement et frottement. *Rev. Eur. Elém. Finis*, (10): 489–504.

Moreau J. J. (1994). Some numerical methods in multi-body dynamics: application to granular media. *Eur. J. of Mechanics*, (A13): 93–114.

Moreau, J. J. (1988). *Unilateral contact and dry friction in finite freedom dynamics.* In Moreau, J. J. and Panagiotopoulos, P.D., editors, *Non Smooth Mechanics and Applications*, CISM Courses and lectures, Number 302, Springer, 1–82.

Moreau, J. J. (1999). Some basics of unilateral dynamics. In Pfeiffer, F. and Glocker, PH., editors, *Unilateral Multibody Contacts*, Solid Mechanics and its applications, Number 72, Kluwer, 1–14.

Raous, M. and Barbarin, S. (1992). Preconditioned conjugate gradient method for a unilateral problem with friction. In Curnier, A. , Editor, *Contact Mechanics*, Presses Polytechn. et Univ. Romandes, 423–432.

Raous, M., Cangémi, L., and Cocou, M. (1997). Un modèle couplant adhérence et frottement pour le contact unilatéral entre deux solides déformables. *C. R. Acad. Sci. Paris*, Série II b (329): 503–509.

Raous, M. (1999). Quasistatic Signorini problem with Coulomb friction and coupling to adhesion. In Wriggers, P. and Panagiotopoulos, P.D., editors, *New developments in contact problems*, CISM Courses and lectures, Number 383, Springer, 101–178.

Raous, M., Cangémi, L., and Cocou, M. (1999). Consistent n pling adhesion, friction and unilateral contact. *Comp. Meth. Mech. and Engrg.*, (177)3-4: 383–399.

Raous, M. and Monerie, Y. (2002). Unilateral contact, friction and ac sion: 3D cracks in composite materials. In Martins, J.A.C. and Mo teiro Marques M.D.P., eds, *Contact mechanics*, Kluwer, 333–346.

Raous, M., Monerie, Y. and Rosu, I. (2002). Dynamic formulation in contact mechanics: flutter instabilities and adhesion problems. Euromech 434 *Contact mechanics of coated bodies*, Moscow, May 2002.

Raous, M., Belloeil, V. and Rosu, I. (2004). Modélisation de l'adhésion par collage, *LCPC contract report*, LMA, 85 pages.

Talon, C. and Curnier, A. (2003) A model of adhesion coupled to contact and friction. *European Journal of Mechanics A/Solids*, (22): 545–565.

Vola, D., Pratt, E., Jean, M., and Raous, M. (1998). Consistent time discretization for dynamical frictional contact problems and complementarity techniques. *Rev. Eur. Eléments Finis* (7): 149–162.

Chapter 10

THE CLAUSIUS-DUHEM INEQUALITY, AN INTERESTING AND PRODUCTIVE INEQUALITY

Michel Frémond

Laboratoire Lagrange, LCPC, 58 boulevard Lefebvre, 75732 Paris Cedex 15, France

fremond@lcpc.fr

Abstract The Clausius-Duhem inequality which appears in continuum mechanics when combining the momentum balance law and the two laws of thermodynamics may be thought as an interesting inequality or as a productive relation to derive constitutive laws. We recall that it gathers most of the quantities which intervene when deriving constitutive laws. In some circumstances it may be much more than a summary: the duality pairings of the inequality give precious information on the quantities which are to be related in predictive theories and to be measured in experiments. Examples are given in various problems of solid mechanics.

Keywords: Clausius-Duhem inequality, constitutive laws, damage, phase change, shape memory alloys, collisions, fracture.

1. Introduction

When building a predictive theory in the continuum mechanics framework, an inequality, the Clausius-Duhem inequality, is derived. It sums up in only one formula most of the elements which intervene in the theory. In this point of view, the list of the quantities which are involve gives a precise idea on the scope and sophistication of both thermal and mechanical aspects of the theory. In some cases, it may be more than interesting, it may be productive by identifying quantities which are to be related in constitutive laws, and by suggesting useful experiments.

An interesting inequality

In classical continuum mechanics, by combining the momentum balance law, the two laws of thermodynamics and using the free energy, one gets an inequality

$$\sigma^d : D(\vec{U}) - \vec{Q} \cdot \overrightarrow{gradT} \geq 0,$$

where σ^d is the dissipative stress, $D(\vec{U})$ is the strain rate, T is the temperature and $T\vec{Q}$ is the heat flux vector, $\vec{Q}$ being the entropy flux vector. This inequality is called the Clausius-Duhem inequality. Scientists do not agree on the status to be given to this inequality. Instead of focusing on the inequality which is the subject of some discussions, let us remark that its expression involves two quantities,

- the mechanical dissipation

  $$\sigma^d : D(\vec{U}),$$

  which is the opposite of the power of the dissipative interior forces. This quantity results directly from the choice of the power of the interior forces;

- and the thermal dissipation

  $$-\vec{Q} \cdot \overrightarrow{gradT},$$

  which results from the first law of thermodynamics.

The inequality gathers quantities which are to be chosen by scientists or engineers to build a predictive theory of some thermo-mechanical phenomenon. It gives the quantities which are significant. Moreover it points out the ones which are to be related: it is wise to have experiments where the dissipative stress σ^d is plotted versus the strain rate $D(\vec{U})$, and to have experiments where the heat flux $T\vec{Q}$ is plotted versus the gradient of temperature $\overrightarrow{gradT}$. All the classical dissipative constitutive laws and the classical thermal Fourier law are obtained in this framework.

In terms of mathematics the various products are duality products between linear spaces, namely products between quantities describing the evolution $D(\vec{U})$, the thermal heterogeneity $\overrightarrow{gradT}$ and quantities describing the interior forces σ^d and $T\vec{Q}$. But this duality structure is not the only one. There are cases where a product is not a duality product between two linear spaces. We give an example down below.

We may remark that it is not the inequality which is the more important, but the expression of the two dissipations which are not fixed once for all, and contain information on the sophistication and pro... the predictive theories. In other words, the important element... Clausius-Duhem inequality are: the power of the interior forces wh... chosen when dealing with the principle of virtual power, and the t... mal power which is chosen when dealing with the energy balance. In th... sequel, we give various examples of such choices. We will see that the choice of the quantities to be related when building predictive theories is not always obvious. In such a situation, the information gathered in the Clausius-Duhem inequality is valuable for the development of theory and experiments.

3. Examples

We give examples of the various form of the inequality:

- inequalities of predictive theories for damage, phase change and shape memory alloys, contact with adhesion and friction, are examples of situation where it is easy to know what are the quantities to appear in the inequality. They show how various it may be;
- inequality for a shock in an adiabatic ideal fluid where the quantity which describes the evolution is unusual and is not an element of a linear space;
- inequality for collisions of two rigid balls, for collisions and fractures of solids where the quantity which describes the evolution is not easy to choose.

3.1 Damage

Damage of material results from microscopic motions creating microfractures and microcavities resulting in the decreasing of the material stiffness. In a predictive theory accounting for damage, the power of these motions has to be taken into account. Thus we assume that the power of interior forces depends on the damage rate, which is clearly related to microscopic motions, and on the gradient of damage rate accounting for local microscopic interactions. In this setting, the Clausius-Duhem inequality is

$$\sigma^d : D(\vec{U}) + B^d \frac{d\beta}{dt} + \vec{H}^d \cdot \overrightarrow{grad} \frac{d\beta}{dt} - \vec{Q} \cdot \overrightarrow{grad T} \geq 0.$$

In this formula the quantity $d\beta/dt$ where β is the volume fraction of microfractures and microcavities, accounts at the macroscopic level, for the velocity of the microscopic motions responsible for the damage. Quantities B and $\vec{H}$ are work and work flux vectors whose physical meaning y an equation of motion in the same way the physical meaning tress σ is given by an equation of motion:

$$\sigma\vec{N},$$

s the force applied by the exterior of a domain with outward normal vector $\vec{N}$ onto this domain and, in the same way

$$\vec{H}.\vec{N},$$

is the amount of work provided, without macroscopic motion, to the domain by the exterior.

The constitutive laws relate the forces σ^d, B^d, $\vec{H}^d$ and $\vec{Q}$ to the quantities describing the evolution $D(\vec{U})$, $d\beta/dt$, $\overrightarrow{grad}(d\beta/dt)$ and $\overrightarrow{grad}T$. The numerical and theoretical results show that the theory describes correctly the damage effects at the macroscopic level, (Frémond and Nedjar, 1993), (Nedjar, 1995). Experiments on samples give properties of B^d and σ^d. Experiments on structures are needed for the properties of $\vec{H}^d$ because they depend on the inhomogeneity of the damage. Of course the thermal effects may be coupled to the damage theory, (Huon et al., 2003). The theory which is based on the distinction between macroscopic and microscopic motions, shows what occurs when a macroscopic motions becomes microscopic: the motion vanishes but its damaging effects remain, (Frémond, 2002), (Bonetti and Frémond, 2004).

3.2 Phase change. Shape memory alloys

Liquid-solid phase change, for instance water-ice phase change, involves microscopic motions. At the macroscopic level, the velocities of these motions may be represented by the liquid water volume fraction velocity $d\beta/dt$. The inequality is

$$B^d\frac{d\beta}{dt} + \vec{H}^d \cdot \overrightarrow{grad}\frac{d\beta}{dt} - \vec{Q} \cdot \overrightarrow{grad}T \geq 0.$$

The elements of this theory allow to describe ice-water phase change, soil freezing with many engineering applications in road and railway track winter maintenance, (Frémond, 2001).

The striking properties of shape memory alloys result from solid-solid phase change. Microscopic motions occur during those phase changes. At the macroscopic level, the volume fractions of the different phases account for the microscopic evolution. Let $\vec{\beta}$ be the vector of the volume fractions of the different phases, the inequality

$$\sigma^d : D(\vec{U}) + \vec{B}^d \cdot \frac{d\vec{\beta}}{dt} + H^d : \overrightarrow{grad}\frac{d\vec{\beta}}{dt} - \vec{Q} \cdot \overrightarrow{grad}T \geq 0,$$

relates the interior forces to the quantities which characterize the evolution. Simple constitutive laws give the main physical properties of shape memory alloys, (Frémond, 2001). By simple constitutive laws, we mean linear constitutive laws derived from quadratic pseudo-potentials of dissipation introduced by Jean Jacques Moreau, (Moreau, 1970) and, reactions to internal constraints, if any, which cannot be avoided. Example of internal constraints are, (Frémond, 2001), (Frémond, 2001): volume fractions have values between 0 and 1, solids cannot interpenetrate, voids cannot appear,...

3.3 Contact with adhesion

Two solids in contact may have microscopic interactions, for instance glue fibers may tighten the two solids. A predictive theory of the contact has to take into account the power of these microscopic interactions and motions. Let β be the active glue fibers surface fraction. The microscopic velocities may be schematized by the velocity $d\beta/dt$. A possible Clausius-Duhem inequality is

$$\vec{R}^d \cdot (\vec{U}_2 - \vec{U}_1) + B^d \frac{d\beta}{dt} + \vec{H}^d \cdot \overrightarrow{grad}_s \frac{d\beta}{dt} - Q_s (T_2 - T_1) \geq 0,$$

where the $\vec{U}_i$ are the velocities of the two contacting solids and the T_i are their temperatures, $\vec{R}^d$ is the interaction force, B^d and $\vec{H}^d$ are the work and surface work vectors within the glue; Q_s is the amount of entropy received by the contact surface, i.e., the difference between the entropy received from side 2 and the entropy received from side 1; $\overrightarrow{grad}_s$ is the surface gradient. The constitutive laws take into account adhesion, Coulomb friction, and other sophisticated contact phenomena involving or not thermal effects, for instance in composite materials, (Point, 1988), (Ghidouche and Point, 1988), (Frémond, 2001), (Raous et al., 1999), (Bruneaux, 2004).

3.4 Shock in an adiabatic ideal fluid. A product which is not a duality pairing

For the sake of simplicity, let us consider a one dimensional adiabatic ideal fluid flow with a shock (see for instance, (Germain, 1973)). The equations of the shock are:

- the mass balance

$$\rho_1(U_1 - W) = \rho_2(U_2 - W) = m,$$

where the $\rho_i, i = 1, 2$ are the densities of the fluid with velocities U_i on the sides 1 and 2 of the discontinuity surface with velocity W. The mass balance defines the mass flow m;

- the momentum balance

$$m[U] = [-p],$$

where $[A] = A_2 - A_1$ denotes the discontinuity of quantity A, and p is the pressure given by the derivative of the fluid free energy $\Psi(\rho, T)$ depending on the density ρ and the temperature T

$$p = -\frac{\partial \Psi}{\partial (1/\rho)},$$

or by using the mass balance

$$m^2[\frac{1}{\rho}] = [-p];$$

- the first law

$$m([e] + \frac{p_1 + p_2}{2}[\frac{1}{\rho}]) = 0,$$

where $e(\rho, T)$ is the internal energy;

- and the second law

$$m[s] \geq 0,$$

where $s(\rho, T)$ is the entropy.

By an easy computation, the second law gives

$$m[s] = \xi F^d \geq 0, \tag{10.1}$$

with

$$F^d = F^d(\rho_1, T_1, \rho_2, T_2, |m|) = \frac{-2\,|m|}{T_1 + T_2}\left([\Psi] + \frac{s_1 + s_2}{2}[T] + \frac{p_1 + p_2}{2}[\frac{1}{\rho}]\right), \tag{10.2}$$

and ξ is a selection of the graph $S(m)$,

$$\xi \in S(m),$$

defined by

$$S(x) = \begin{cases} if\ x > 1,\ S(x) = \{1\}, \\ if\ x = 0,\ S(0) = [-1, 1], \\ if\ x < 1,\ S(x) = \{-1\}. \end{cases}$$

By abuse of notation, we denote ξ, the sign of m, by $sg(m)$. The constitutive law is a relation between $sg(m)$ and F^d such that inequality (10.1) is satisfied. Let us remark that formula (10.2) gives

$$\begin{aligned} F^d > 0 &\Longrightarrow |m| \neq 0, \\ F^d = 0 &\Longrightarrow |m| \in [0, \infty), \\ F^d < 0 &\Longrightarrow |m| \neq 0. \end{aligned}$$

Thus it is obvious that there is one and only one possible constitutive law

$$\begin{aligned} F^d > 0 &\Longrightarrow sg(m) = 1, \\ F^d = 0 &\Longrightarrow sg(m) \in [-1, 1], \\ F^d < 0 &\Longrightarrow sg(m) = -1, \end{aligned}$$

which may be written

$$F^d \in \partial I(sg(m)), \ or \ sg(m) \in \partial I^*(F^d), \tag{10.3}$$

where $\partial I(sg(m))$ is the subdifferential set of the indicator function I of the segment $[-1, 1]$ and, ∂I^* is the subdifferential set of the dual function I^* of I. They are elements of convex functions theory developed by Jean Jacques Moreau, (Moreau, 1966).

We conclude that the adequate quantity to describe the evolution is the sign of m: $sg(m)$. Clearly, being an element of a graph, it is not an element of a linear space. The effect of the constitutive law (10.3) is only to discriminate between the two possible directions of the flow on a discontinuity line. Then the equations for a shock are

$$\rho_1(U_1 - W) = \rho_2(U_2 - W) = sg(m)\,|m|\,, \tag{10.4}$$

$$|m|^2\,[\frac{1}{\rho}] = [-p(\rho, T)], \tag{10.5}$$

$$|m|\,([e(\rho, T)] + \frac{p(\rho_1, T_1) + p(\rho_2, T_2)}{2}[\frac{1}{\rho}]) = 0, \tag{10.6}$$

$$sg(m) \in \partial I^*(F^d(\rho_1, T_1, \rho_2, T_2, |m|)). \tag{10.7}$$

A classical problem is to compute the quantities ρ_2, T_2, U_2 on the side 2 and the direction of the flow $sg(m)$ depending on the quantities on the other side ρ_1, T_1, U_1 and on the mass outflow $|m|$. The solution is straightforward:

- when $|m| \neq 0$, equations (10.5) and (10.6) give ρ_2 and T_2. Then constitutive law (10.7) gives $sg(m)$, and the velocities U_2 and W are given by (10.4);

- when $|m| = 0$, the discontinuity surface is a contact surface of two identical fluids with equal pressure (there is only equation (10.5) expressing that the pressures of the two fluids are equal).

This example is very specific but it shows that the theory exhibits the pertinent quantity, $sg(m)$, to describe the evolution, (Frémond, 2001).

3.5 Collisions of solids and thermal effects

Consider two balls moving on a line. Both of them are schematized by points with velocity U_i, $i = 1, 2$. The system made of the two balls is *deformable* because the distance of the two balls may change, (Frémond, 2001). The deformation of the system may be measured by the velocity of deformation

$$D(U) = U_2 - U_1,$$

where $U = (U_1, U_2)$. We assume that the collisions are instantaneous. Thus there are velocities U^- before a collision and U^+ after. Because the system is deformable, there are interior forces: in this example, it is an interior percussion P^{int}. We account also for the thermal phenomena due to collisions with temperatures which are also discontinuous: T_i^- are the temperatures before the collision and T_i^+ are the temperatures after. The ball 1 receives from ball 2, heat impulse $T_1^+ B_{12}^+ + T_1^- B_{12}^-$ and entropy impulse $B_{12} = B_{12}^+ + B_{12}^-$. Note that besides these interior thermal impulses, the two balls may exchange thermal impulses with the exterior which appear in the energy balance of the system. Let us define the difference of the average temperatures

$$\delta\bar{T} = \frac{T_2^- + T_2^+}{2} - \frac{T_1^- + T_1^+}{2},$$

which is the non-smooth equivalent of the gradient of temperature. The Clausius-Duhem inequality which results from the two laws of thermodynamics is

$$P^{int} D\left(\frac{U^+ + U^-}{2}\right) - \delta\bar{T}\frac{B_{21} - B_{12}}{2} \geq 0. \tag{10.8}$$

To describe the collisions of the two balls we need constitutive laws for the percussion P^{int} and for the interior entropy impulse $B_{21} - B_{12}$. In this perspective the inequality is useful: it is clear that the percussion is going to depend on the velocity of deformation but it is not easy to guess that it depends actually on the average relative velocities before and after the collision. Thus the inequality is a guide for theory and experiments. The percussion P^{int} is given by the equation of motion

$$m_1\left(U_1^+ - U_1^-\right) = P^{int},$$

where m_1 is the mass of ball 1. In experiments, velocities of the b. U^+ and U^- are measured. Thus percussion P^{int} may be plotted vers the average velocity of deformation $D((U^+ + U^-)/2)$ in order to get constitutive laws. For collisions of a small solid with a marble table this procedure gives excellent results: a Coulomb type constitutive law is identified, (Cholet, 1998), (Dimnet, 2001). Plotting the percussion versus either the velocity of deformation before collision or the velocity of deformation after, gives actually very bad results: the plots show that there is no functional dependence between these quantities. In this situation the inequality is useful because it is not clear what are the quantities which are to be related and measured.

Notice that we do not infer that P^{int} depends only on $D(U^+ + U^-)$. It may also depend on other quantities χ but in any case it has to satisfy inequality (10.8). For instance, P^{int} involves a reaction to the internal constraint of impenetrability of the two balls

$$D(U^+) = U_2^+ - U_1^+ \geq 0. \tag{10.9}$$

This percussion is

$$P^{reac} \in \partial I_+(D(U^+), \tag{10.10}$$

where ∂I_+ is the subdifferential set of the indicator function of $\mathbb{R}^+$. It depends on $D(U^+) = D(U^+ + U^-) - D(U^-)$. Thus P^{int} depends on $\chi = D(U^-)$ and

$$P^{int} = P^{int}\left(D\left(U^+ + U^-\right), \chi\right) = P^{reac} + P^{dis}\left(D\left(U^+ + U^-\right)\right), \tag{10.11}$$

where $P^{dis}(D(U^+ + U^-))$ takes into account all the other dissipative properties of the colliding balls.

Concerning the thermal results one may think that it is rather difficult to measure the temperature discontinuities. But experimental results are now available: measurements of temperatures during collisions of steel balls with steel planes show jumps of temperature up to 7^oC in less than a tenth of a second, (Pron, 2000). Thermal measurements are actually an experimental tool to improve the determination of collisions constitutive laws.

The idea that systems made of solids either rigid or deformable and that systems made of solids and fluids, are deformable is productive, (Cholet, 1998), (Dimnet, 2001), (Frémond, 2001), (Frémond et al., 2003). The related inequalities contain different works involving different velocities of deformations of the system and of the elements of the system.

Collisions and fractures of solids

When a plate falls on the floor, it breaks. Consider a rock avalanching a mountain slope and colliding a concrete protective wall: depending on the circumstances, the rock and the wall break, or only one of them breaks or none of them breaks, (Frémond, 2002a), (Bonetti and Frémond, 2004a), (Bonetti and Frémond, 2004b), (Bonetti and Frémond, 2004c). Dealing with this problem at the macroscopic or engineering level, we assume the collisions to be instantaneous: at collision time t, the velocity field is discontinuous with respect to time: $\vec{U}^-(\vec{x},t)$ is the smooth velocity field before collision and $\vec{U}^+(\vec{x},t)$ is the velocity field after collision. A fracture is characterized by a spatial discontinuity of $\vec{U}^+(\vec{x},t)$. The predictive theory introduces new interior forces: percussion stresses Σ together with surface percussion $\vec{R}$ on the fractures. For the sake of simplicity, we assume that a solid collides an indestructible plane. At collision time, the solid occupies domain Ω and impacts the plane on part $\partial\Omega_0$ of its boundary. The fractures are unknown surfaces Γ. The inequalities are

$$\Sigma : D(\frac{\vec{U}^+ + \vec{U}^-}{2}) \geq 0,$$

in the volume $\Omega\backslash\Gamma$;

$$-\vec{R} \cdot \left[\frac{\vec{U}^+ + \vec{U}^-}{2}\right] \geq 0, \tag{10.12}$$

on the fractures Γ, where $[A]$ is the spatial discontinuity of A on the oriented fracture;

$$\vec{R} \cdot \frac{\vec{U}^+ + \vec{U}^-}{2} \geq 0,$$

on the contact surface $\partial\Omega_0$.

The three inequalities concern: the volume where the percussion stress is related to the average strain rate, the fracture where the percussion is related to the average discontinuity of velocity and the contact surface with the plane obstacle where the percussion, the action of the solid on the obstacle, is related to the average velocity. This average velocity becomes a spatial discontinuity if the velocity of the obstacle, which is zero in this example, is taken into account. By introducing the obstacle in the system and taking into account its velocity $\vec{U}_{obs}$, the inequality on the contact surfaces has the same structure that the inequality on fractures, for instance inequality (10.12)

$$-\vec{R} \cdot \left[\frac{\vec{U}^+ + \vec{U}^-}{2}\right] \geq 0,$$

with

$$\left[\vec{U}\right] = \vec{U}_{obs} - \vec{U}.$$

This remark is useful both for choosing constitutive laws which involve impenetrability condition on fractures, and for mathematics which deal easily with discontinuities of velocity and not so easily with traces, (Bonetti and Frémond, 2004a), (Bonetti and Frémond, 2004b), (Bonetti and Frémond, 2004c).

4. Conclusion

The Clausius-Duhem inequality is interesting and useful because it gathers physical information which has been introduced in the power of the interior forces and in the first law of thermodynamics. It is productive in situations where it is not clear and precise what are the pertinent quantities to relate in constitutive laws.

Let us also recall that Clausius-Duhem is not very restricting and that it leaves marvelous opportunities to fancy and invention of scientists and engineers to build constitutive laws.

References

E. Bonetti, M. Frémond, 2004, Damage theory: microscopic effects of vanishing macroscopic motions, Comp. Appl. Math., 22 (3), 1-21.

E. Bonetti, M. Frémond, 2004a, Collisions and fractures, to appear in the Vietnam Journal of Mathematics.

E. Bonetti, M. Frémond, 2004b, Collisions and fractures: a model in SBD, Rend. Mat. Acc. Lincei, 9 (15), 47-57.

E. Bonetti and M. Frémond, 2004c, Collisions and fractures: a 1-D theory. How to tear off a chandelier from the ceiling, Journal of Elasticity, 74 (1), 47-66.

M. A. Bruneaux, 2004, Durabilité des assemblages collés : modélisation mécanique et physico-chimique, thèse, Ecole nationale des Ponts et Chaussées, Paris.

C. Cholet, 1998, Chocs de solides rigides, thèse, Université Pierre et Marie Curie, Paris.

E. Dimnet, 2001, Collisions de solides déformables, thèse, Ecole nationale des Ponts et Chaussées, Paris.

M. Frémond, 2001, Non-smooth Thermomechanics, Springer-Verlag, Heidelberg.

M. Frémond, 2001, Internal constraints in mechanics, Philosophical Transactions of the Royal Society, vol 359, n° 1789, 2309-2326.

M. Frémond, 2002, Damage theory. A macroscopic motion vanishes but its effects remain, *Comp. Appl. Math.*, 21 (2), 1-14.

M. Frémond, 2002a, Collisions and damage, Tendencias em Matematica Aplicada e Computacional, Vol 3, n°1, SBMAC, Sao Carlos-SP, ISBN 85-86883-06-9, Brazil.

M. Frémond, R. Gormaz, J. San martin, 2003, Collision of a solid with an incompressible fluid, Theoretical and Computational Fluid Dynamics, 16, 405-420.

M. Frémond, B. Nedjar, 1993, Endommagement et principe des puissances virtuelles, C. R. Acad. Sci., Paris, 317, II, n°7, 857-864.

P. Germain, 1973, Mécanique des milieux continus, Masson, Paris.

H. Ghidouche, N. Point, 1988, Unilateral contact with adherence, in Free boundary problems: Theory and application, K. H. Hoffmann, J. Spreckels, eds, Pittman, Longman, Harlow.

V. Huon, B. Cousin, O. Maisonneuve, 2003, Study of thermal and kinematic phenomena associated with quasi-static deformation and damage process of some concretes, in Novel approaches in civil engineering, M. Frémond, F. Maceri, eds. Springer, Heidelberg, 187-201.

J. J. Moreau, 1966, Fonctionnelles convexes, Séminaire sur les équations aux dérivées partielles, Collège de France, and 2003, Dipartimento di Ingegneria Civile, Tor Vergata University, Roma.

J. J. Moreau, 1970, Sur les lois de frottement, de viscosité et de plasticité, C. R. Acad. Sci., Paris, 271, 608-611.

B. Nedjar, 1995, Mécanique de l'endommagement. Théorie du premier gradient et application au béton, thèse, Ecole nationale des Ponts et Chaussées, Paris.

N. Point, 1988, Unilateral contact with adhesion, Math. Methods Appl. Sci., 10, 367-381.

H. Pron, 2000, Application des effets photothermiques et thermomécaniques à l'analyse des contraintes appliquées et résiduelles, thèse, Université de Reims Champagne-Ardenne.

M. Raous, L. Cangemi, M. Cocu, 1999, A consistent model coupling adhesion, friction and unilateral contact, Computer Methods in Applied Mechanics and Engineering, 177, n°3-4, 383-399.

Chapter 11

UNILATERAL CRACK IDENTIFICATION

Elastodynamic data, BEM and filter algorithm

Georgios E. Stavroulakis

*University of Ioannina, Dept. of Applied Mathematics and Mechanics, GR-73100 Ioannina, Greece**

gestavr@cc.uoi.gr

Marek Engelhardt
and Heinz Antes

Technical University of Braunshweig, Department of Civil Engineering, Institute of Applied Mechanics, Braunschweig, Germany†

marek.engelhardt@tu-bs.de

heinz.antes@tu-bs.de

Abstract Wave propagation and reflection from unilateral cracks in two-dimensional elasticity can be numerically modeled by suitable boundary element methods and nonsmooth mechanics techniques. This tool is used, in connection with Kalman-filter based optimization, for the study of the inverse crack identification problem.

Keywords: unilateral contact, cracks, boundary elements, inverse problems, Kalman filters

Introduction

Certain classes of inverse problems are related to optimal shape design of structures. This is the case of the crack identification problems

*Partial funding provided by the Greek-German Cooperation program IKYDA.

†Partial funding provided by the German Research Foundation DFG and the German-Greek Cooperation program IKYDA.

discussed here. Unilateral and more general nonsmooth effects of the mechanical problem, make the formulation and the solution of inverse problems a considerably more complicated task. Nonsmooth effects are caused by partial or total closure of cracks and related frictional effects, which are neglected in the majority of publications in fracture mechanics and inverse analysis.

The problem of unilateral crack identification is studied numerically in this paper. Previous publications of our group have been restricted either to static problems or to dynamic problems for relatively simple geometries (the so-called *impact-echo* nondestructive evaluation test), see (Stavroulakis, 2000, Stavroulakis, 1999). The mechanical problem of wave propagation and reflection from unilaterally working cracks has been solved numerically by means of boundary element techniques combined with a linear complementarity solver (we call it briefly *BEM-LCP* approach, see (Antes and Panagiotopoulos, 1992, Stavroulakis, Antes and Panagiotopoulos, 1999)). The inverse problem is solved by a filter-driven optimization algorithm, analogously to our previous work in statics presented in (Stavroulakis and Antes (2000)). Alternative methods to solve the inverse problem include *neural networks* and other *soft computing* tools which have been presented elsewhere (Stavroulakis et al., 2004). It should be emphasized here that classical optimization is the less effective method to solve crack identification problems of this kind and often fails to provide results, cf. (Stavroulakis, 2000).

1. Mechanical Modeling

Let us consider a two-dimensional elastodynamical problem for a linearly elastic medium. Using *Betti*'s reciprocal equation in statics, or it's generalization in dynamics, the so-called *Graffi*'s lemma, and the explicit solution of the mechanical problem for unitary boundary loads (impulses) in the form of the fundamental solutions, one formulates the boundary integral equation:

$$c_{ij}(\boldsymbol{\xi})u_j(\boldsymbol{\xi},t) = \int_\Gamma t_i(\mathbf{x},\tau) * u_{ij}^*(\mathbf{x},\boldsymbol{\xi},t,\tau)d\Gamma - \oint_\Gamma t_{ij}^*(\mathbf{x},\boldsymbol{\xi},t,\tau) * u_i(\mathbf{x},\tau)d\Gamma \qquad (11.1)$$

Here volumic forces and initial conditions are not considered, for simplicity. Furthermore c_{ij} is an integration constant, C in the second integral denotes the *Cauchy principal value*, $\beta^*(\mathbf{x},\boldsymbol{\xi},t,\tau), \beta = u_{ij}, t_{ij}$ denotes the fundamental solution which represents at point $\mathbf{x}$ and time-instant t the solution of the homogeneous equation due to an impulse at the source point $\boldsymbol{\xi}$ applied at time τ and $*$ denotes the convolution operator.

This relation is the basis for the construction of the boundary element method for the case of classical external and internal boundaries (holes). For cracks the equations (11.1) written for the points lying on the two adjacent sides of the crack, Γ^+ and Γ^- of Figure 11.1, are linearly dependent.

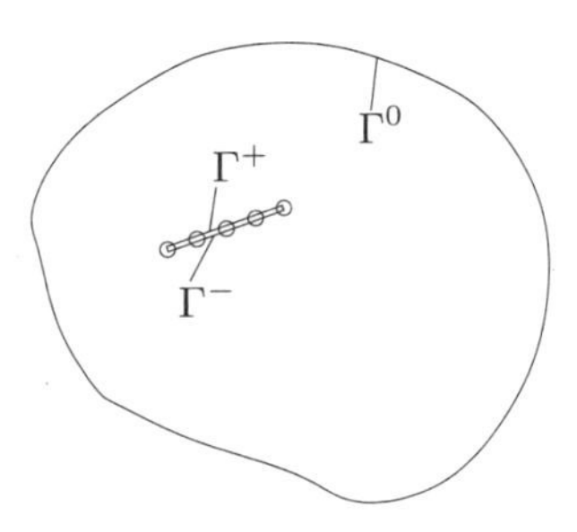

Figure 11.1. Discretization of a crack with similar discretization points

One possibility to overcome this problem is to use another boundary equation at the second face of the crack. This leads to a *hypersingular boundary integral equations*, where instead of the displacements u_j one uses the boundary tractions t_j as the basis of the calculations:

$$t_j(\boldsymbol{\xi}, t) = \int_\Gamma t_i * d_{ij}^* d\Gamma - \int_\Gamma s_{ij}^* * u_i d\Gamma \tag{11.2}$$

1.1 Boundary Element Method

After appropriate discretization of boundary values with boundary elements and a suitable time-discretization, e.g. by *point collocation*, one gets for each time step n a procedure which leads to a recursive formula for the discrete elastic boundary equations in time domain:

$$\mathbf{C}^1 \mathbf{y}^n = \mathbf{D}^1 \bar{\mathbf{y}}^n + \mathbf{R}^n \tag{11.3}$$

Here vector $\mathbf{y}^n$ denotes the unknowns and vector $\bar{\mathbf{y}}^n$ the given boundary quantities, which are boundary displacements $\mathbf{u}$ and tractions $\mathbf{t}$ at the corresponding part of the boundary Γ, for the time instant n:

$$\mathbf{y}^n = \begin{bmatrix} \mathbf{t}_{\Gamma_u} \\ \mathbf{u}_{\Gamma_t} \end{bmatrix}^n, \qquad \bar{\mathbf{y}}^n = \begin{bmatrix} \bar{\mathbf{u}}_{\Gamma_u} \\ \bar{\mathbf{t}}_{\Gamma_t} \end{bmatrix}^n \tag{11.4}$$

In the matrices $\mathbf{C}^1$ and $\mathbf{D}^1$ of equation (11.3) are included the stress and displacement fundamental solutions of the first (due to the translation) time step. In fact, the fundamental solutions are included in matrices $\mathbf{U}$ and $\check{\mathbf{T}}$, which in turn are rearranged according to the boundary conditions of the studied problem in order to form the matrices of equation (11.3):

$$\mathbf{C}^1 = \left[-\mathbf{U}_{\Gamma_u} | \, \check{\mathbf{T}}_{\Gamma_t}\right]^1, \qquad \mathbf{D}^1 = \left[-\check{\mathbf{T}}_{\Gamma_u} | \, \mathbf{U}_{\Gamma_t}\right]^1 \tag{11.5}$$

Finally, the vector $\mathbf{R}^n$ introduces in the equation (11.3) the influence of all previous time steps, up to the time step n, as follows:

$$\mathbf{R}^n = \sum_{k=2}^{\bar{n}} \left(\mathbf{U}^k \mathbf{t}^{n-k+1} - \mathbf{T}^k \mathbf{u}^{n-k+1}\right) \tag{11.6}$$

1.2 Unilateral Contact for Cracks

The classical unilateral contact behavior describes, at each point of a unilateral boundary, a highly nonlinear relation between a boundary traction normal to the considered boundary t_n and the corresponding displacement u_n: both quantities obey to certain inequalities, expressing the inadmissibility of tension stresses and penetration, respectively, and are complementary quantities, which means that at most one of them may attain a nonzero value. Schematically we have the *complementarity relations*

$$\Delta\mathbf{u}_n \geq \mathbf{0}, \mathbf{t}_n \geq \mathbf{0}, \Delta\mathbf{u}_n\mathbf{t}_n = \mathbf{0} \tag{11.7}$$

This formulation, for the frictionless unilateral contact problem, and analogous relations for frictional models, are coupled with the boundary equations outlined in the previous section and lead to the formulation of a *linear complementarity problem* within each time step. This is solved here, for example, by *Lemke*'s algorithm.

2. Inverse Problem

2.1 Least Output Error Formulation

Let us assume that our mechanical model is suitably parametrized, such that cracks, holes and other defects are described by the values of the parameters $\mathbf{z}$. This model provides, for each value of the parameter vector, an *estimate* of the solution. Obviously some of the boundary values of this estimate, let us say displacements, will be compared with the *measurement*. This leads to a least square formulation of the inverse problem, which requires the solution of the following problem:

$$F := \sum_{i=1}^{m} \left(u_{i,\text{measurement}} - u_{i,\text{estimate}}(\mathbf{z})\right)^2 \tag{11.8}$$

2.2 Filter Algorithm

Let us consider an extended *Kalman* filter algorithm for the estimation of a nonlinear dynamical system. The variables to be estimated $\mathbf{z}_k$ correspond to the parameters of the unknown defect, which do not change during the considered dynamical test loading. In other words one has a stationary, time-invariant process. Therefore the transitions matrix $\mathbf{\Phi}_k$, which describes the transition of the system from time step t_k to time step t_{k+1}, is the unity matrix $\mathbf{I}$. Accordingly we set the error in the system $\mathbf{w}_k$ equal to zero. Finally the equation of the considered process reads:

$$\mathbf{z}_{k+1} = \mathbf{z}_k \; . \tag{11.9}$$

The measurement vector $\mathbf{u}_k$ corresponds to the displacements at the external boundary of the plate. They depend nonlinearly on the defect parameters through the mechanical model, which is done by the previously outlined boundary element model

$$\mathbf{u}_k = \mathbf{h}(\mathbf{z}_k) + \mathbf{v}_k \tag{11.10}$$

where, $\mathbf{v}$ is the measurement error. The application of Kalman filter technique for the solution of the defect identification problem requires within iteration step the minimization of the linear approximation of relation (11.10). The theory of extended Kalman filter is used, where the linearized measurement matrix $\mathbf{H}_k$ is defined within each iteration step by means of a sensitivity analysis. The vector $\mathbf{z}_k$ is covered with the experimentally obtained (or calculated) displacements of each discretization point of the external boundary for each considered time step.

The steps of the modified Kalman filter identification algorithm for the solution of the inverse, defect identification problem, are:

Kalman filter identification

0. Initialization ($k = 0$)
 $\mathbf{\Phi}_k = \mathbf{I}$, $\hat{\mathbf{z}}_0^-$ and $\mathbf{P}_0^-$ take random initial values
1. Calculation of displacements $\hat{\mathbf{u}}_k$ with BEM
2. Calculation of $\mathbf{H}_k$ with a sensitivity analysis of $\hat{\mathbf{u}}_k$
3. Correction of the error covariance matrix
 $\mathbf{P}_k^{-1} = \left(\mathbf{P}_k^-\right)^{-1} + \mathbf{H}_k^T \mathbf{R}_k^{-1} \mathbf{H}_k$
4. Calculation of Kalman gain matrix,

for the optimization of the defect parameter estimate $\hat{\mathbf{z}}_k$
$\mathbf{K}_k = \mathbf{P}_k \mathbf{H}_k^T \mathbf{R}_k^{-1}$

5. Update of estimated parameters $\hat{\mathbf{z}}_k$
$\hat{\mathbf{z}}_{k+1} = \hat{\mathbf{z}}_k + \mathbf{K}_k \left(\mathbf{u}_k - \hat{\mathbf{u}}_k^- \right)$
6. Update of error covariance matrix
$\mathbf{P}_{k+1}^- = \mathbf{P}_k + \mathbf{Q}_k$
7. Continue to next step, set $k = k + 1$ and continue with step 1.

3. Numerical Examples

3.1 Crack identification in a quadrilateral plate

Let us consider the quadrilateral example of Figure 11.2. Results for more complicated, $L-$ and $U-$ shaped plates, are presented in (Engelhardt, 2004). For the constitutive modeling of the elastic material, the following data are used: elastic constant $E = 1 \cdot 10^6$ and Poisson's ratio $\nu = 0.3$. The steps of the algorithm for the calculation of the center

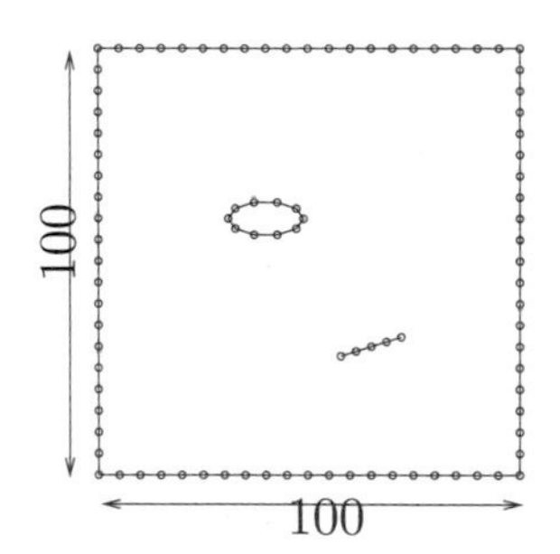

Figure 11.2. Discretization of boundary and defects

coordinates, the length and the direction of a rectilinear crack described by $\mathbf{z} = [50\ 50\ 10\ 0]$, starting from the initial estimate $\hat{\mathbf{z}}_0^- = [70\ 20\ 20\ 1]$, are documented in Figure 11.3.

3.2 The Effect of Unilateral Behavior

Unilateral contact along the crack is activated if one considers a compressive Heaviside loading at the upper boundary of the plate shown in Figure 11.2 As it is shown in Figure 11.4 the algorithm does not converge. This should be compared with the results of the same loading in the opposite direction (tension), which is documented in Figure 11.5. Better results can be obtained if one considers a non-horizontal crack,

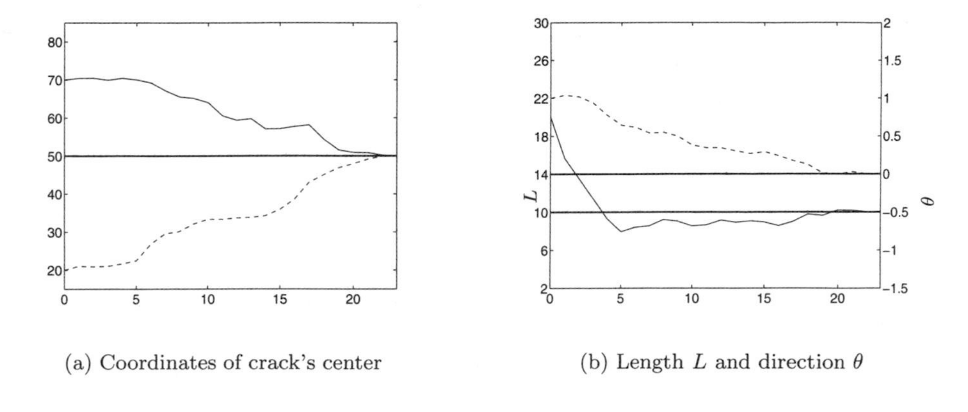

(a) Coordinates of crack's center

(b) Length L and direction θ

Figure 11.3. Convergence history, from starting point [70 20 20 1]

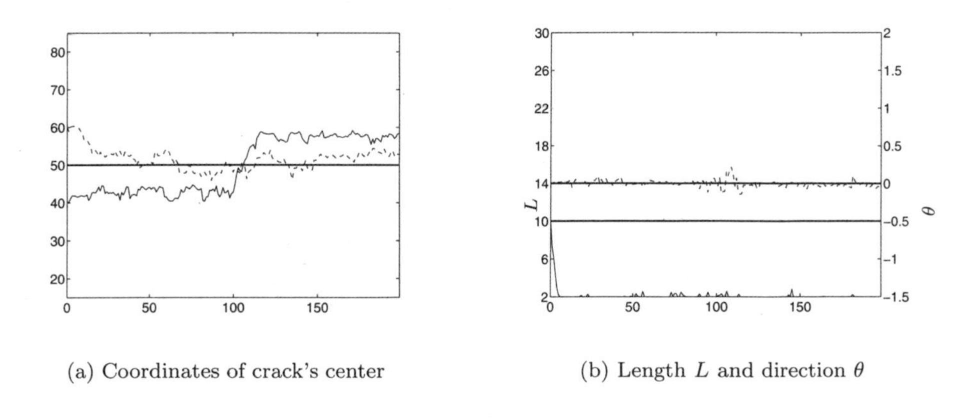

(a) Coordinates of crack's center

(b) Length L and direction θ

Figure 11.4. Compressive loading, iterations from starting point [40 60 10 0]

where the activation of the contact mechanism by the incoming wave is less possible, or by using of different wave-type impulsive loadings.

3.3 The Effect of Loading

Beyond the Heaviside loading, we considered other types of impulsive loadings: a modified Ricker-impulse

$$R^*(t) = e^{-\frac{(t+1)^2}{2}} \left((t+1)^2 - 1\right) \tag{11.11}$$

and sinusoidal loadings.
The steps of the algorithm for a Ricker and a sinusoidal impulse are shown in Figures 11.6 and 11.7, respectively. Initial estimate of the solu-

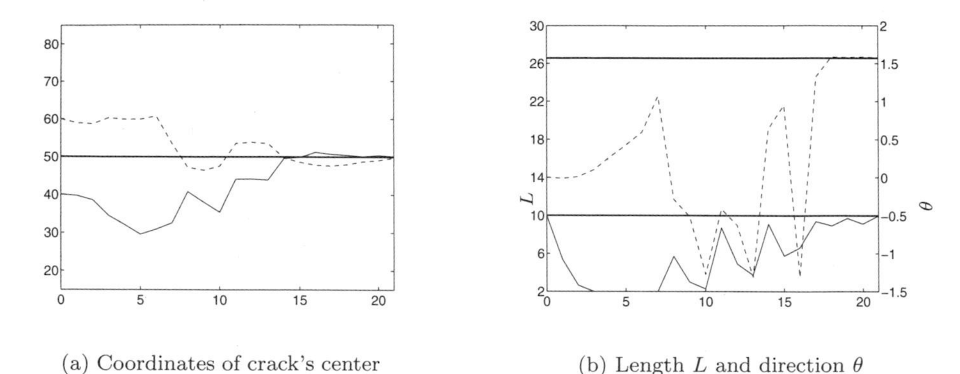

(a) Coordinates of crack's center

(b) Length L and direction θ

Figure 11.5. Convergence history, from starting point [40 60 10 0]

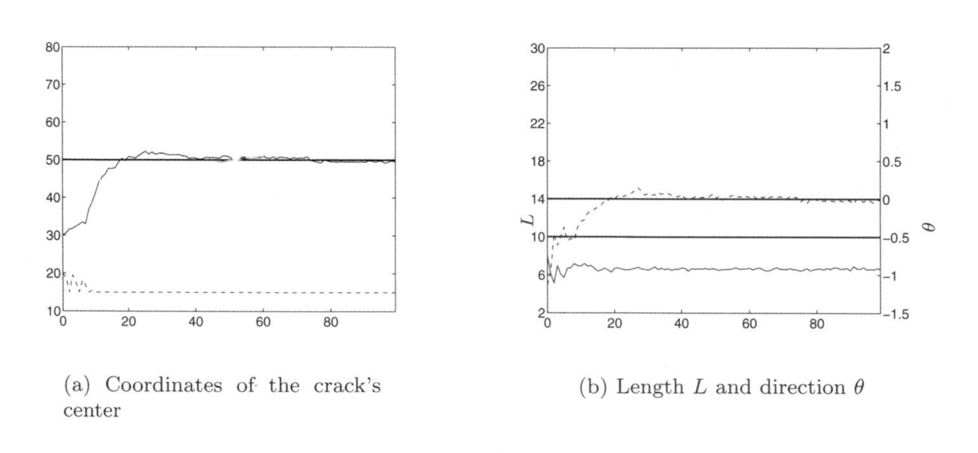

(a) Coordinates of the crack's center

(b) Length L and direction θ

Figure 11.6. Ricker-Impulse: Iterations from starting point [30 20 8 2]

tion vector is $\hat{\mathbf{z}}_0^- = [30\ 20\ 8\ 2]$, while the solution reads $\mathbf{z} = [50\ 50\ 10\ 0]$. In both cases the algorithm does not solve the problem satisfactorily. A thorough investigation has shown that the error function considered here for the inverse problem is a nonconvex function and that the algorithm stops in a local minimum of this error function. One should emphasize that in inverse analysis a local minimum provides no useful information (Stavroulakis, 2000), while in classical optimal design one may be satisfied with the calculation of some *suboptimal* solution.

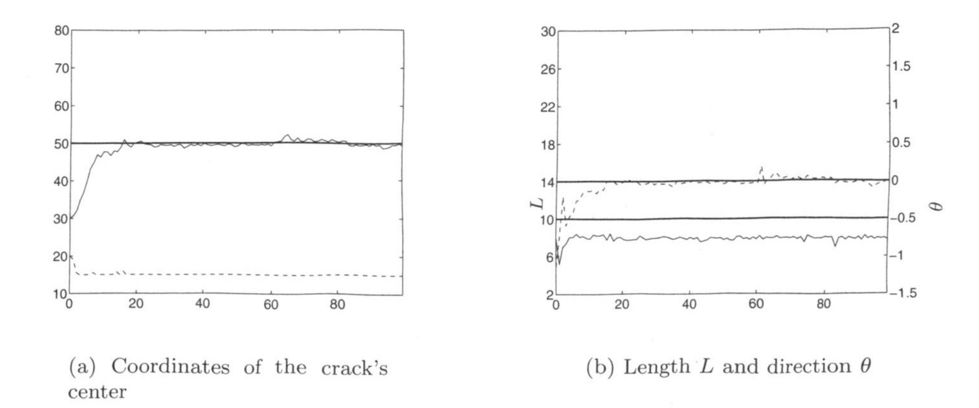

(a) Coordinates of the crack's center

(b) Length L and direction θ

Figure 11.7. Sinus-Function: Iterations from starting point [30 20 8 2]

4. Conclusions

A unilateral crack identification problem in elastodynamics may be quite complicated. The problem may become nonsmooth and nonconvex (see also the discussion in (Stavroulakis, 2000)). A thorough investigation has shown that the ability to solve this problem, by following a classical output error minimization formulation, depends on the orientation of the crack, the loading, the starting point (for iterative algorithms) and the shape of the structure. In some cases we may explain the inability of a chosen algorithm to find a solution. Usually this happens if the error function has local minima and the algorithm stops there. The proposed filter-driven optimization algorithm is much more effective than classical optimization methods, but still may fail by stopping in local minima of the error function. An alternative method of comparable performance but different advantages and disadvantages is the use of neural networks. Finally, the use of gradient-free, genetic optimization algorithms seems to be the most powerful method for the numerical solution of the inverse problem, able to avoid local minima. Nevertheless this last method, especially in dynamics, is connected with enormous demans on computing time.

References

Antes, H. and Panagiotopoulos, P.D. (1992). *The Boundary Integ proach to Static and Dynamic Contact Problems. Equality equality Methods.* Basel, Boston: Birkhäuser Verlag.

Engelhardt, M. (2004). Numerische Verfahren zur Identifizierung von Fehlstellen aus Randdaten. PhD Thesis, Department of Civil Engineering, Carolo Wilhelmina Technical University, Braunschweig, Germany, to appear.

Stavroulakis, G.E. (1999). Impact-echo from a unilateral interlayer crack. LCP-BEM modelling and neural identification. *Engineering Fracture Mechanics*, Vol. 62(2-3), pp. 165-184.

Stavroulakis, G.E., Antes, H. and Panagiotopoulos, P.D. (1999). Transient elastodynamics around cracks including contact and friction. *Computer Methods in Applied Mechanics and Engineering*, Vol. 177(3-4), pp. 427-440.

Stavroulakis, G.E. and Antes, H. (2000). Unilateral crack identification. A filter-driven, iterative, boundary element approach. *Journal of Global Optimization*, Vol. 17(1-4), pp. 339-352.

Stavroulakis, G.E. (2000). *Inverse and Crack Identification Problems in Engineering Mechanics.* Dordrecht, The Netherlands: Kluwer Academic Publishers.

Stavroulakis, G.E., Engelhardt, M., Likas, A., Gallego, R., Antes, H. (2004). Neural network assisted crack and flaw identification in transient dynamics. *Journal of Theoretical and Applied Mechanics, Polish Academy of Sciences* in press.

Chapter 12

PENALTY APPROXIMATION OF PAINLEVÉ PROBLEM

Michelle Schatzman

MAPLY (Laboratoire de Mathématiques Appliquées de Lyon), CNRS et UCBL, 69622 Villeurbanne Cedex, France

schatz@maply.univ-lyon1.fr

Abstract An elongated solid bar drops onto a rigid foundation, and Coulomb friction acts at the contact. A penalty approximation is defined, and is proved to converge to a solution of the problem.

Keywords: Penalty, Coulomb friction, impact

1. Introduction

The mathematical status of many problems in non-smooth mechanics has been greatly advanced by the work of Jean Jacques Moreau.

I have worked independently of Jean Jacques' school and set up my own techniques and approaches to problems of non-smooth mechanics; see for instance (Schatzman, 1977) and (Schatzman, 1978) for a first appearance of vector-valued measures in the context of impact mechanics. It was a great honor to be present for the celebration of Jean Jacques Moreau's eightieth birthday, and to have the opportunity to praise the immense impact he had on non smooth mechanics.

Therefore, dear Jean Jacques, I hope that you will stay with us many more years in good health and good spirits, with the sharpness of mind and the creativity that are your trademark, and may you reach your hundred and twentieth birthday.

In this article, I will report on recent results on the penalty approach to combined friction and impact. I will concentrate on a very specific problem, namely the case of a solid bar dropping onto a rigid foundation. This problem has been solved mathematically through a time-stepping approach by David Stewart (Stewart, 1998), which is a very difficult paper.

Here, I propose to use a penalty approach, which is applied only to the non-interpenetration condition. I prove that as the stiffness in the penalty approximation tends to infinity, the solution of the system of differential equations tends to a solution of the limiting problem, in a slightly generalized sense. There is a substantial mathematical difficulty here: the passage to the limit in the Coulomb condition seems to involve the multiplication of a measure by a function which is discontinuous precisely where the measure has a Dirac mass. There is a way around this difficulty, since the details of the convergence are accessible, and therefore a relation between the velocities after impact can be deduced from the velocities before impact.

The paper is organized as follows. In Section 2, the notations are introduced, the system of differential equations and its approximation are described.

In Section 3, an auxiliary problem is stated: it is a multivalued equation of first order, involving the multivalued sign function multiplied by a variable coefficient. Complete proofs will be given elsewhere.

Section 4 gives an existence result for the approximate problem; it uses a local existence theorem through the contraction principle and an energy estimate in order to go global.

The convergence of the approximate solution is sketched in Section 5. Many interesting and intricate technical steps have been skipped, and will be given in detail in a later publication.

Finally, an appendix contains the statement of some technical results relative to the weak convergence of measures.

2. Notations and position of the problem

Let 2ℓ be the length of the bar, (X, Y) the coordinates of the center of mass, θ the angle of the bar with the horizontal; then the coordinates of the possible contact points are (x, y) given by

$$x = X - \ell \cos\theta \text{ and } y = Y - \ell \sin\theta \tag{12.1}$$

or

$$x = X + \ell \cos\theta \text{ and } y = Y + \ell \sin\theta. \tag{12.2}$$

We concentrate on the case (12.1); both cases can occur at the same time only if θ vanishes modulo π. This situation creates difficulties, which are not yet solved, and we exclude it *a priori*.

Denote by m the mass of the bar and by μ the friction coefficient. Then, the moment of inertia of the bar about its center of mass is $J =$

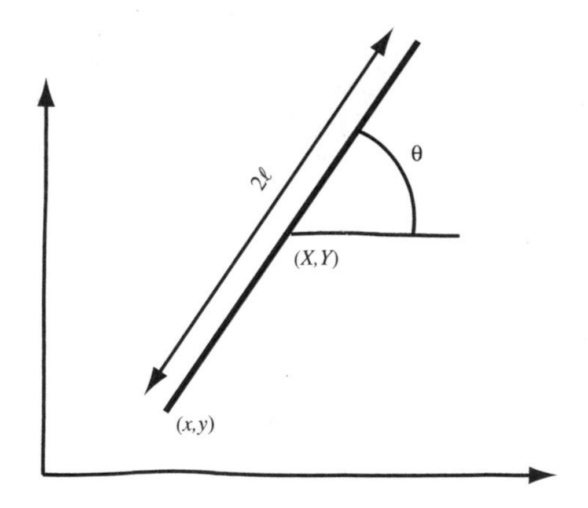

Figure 12.1. The geometry of the bar.

$m\ell^2/3$. If $\mathbf{R}_T$ and $\mathbf{R}_N$ denote respectively the tangential and the normal reactions, and g denotes the gravity, the equations of motion can be written

$$m\ddot{X} = \mathbf{R}_T, \tag{12.3}$$

$$m\ddot{Y} = \mathbf{R}_N - mg, \tag{12.4}$$

$$J\ddot{\theta} = -\ell\mathbf{R}_N \cos\theta + \ell\mathbf{R}_T \sin\theta. \tag{12.5}$$

The condition of interpenetration is a Signorini condition:

$$0 \leq \mathbf{R}_N \perp y \geq 0, \tag{12.6}$$

and friction is modeled through a Coulomb law:

$$-\mathbf{R}_T \in \mu\mathbf{R}_N \,\mathrm{Sign}(\dot{x}). \tag{12.7}$$

The multivalued operator Sign is given by

$$\mathrm{Sign}(r) = \begin{cases} \{-1\} & \text{if } t < 0, \\ [-1,+1] & \text{if } t = 0, \\ \{+1\} & \text{if } t > 0. \end{cases} \tag{12.8}$$

The purpose of this article is to show the existence of a solutio
the system (12.3), (12.4), (12.5), (12.6), (12.7) in an appropriate
We construct an approximate solution by a penalty method
the Signorini condition, that is, we replace (12.6) by

$$\mathbf{R}_N = -m\frac{y^-}{\tau^2};$$

here y^-, the negative part of y is defined to be $\max(-y, 0)$ and τ is a positive parameter having the dimension of time, which will eventually tend to 0.

We consider now the multivalued differential system

$$\ddot{X} + \mu\sigma\frac{y^-}{\tau^2} = 0, \tag{12.10}$$

$$\ddot{Y} - \frac{y^-}{\tau^2} + g = 0, \tag{12.11}$$

$$\frac{\ell}{3}\ddot{\theta} + \frac{y^-}{\tau^2}\cos\theta + \mu\sigma\frac{y^-}{\tau^2}\sin\theta = 0, \tag{12.12}$$

$$\sigma \in \operatorname{Sign}(\dot{x}). \tag{12.13}$$

This system is supplemented by Cauchy conditions on X, Y, θ and their first derivatives at time 0:

$$X(0), \quad Y(0), \quad \theta(0), \quad \dot{X}(0), \quad \dot{Y}(0), \quad \dot{\theta}(0) \text{ are given.} \tag{12.14}$$

3. An auxiliary problem

In order to solve (12.10)–(12.14), we need to solve an auxiliary problem relative to a class of multivalued scalar equations:

LEMMA 12.1 *Let ϕ and ψ belong to $L^1(0,T)$, and assume moreover that ϕ is almost everywhere non negative. Then, there exists a unique absolutely continuous solution of*

$$\dot{\xi} + \phi \operatorname{Sign}(\xi) \ni \psi \tag{12.15}$$

for all initial condition $\xi(0) = \xi^0$. Moreover, given $\hat{\phi}$ and $\hat{\psi}$ satisfying the same conditions, we denote by $\hat{\xi}$ the solution of the analogous problem; then the following estimate holds:

$$|\xi(t) - \hat{\xi}(t)| \leq |\xi^0 - \hat{\xi}^0| + \int_0^t \left(|\phi(s) - \hat{\phi}(s)| + |\psi(s) - \hat{\psi}(s)|\right) ds \tag{12.16}$$

iple of the proof. The existence is clear for piecewise constant func- and ψ; one obtains the existence in general by a density argu- e uniqueness is a straightforward consequence of monotonicity, n (12.16) is also a consequence of monotonicity. Details will sewhere.

4. Existence for the approximate problem

We will now perform some algebraic manipulations on the system (12.10)–(12.13); indeed, we infer from the definition $x = X - \ell \cos\theta$ the identity

$$\ddot{X} = \ddot{x} - \ell\ddot{\theta}\sin\theta - \ell\dot{\theta}^2\cos\theta. \tag{12.17}$$

We substitute now the value of $\ell\ddot{\theta}$ given by (12.12) into (12.17), and we obtain the following equation for $\dot{x}$:

$$\frac{\mathrm{d}}{\mathrm{d}t}\dot{x} + \frac{\mu(1+3\sin^2\theta)\sigma y^-}{\tau^2} = \ell\dot{\theta}^2\cos\theta - \frac{3y^-\sin\theta\cos\theta}{\tau^2}. \tag{12.18}$$

Therefore, if we assume that Y, θ and $\dot{\theta}$ are given continuous functions, according to Lemma 12.1 there exists a unique solution of Equ. (12.18), which satisfies the Cauchy data

$$\dot{x}(0) = \dot{X}_0 + \ell\dot{\theta}_0\sin\theta_0.$$

We denote this solution by

$$\dot{x}(t) = \mathcal{X}(Y, \theta, \dot{\theta})(t), \tag{12.19}$$

the initial data being understated, for simplicity of notation.

We are now able to prove an existence and uniqueness theorem for the penalized problem:

THEOREM 12.2 *Given initial data* (12.14), *there exists a unique solution on* $[0,T]$ *of* (12.10)–(12.13), *such that* $X, \dot{X}, Y, \dot{Y}, \theta$ *and* $\dot{\theta}$ *are continuous, while the second derivatives* $\ddot{X}$, $\ddot{y}$ *and* $\ddot{\theta}$ *are bounded.*

Proof.

Step 1. Existence of a local solution. Assuming that there exists a solution, we may rewrite Equ. (12.12) by expressing $\sigma y^-/\tau^2$ in terms of $\ddot{x}$, θ and Y; indeed, we infer from Equ. (12.18) the following relation:

$$\mu\frac{\sigma y^-}{\tau^2} = \left(\ell\dot{\theta}^2\cos\theta - \frac{3y^-}{\tau^2}\cos\theta\sin\theta - \ddot{x}\right)\frac{1}{1+3\sin^2\theta}.$$

We substitute this expression into Equ. (12.12), and we obtain the relation

$$\ddot{\theta} = -\frac{3y^-}{\ell\tau^2}\frac{\cos\theta}{1+3\sin^2\theta} - \frac{3\dot{\theta}^2\cos\theta\sin\theta}{1+3\sin^2\theta} + \frac{3}{\ell}\frac{\sin\theta}{1+3\sin^2\theta}\ddot{x}. \tag{12}$$

We transform the differential system satisfied by U into an integral equation; we treat the term containing $\ddot{x}$ through an integration by parts.

Let us define now

$$U = \begin{pmatrix} Y \\ Z \\ \theta \\ \eta \end{pmatrix}, \quad U_0 = \begin{pmatrix} Y_0 \\ \dot{Y}_0 \\ \theta_0 \\ \dot{\theta}_0 \end{pmatrix}, \quad e_4 = \begin{pmatrix} 0 \\ 0 \\ 0 \\ 1 \end{pmatrix} \text{ and } F(U) = \begin{pmatrix} F_1(U) \\ F_2(U) \\ F_3(U) \\ F_4(U) \end{pmatrix},$$

with F_i given as follows:

$$\begin{aligned} F_1(U) &= Z, \\ F_2(U) &= \frac{(Y - \ell \sin\theta)^-}{\tau^2} - g, \\ F_3(U) &= \eta, \\ F_4(U) &= -\frac{3(Y - \ell\sin\theta)^-}{\ell\tau^2} \frac{\cos\theta}{1 + 3\sin^2\theta} - \frac{3\eta^2 \cos\theta \sin\theta}{1 + 3\sin^2\theta}. \end{aligned}$$

When we integrate by parts oder $[0, t]$ the $\ddot{x}$ term in Equ. (12.20), we obtain the expression

$$\dot{x}(t)\frac{3}{\ell}\frac{\sin\theta}{1 + 3\sin^2\theta}(t) - \dot{x}(0)\frac{3}{\ell}\frac{\sin\theta}{1 + 3\sin^2\theta}(0) - \int_0^t \dot{x}(s)\frac{3}{\ell}\eta(s)\frac{\mathrm{d}}{\mathrm{d}\theta}\frac{\sin\theta}{1 + 3\sin^2\theta}\,\mathrm{d}s.$$

Therefore, it is convenient to define the functions

$$\begin{aligned} \chi_0(r) &= \frac{3}{\ell}\frac{\sin r}{1 + 3\sin^2 r}, \\ \chi(U)(t) &= \chi_0\left(\theta(0) + \int_0^t \eta(s)\,\mathrm{d}s\right), \\ \omega(U) &= -\chi_0'(\theta)\eta. \end{aligned}$$

We also need the functions, which appear naturally in the definition of $\mathcal{X}$:

$$\begin{aligned} \phi(U) &= \mu\frac{(1 + 3\sin^2\theta)(Y - \ell\sin\theta)^-}{\tau^2}, \\ \psi(U) &= \ell\eta^2\cos\theta - \frac{3(Y - \ell\sin\theta)^- \cos\theta\sin\theta}{\tau^2}. \end{aligned}$$

bserve that χ maps continuous functions to continuous functions, while naps $\mathbb{R}^4$ to itself and ω, ϕ and ψ map $\mathbb{R}^4$ to $\mathbb{R}$.

nally, it is convenient to write $\mathcal{X}(U)$ instead of $\mathcal{X}(Y, \theta, \eta)$.

Equ. (12.20) together with Cauchy data is equivalent to

$$\eta(t) = \dot\theta(0) + \int_0^t F_4(U(s))\,\mathrm{d}s + \chi(U)(t)\mathcal{X}(U)(t) - \chi_0(\theta(0))\dot x(0) + \int_0^t \mathcal{X}(U)(s)\omega(U(s))\,\mathrm{d}s,$$

provided that we let $\dot\theta = \eta$.

At this point, we define

$$G(U)(t) = \chi(U)(t)\mathcal{X}(U)(t) - \chi_0(\theta(0))\dot x(0) + \int_0^t \omega(U)\mathcal{X}(U)\,\mathrm{d}s.$$

It is equivalent to solve (12.10)–(12.13) with Cauchy data (12.14) and to find a fixed point of the operator $\mathcal{T}$ defined by

$$(\mathcal{T}U)(t) = U_0 + \int_0^t F(U(s))\,ds + e_4 G(U)(t).$$

Let B be the closed ball of radius $r > 0$ about U_0 in $\mathbb{R}^4$, and let $\tilde B$ be $B \times B$ deprived from its diagonal. Define for all $r > 0$ the quantities:

$$\begin{aligned}
R_1 &= \max_{V\in B} \max\big(|F(V)|, |\phi(V)| + |\psi(V)|, |\omega(V)|\big),\\
R_2 &= \max|\chi_0|,\\
L_1 &= \sup_{(V,V')\in B'} \frac{1}{|V - V'|}\big(|F(V) - F(V')|, |\phi(V) - \phi(V')|\\
&\qquad + |\psi(V) - \psi(V')|, |\omega(V) - \omega(V')|\big),\\
L_2 &= \max|\chi_0'|,\\
R &= \max(R_1, R_2), \quad L = \max(L_1, L_2).
\end{aligned}$$

The numbers R and L are finite, for all finite values of r.

Equip the space $C^0([0, t_0]; \mathbb{R}^4)$ with the maximum of the sup norm in each component; the number t_0 is a positive number that will be chosen later; let the set $\mathcal{B}$ be the ball of radius r about the constant function U_0 in $C^0([0, t_0]; \mathbb{R}^4)$. We show now that for small enough values of t_0, $\mathcal{T}$ is a strict contraction from $\mathcal{B}$ to itself.

If U belongs to $\mathcal{B}$, we have the estimate

$$|(\mathcal{T}U)(t) - U_0| \le tR + |G(U)(t)|;$$

Decompose G into two pieces:

$$G_1(U) = \chi(U)\mathcal{X}(U) - \chi_0(\theta_0)\dot x_0 \text{ and } G_2(U)(t) = \int_0^t \omega(U)\mathcal{X}($$

Thanks to Equ. (12.15), and by definition of R, we have

$$|\mathcal{X}(U)(t) - \dot{x}_0| \leq Rt, \tag{12.22}$$

which implies

$$|\mathcal{X}(U)(t)| \leq |\dot{x}_0| + Rt. \tag{12.23}$$

We also have, thanks to Equ. (12.16) and the definition of L

$$|\mathcal{X}(U^*)(t) - \mathcal{X}(U)(t)| \leq Lt\|U^* - U\|. \tag{12.24}$$

Let us estimate now $\chi(U)(t)$: by definition of R,

$$|\chi(U)(t)| \leq R \tag{12.25}$$

and by definition of L:

$$|\chi(U^*)(t) - \chi(U)(t)| \leq Lt\|U^* - U\| \text{ and } |\chi(U)(t) - \chi_0(\theta_0)| \leq Ltr. \tag{12.26}$$

We use the triangle inequality to obtain the estimates

$$\begin{aligned}|G_1(U)(t)| &\leq |\mathcal{X}(U)(t) - \dot{x}_0||\chi(U)(t)| + |\dot{x}_0||\chi(U)(t) - \chi_0(\theta_0)| \\ &\leq R^2 t + |\dot{x}_0|Ltr,\end{aligned}$$

thanks to (12.22), (12.25), (12.26) and

$$|G_2(U)(t)| \leq R\int_0^t (|\dot{x}_0| + Rs)\,\mathrm{d}s = R\big(|\dot{x}_0| + Rt^2/2\big).$$

thanks to (12.23). Therefore, if

$$t_0(R + R^2 + |\dot{x}_0|Lr + R|\dot{x}_0|) + R^2 t_0^2/2 \leq r,$$

$\mathcal{T}$ maps $\mathcal{B}$ to itself.

We move on to contractivity. It is plain that

$$|(\mathcal{T}U^* - \mathcal{T}U)(t)| \leq Lt\|U^* - U\| + |G(U^*)(t) - G(U)(t)|$$

and we reuse estimates (12.22) to (12.26) to treat the G terms:

$$\begin{aligned}&|G_1(U^*)(t) - G_1(U)(t)| \\ &\quad\leq |\mathcal{X}(U^*)||\chi(U^*)(t) - \chi(U)(t)| + |\chi(U)(t)||\mathcal{X}(U^*) - \mathcal{X}(U)| \\ &\quad\leq (|\dot{x}_0| + Rt)Lt\|U^* - U\| + RLt\|U^* - U\|\end{aligned}$$

and very similarly

$$|G_2(U^*)(t) - G_2(U)(t)| \leq \int_0^t \big(RLs\|U^* - U\| + L(|\dot{x}_0| + Rs)\big)\,\mathrm{d}s.$$

fore, if

$$Lt_0\big(2|\dot{x}_0| + 2Rt_0 + R\big) < 1,$$

ict contraction, so that the system (12.10)–(12.14) possesses a
ition.

Step 2. Global solution. We obtain an energy identity: if we multiply (12.10) by $\dot{X}$, (12.11) by $\dot{Y}$ and (12.12) by $\ell^2\dot{\theta}$, and we integrate over $[0,t]$ for $t \leq t_0$, we get

$$\frac{1}{2}\left(\dot{X}^2 + \dot{Y}^2 + \frac{\ell^2\dot{\theta}^2}{3} + \frac{(y^-)^2}{2\tau^2} + gY\right)\Bigg|_0^t + \int_0^t \mu|\dot{x}|\frac{y^-}{\tau^2}\,\mathrm{d}s = 0. \tag{12.27}$$

A Gronwall inequality argument gives then

$$\max_{[0,t_0]} \max(|\dot{X}|, |\dot{Y}|, |\dot{\theta}|, |y^-/\tau|) \leq C, \tag{12.28}$$

where the number C depends neither on t nor on τ, provided that $Y_0 - \ell\sin\theta_0 = y(0)$ is non negative.

Therefore, the standard continuation argument shows that the solution exists up to time T.

Moreover, if we integrate (12.10) over $[0,T]$, we obtain immediately from the identity

$$\dot{Y}(T) - \dot{Y}_0 + gT = \int_0^T \frac{y^-}{\tau^2}\,\mathrm{d}t$$

the bound

$$\max_{\tau\geq 0} \int_0^T \frac{y^-}{\tau^2}\,\mathrm{d}t \leq C. \tag{12.29}$$

□

5. Convergence of the approximate solution

In this section, we put back the index τ on all the quantities, which depend on τ.

THEOREM 12.3 *For all Cauchy data (12.14), and as long as the second end of the bar does not hit the rigid foundation, there exist three functions X, Y and θ, whose derivatives are of bounded variation, and which satisfy the following equations:*

$$\ddot{X}_0 + \mu\xi_0 = 0, \tag{12.30}$$

$$\ddot{Y}_0 - \eta_0 + g = 0, \tag{12.31}$$

$$\frac{\ell}{3}\ddot{\theta}_0 + \eta_0\cos\theta_0 + \mu\xi_0\sin\theta_0 = 0, \tag{12.32}$$

$$0 \leq Y_0 - \ell\sin\theta_0 \perp \eta_0 \geq 0, \tag{12.33}$$

$$\xi_0 = \sigma_0\eta_0. \tag{12.34}$$

The function σ_0 and the measures η_0 and ξ_0 satisfies the following conditions: η_0 is non negative, according to (12.33), $|\sigma_0| \leq 1$, η_0-almost everywhere; moreover, if the decomposition of η_0 and ξ_0 into their atomic and continuous parts are respectively

$$\eta_0 = \eta_0^{\mathrm{a}} + \eta_0^{\mathrm{c}}, \quad \xi_0 = \xi_0^{\mathrm{a}} + \xi_0^{\mathrm{c}},$$

then

$$\xi_0^{\mathrm{c}} = \sigma_0^{\mathrm{c}} \eta_0^{\mathrm{c}}, \quad \xi_0^{\mathrm{a}} = \sigma_0^{\mathrm{a}} \eta_0^{\mathrm{a}}$$

and we have the relations

1 *the continuous part σ_0^{c} of σ_0 satisfies*

$$\sigma_0^{\mathrm{c}} \in \mathrm{Sign}(\dot{x}_0), \ \eta_0^{\mathrm{c}}\text{-almost everywhere;}$$

2 *let t_i be an atom of η_0; if $\dot{x}_0(t_i + 0)$ and $\dot{x}_0(t_i - 0)$ have the same sign, then $\sigma^{\mathrm{a}}(t_i) = \mathrm{sign}(\dot{x}_0(t_i \pm 0))$;*

3 *if $\dot{x}_0(t_i - 0)$ does not vanish and $\dot{x}_0(t_i + 0)$ vanishes or has the opposite sign to $\dot{x}_0(t_i - 0)$, then $\sigma_0^{\mathrm{a}}(t_i) \in [-1, 1]$; moreover, a precise expression of $\sigma_0^{\mathrm{a}}(t_i)$ can be given in terms of the values on the left and on the right of all the derivatives of X_0, Y_0 and θ_0 at t_i.*

Proof. Let us write

$$\eta_\tau = \frac{y_\tau^-}{\tau^2} \text{ and } \xi_\tau = \sigma_\tau \frac{y_\tau^-}{\tau_2}.$$

We may always extract subsequences, which will still be denoted by ξ_τ and η_τ, such that ξ_τ and η_τ converge weakly $*$ in the space $M^1([0,T])$ of Radon measures to their respective limits ξ_0 and η_0.

We will identify η_τ with $\eta_\tau \mathrm{d}t$, and similarly for ξ_τ. Therefore, any integral that contains η_τ or ξ_τ will contain its element of integration. This choice of notations makes life easier in the limit, when we have to integrate with respect to more or less arbitrary Radon measures.

We have to characterize now the relation between η_0 and ξ_0. This cannot be a trivial matter since we expect that the limit of $\dot{x}_\tau$ will change sign, and if it changes sign where η_τ has an atom, then we have to pass to the limit in relation (12.13), which amounts to multiplying an atom by a function, which is discontinuous at this atom.

Therefore, it is better to use the weak formulation of (12.13). Indeed, any given $\tau > 0$, Equ. (12.13) has an equivalent weak form

$$\forall z \in C^0([0,T]), \qquad \int_0^T \big(|z + \dot{x}_\tau| - |\dot{x}_\tau|\big)\eta_\tau \geq \int_0^T \xi_\tau z. \tag{12.35}$$

There will be an easy part in the passage to the limit and a difficult part. Let η_0 be decomposed into its atomic and continuous parts:

$$\eta_0 = \eta_0^{\mathrm{a}} + \eta_0^{\mathrm{c}}.$$

Then, according to Theorem A.3, there exists a sequence of continuous functions Ψ_τ with values in $[0,1]$ such that

$$\eta_\tau \Psi_\tau \stackrel{*}{\rightharpoonup} \eta_0^{\mathrm{a}}, \quad \eta_\tau(1-\Psi_\tau) \stackrel{*}{\rightharpoonup} \eta_0^{\mathrm{c}}.$$

Moreover, the support of Ψ_τ tends to the set of atoms of ξ_0. If we consider now $\xi_\tau \Psi_\tau$ and $\xi_\tau(1-\Psi_\tau)$, we may extract subsequences such that both measures converge weakly $*$ to their respective limits ξ_0^{a} and ξ_0^{c}.

Let us prove that these are precisely the atomic and continuous parts of ξ_0. The support of ξ_0^{a} is included in the set of atoms of η_0; we also have the inequality

$$|\xi_0^{\mathrm{c}}| \le |\eta_0^{\mathrm{c}}|; \tag{12.36}$$

therefore, ξ_0^{c} is continuous and the support of ξ_0^{c} is at most countable, which gives the desired conclusion.

We will write henceforth

$$\eta_\tau^{\mathrm{c}} = (1-\Psi_\tau)\eta_\tau, \qquad \eta_\tau^{\mathrm{a}} = \Psi_\tau \eta_\tau,$$
$$\xi_\tau^{\mathrm{c}} = (1-\Psi_\tau)\xi_\tau, \qquad \xi_\tau^{\mathrm{a}} = \Psi_\tau \xi_\tau;$$

the reader should beware: these four measures are identified with the product of an essentially bounded function by Lebesgue's measure, and therefore, they are continuous measures. It is plainly possible for a sequence of continuous measures to converge to an atomic measure, and conversely.

The following inequality holds for all function ϕ with values in $[0,1]$:

$$|\dot{x}_\tau + z\phi| - |\dot{x}_\tau| \le \phi\big(|\dot{x}_\tau + z| - |\dot{x}_\tau|\big). \tag{12.37}$$

Therefore, if we substitute z by $1-\Psi_\tau$ in (12.37), we can pass to the limit; indeed, let us decompose $\dot{x}_\tau$ into two pieces:

$$\dot{x}_\tau^{\mathrm{c}}(t) = \dot{x}(0) + \int_0^t \left(\ell\dot{\theta}_\tau^2 - 3\eta_\tau^{\mathrm{c}} \sin\theta_\tau \cos\theta_\tau - \mu\big(1 + 3\sin^2\theta_\tau\big)\xi_\tau^{\mathrm{c}}\right) \mathrm{d}t,$$
$$\dot{x}_\tau^{\mathrm{a}} = -\int_0^t \big(\mu(1+3\sin^2\theta_\tau)\xi_\tau^{\mathrm{a}} + 3\sin\theta_\tau\cos\theta_\tau \eta_\tau^{\mathrm{a}}\big).$$

Thanks to Lemma A.2, $\dot{x}_\tau^{\mathrm{c}}$ converges uniformly to its limit

$$\dot{x}_0^{\mathrm{c}}(t) = \dot{x}(0) + \int_0^t \left(\ell\dot{\theta}_0^2 \,\mathrm{d}t - 3\eta_0^{\mathrm{c}} \sin\theta_0 \cos\theta_0 - \mu(1+3\sin^2\theta_0)\xi_0^{\mathrm{c}}\right).$$

This proves the following convergence:

$$\int_0^T \big(|\dot{x}_\tau^{\mathrm{c}} + z| - |\dot{x}_\tau^{\mathrm{c}}|\big)\eta_\tau^{\mathrm{c}} \to \int_0^T \big(|\dot{x}_0^{\mathrm{c}} + z| - |\dot{x}_0^{\mathrm{c}}|\big)\eta_0^{\mathrm{c}}.$$

We have to estimate now

$$\int_0^T \big(|\dot{x}_\tau + z| - |\dot{x}_\tau| - |\dot{x}_\tau^{\mathrm{c}}| + |\dot{x}_\tau^{\mathrm{c}}|\big)\eta_\tau^{\mathrm{c}}. \tag{12.38}$$

But the expression inside parentheses in (12.38) vanishes wherever $\dot{x}_\tau^{\mathrm{c}}$ is equal to $\dot{x}_\tau$, that is outside of the support of Ψ_τ. Thus, we have the estimate

$$\left|\int_0^T \big(|\dot{x}_\tau + z| - |\dot{x}_\tau| - |\dot{x}_\tau^{\mathrm{c}}| + |\dot{x}_\tau^{\mathrm{c}}|\big)\eta_\tau^{\mathrm{c}}\right| \leq \sup_t |\dot{x}_\tau^{\mathrm{a}}(t)| \int 1_{\{\Psi_\tau > 0\}}\, \eta_\tau^{\mathrm{a}} \tag{12.39}$$

and by construction the integral factor in the right hand side of (12.39) tends to 0 as τ tends to 0. Also, the supremum of $|\dot{x}_\tau^{\mathrm{a}}|$ is bounded independently of τ. Therefore, we have in the limit the inequality

$$\int_0^T \big(|\dot{x}_0^{\mathrm{c}} + z| - |\dot{x}_0^{\mathrm{c}}|\big)\,\eta_0^{\mathrm{c}} \geq \int_0^T z\xi_0^{\mathrm{c}}, \tag{12.40}$$

for all continuous function z on $[0,T]$. This is the weak formulation of the multivalued relation (in the sense of measure-valued differential inclusions)

$$\xi_0^{\mathrm{c}} \in \eta_0^{\mathrm{c}}\,\mathrm{Sign}(\dot{x}_0^{\mathrm{c}}). \tag{12.41}$$

There is a slightly different way of writing things; as the measure ξ_0^{c} is absolutely continuous with respect to the measure ξ_0^{c}, there exists an η_0^{c}-measurable function σ^{c}, such that

$$\xi_0^{\mathrm{c}} = \sigma^{\mathrm{c}}\eta_0^{\mathrm{c}}.$$

In fact, σ^{c} is η_0^{c}-essentially bounded, thanks to relation (12.36) Then relation (12.41) is equivalent to

$$\sigma^{\mathrm{c}} \in \mathrm{Sign}(\dot{x}_0^{\mathrm{c}}),\ \ \eta_0^{\mathrm{c}}\text{–almost everywhere.} \tag{12.42}$$

Let us move on to the atomic part of the measure ξ_0; of course this will be the difficult part of the proof. Since it is highly technical, it will be only sketched.

We need more notations: define

$$\kappa = \mu(1 + 3\sin^2\theta), \quad \lambda = 3\sin\theta\cos\theta. \tag{12.43}$$

In order to emphasize the dependence of κ and λ over τ, we will affix an index τ whenever necessary.

Then, close to an atom of η at t_i, the equation for $\dot{x}_\tau$ is asymptotic to the equation

$$\frac{\mathrm{d}\dot{x}_\tau}{\mathrm{d}t} + (\kappa_0(t_i)\,\mathrm{Sign}(\dot{x}_\tau) + \lambda_0(t_i))\eta_\tau \ni 0. \tag{12.44}$$

The comparison between (12.44) and (12.17) can be performed with the help of Lemma 12.1. The nice feature of (12.44) is that the passage to the limit is easy here, provided, of course, that we use Theorem A.3.

Then, we find the following relations between the velocity on the left of t_i and the velocity on the right of t_i, where we have deonted by m_i the mass of the atom of η_0 at t_i:

- if $\dot{x}_0(t_i-0) > 0$ and $v = \dot{x}_0(t_i-0) - \mu m_i(\kappa_0(t_i)+\lambda_0(t_i)) > 0$, then $\dot{x}_0(t_i+0) = v$;
- if $\dot{x}_0(t_i-0) > 0$, $\dot{x}_0(t_i-0) - \mu m_i(\kappa_0(t_i) + \lambda_0(t_i), 0$ and $\lambda_0(t_i) - \kappa_0(t_i) > 0$, then $\dot{x}_0(t_i+0) > 0$ and its value is given by

  $$\dot{x}_0(t_i+0) = \frac{(\kappa_0(t_i) - \lambda_0(t_i))(m_i + n_i)}{(\kappa_0(t_i) + \lambda_0(t_i))(m_i - n_i)}\dot{x}_0(t_i-0),$$

  where n_i denotes the mass of ξ_0 at t_i;
- if $\dot{x}_0(t_i-0) > 0$, $\dot{x}_0(t_i-0) - \mu m_i(\kappa_0(t_i) + \lambda_0(t_i), 0$ and $\lambda_0(t_i) - \kappa_0(t_i) \le 0$, then $\dot{x}_0(t_i+0) = 0$.

For $\dot{x}_o(t_i-0)$ there are analogous expressions, which are not given here for simplicity.

The question now is to define m_i and n_i from the data of the problem. In the case when $\dot{x}$ does not change sign at t_i, we pass to the limit in the energy equality (12.27), and we find the formula

$$m_i = \frac{2\dot{Y}_0(t_i-0) - 2\ell\dot{\theta}_0(t_i-0)(\cos\theta(t_i) + \mu\sin\theta(t_i))}{1 + (\cos\theta(t_i) + \mu\sin\theta(t_i))^2}.$$

There are analogous, but much more complicated formulas in the two other cases. □

Appendix: Some results on the weak * convergence of measures

The construction of Radon measure is classical and is performed from two slig[illegible] different points of view in Dieudonné's treatise (Dieudonné, 1970) or the french [illegible]nal (Dieudonné, 1968) and in Malliavin's superb course book (Malliavin, 1995[illegible] french original (Malliavin and Airault, 1994).

I will assume that the reader if familiar with the weak convergence of measures and the definition of the absolute value of a measure.

For a μ integrable set A, we use the standard notation

$$\int_A f\mu = \int f 1_A \mu.$$

A measure μ has an atom at ξ iff $\langle \mu, 1_{\{\xi\}} \rangle$ does not vanish. We say that the mass of this atom is $\langle \mu, 1_{\{\xi\}} \rangle$. Then $|\mu|$ also has an atom at x, and the mass of this atom is $|\langle \mu, 1_{\{\xi\}} \rangle|$. The basic case of a measure, which has an atom is the Dirac mass $\delta(\cdot - \xi)$ at $\xi \in [0, T]$, which is defined by

$$\langle \delta(\cdot - \xi), \phi \rangle = \phi(\xi).$$

It is a classical fact that every Radon measure can be decomposed into the sum of its atomic part, of the form

$$\mu^{\mathrm{a}} = \sum_\alpha m_\alpha \delta(\cdot - \xi_\alpha).$$

and of its continuous part, which has no atoms and is defined as

$$\mu^{\mathrm{c}} = \mu - \mu^{\mathrm{a}}.$$

Let us now recall Helly's theorem:

THEOREM A.1 *Let $(\mu_j)_{j \geq 1}$ be a weakly $*$ convergent sequence of Radon measures, denote its limit by μ_∞ and define*

$$\forall j = 1, 2, \ldots, \infty, \quad F_j(x) = \int_{[0,x]} \mu_j. \tag{12.A.1}$$

Then, there exists an at most countable set $D \in [0, T]$ such that for all $x \in [0, T] \setminus D$, the sequence $(F_j(x))_{j \in \mathbb{N}^*}$ converges to $F_\infty(x)$.

There is a refinement of the above result when the limit μ_∞ has no atomic part:

LEMMA A.2 *Let the sequence of Radon measures $(\mu_j)_j$ converge weakly $*$ to a limit μ_∞ without atomic part. Let F_j be defined by (12.A.1); then the convergence of F_j to F_∞ is uniform.*

The proof of this result will be given in a later, developed version, of this article. Its statement makes sensible the main result of the present appendix:

THEOREM A.3 *Let $(\mu_j)_{j \in \mathbb{N}^*}$ be a sequence of measures over $[0, T]$, which converges weakly $*$ to μ_∞. Assume that μ decomposes into its atomic part μ^{a} and its continuous part μ^{c}. There exists a sequence of continuous functions Ψ_j with values in $[0, 1]$ such that the following assertions hold:*

1 *$\Psi_j \mu_j$ converges weakly $*$ to μ^{a};*

2 *$(1 - \Psi_j)\mu_j$ converges weakly $*$ to μ^{c}, and in particular the integral $\int_{[0,x]} (1 - \Psi_j)\mu_j$ converges uniformly to the integral $\int_{[0,x]} \mu_\infty^{\mathrm{c}}$;*

he measure of the support of Ψ_j by $|\mu_j|(1 - \Psi_j)$ tends to 0 as j tends to infinity.

References

H. Brézis. *Opérateurs maximaux monotones et semi-groupes de contractions dans les espaces de Hilbert.* North-Holland Publishing Co., Amsterdam, 1973. North-Holland Mathematics Studies, No. 5. Notas de Matemática (50).

J. Dieudonné. *Éléments d'analyse. Tome II: Chapitres XII à XV.* Cahiers Scientifiques, Fasc. XXXI. Gauthier-Villars, Éditeur, Paris, 1968.

J. Dieudonné. *Treatise on analysis. Vol. II.* Translated from the French by I. G. Macdonald. Pure and Applied Mathematics, Vol. 10-II. Academic Press, New York, 1970.

P. Malliavin and H. Airault. *Intégration et analyse de Fourier. Probabilités et analyse gaussienne.* Collection Maîtrise de Mathématiques Pures. [Collection of Pure Mathematics for the Master's Degree]. Masson, Paris, second edition, 1994.

Paul Malliavin. *Integration and probability*, volume 157 of *Graduate Texts in Mathematics.* Springer-Verlag, New York, 1995. With the collaboration of Hélène Airault, Leslie Kay and Gérard Letac, Edited and translated from the French by Kay, With a foreword by Mark Pinsky.

Michelle Schatzman. Le système différentiel $(d^2u/dt^2) + \partial\varphi(u) \ni f$ avec conditions initiales. *C. R. Acad. Sci. Paris Sér. A-B*, 284(11):A603–A606, 1977.

Michelle Schatzman. A class of nonlinear differential equations of second order in time. *Nonlinear Anal., Theory, Methods and Applications*, 2:355–373, 1978.

David E. Stewart. Convergence of a time-stepping scheme for rigid-body dynamics and resolution of Painlevé's problem. *Arch. Ration. Mech. Anal.*, 145(3):215–260, 1998.

Chapter 13

DISCRETE CONTACT PROBLEMS WITH FRICTION: A STRESS-BASED APPROACH

Franco Maceri and Paolo Bisegna

Department of Civil Engineering, University of Rome "Tor Vergata", 00133 Rome, Italy

franco.maceri@lagrange.it

Abstract Many three-dimensional contact problems with friction between linearly elastic bodies and rigid supports can be successfully modelled by using Signorini's law of unilateral contact and generalized Coulomb's law of anisotropic friction. The discrete stress-based static formulation of such a class of nonlinear problems leads to quasi-variational inequalities, whose unknowns, after condensation on the initial contact area, are the normal and tangential contact forces. In this paper a block-relaxation solution algorithm of these inequalities is studied, at the typical step of which two subproblems are solved one after the other. The former is a problem of friction with given normal forces, the latter is a problem of unilateral contact with prescribed tangential forces. For friction coefficient values smaller than an explicitly-computable limit value, every step of the iteration is shown to be a contraction. The contraction principle is used to establish the well-posedness of the discrete formulation, to prove the convergence of the algorithm, and to obtain an estimate of the convergence rate. An example shows the sub-optimality of the obtained limit value of the friction coefficient.

Keywords: Friction, unilateral contact, computational solid mechanics

1. Introduction

The three-dimensional quasi-static problem of a deformable body monotonically loaded, clamped along a part of its boundary and in receding unilateral frictional contact with a rigid support along another part of its boundary, is analyzed here in the linear elasticity framework. The nonlinear constraint is modelled according to Signorini's law of unilateral contact and anisotropic Coulomb's law of friction with associated slip rule.

In the isotropic case (i.e. if the classical Coulomb's law prevails) this contact problem was extensively studied both in the dynamic context (Jean, 1999, Moreau, 1988) and in elastostatics (Kalker, 1983). Variational (Duvaut and Lions, 1976, Panagiotopoulos, 1985, Panagiotopoulos, 1993) and mathematical-programming (Ciarlet, 1982, Cryer, 1971, Duvaut and Lions, 1976, Glowinski et al, 1975, Klarbring and Bjorkman, 1988, Klarbring, 1986, Kikuchi and Oden, 1988, Panagiotopoulos, 1975) concepts were applied to investigate the well-posedness and computational aspects. Recently some studies were devoted to simple models (Giambanco and Mróz, 2001, Hjiaj et al, 2002, Mróz, 2002, Zmitrowicz, 1999) of anisotropic friction, introducing an elliptic Coulomb's type cone.

The analysis of the isotropic problem is usually based on primal formulations (displacements) or mixed formulations (displacements-multipliers), and a large class of related numerical solution methods has been developed (Alart and Curnier, 1991, Chabrand et al, 1998, Haslinger, 1983, Haslinger et al, 1996, Lebon and Raous, 1992, Maceri and Bisegna, 1998, Simo and Laursen, 1992, Wriggers, 1995). On the contrary, dual (stress-based) formulations have not been given such a quantity of strategies (Dostal and Vondrak, 1997, Haslinger and Panagiotopoulos, 1984, Panagiotopoulos and Haslinger, 1992, Telega, 1988, Telega, 1991), despite the importance of stress from a mechanical point of view. Indeed, the stress based approach gives directly, as a result, the directions of the tangential force on the contact area and of the slip, if present.

The aim of this paper is to present and to state the convergence of a new stress-based algorithm aimed at computing the contact forces, on the basis of a discrete dual formulation of the contact problem, condensed on the initial contact area in order to obtain a small-sized numerical problem.

The highly nonlinear problem arising from the simultaneous presence of the Signorini and generalized Coulomb laws is split into two nested subproblems and solved by block-iteration. At each iteration step, first a friction problem with given normal forces, resulting from the previous iteration step, is solved. This supplies new tangential forces which, in turn, are prescribed in the solution of a unilateral contact problem.

This algorithmic scheme is the dual version of the celebrated PANA algorithm due to P. D. Panagiotopoulos (Panagiotopoulos, 1975), based on the primal formulation of the contact problem.

In this paper the well-posedness of the discrete dual condensed formulation, the convergence of the proposed algorithm and an error estimate of the k-th iterate are established for sufficiently small values of the friction coefficients.

In very special cases the algorithm does not converge, for large enough values of the friction coefficients, as it is shown by a counterexample.

2. Statement of the algorithm

The following discrete frictional three-dimensional contact problem, in dual formulation, condensed on the contact area, is considered:

$$\text{Find } (N,T) \in I\!R^n_- \times K_{-\mu N} \text{ such that, for all } (X,Y) \in I\!R^n_- \times K_{-\mu N},$$
$$(X-N)^t(A_{NN}N+A_{NT}T+D_N)+(Y-T)^t(A_{TN}N+A_{TT}T+D_T) \geq 0 \tag{13.1}$$

where, for any $Z \in I\!R^n_+$

$$K_Z = \Big\{(Y_1,Y_2) \in \mathcal{A}\colon \ \Big[(\Delta^{11}Y_1+\Delta^{12}Y_2)^2+(\Delta^{21}Y_1+\Delta^{22}Y_2)^2\Big]^{1/2} \leq Z\Big\}. \tag{13.2}$$

Here the superscript t denotes transposition; $I\!R^n$ is the space of n-components real vectors; $I\!R^n_+$ (respectively, $I\!R^n_-$) is the cone of $I\!R^n$ composed by vectors with nonnegative (respectively, nonpositive) components; $\mathcal{A} = I\!R^n \times I\!R^n$; squares, square roots and inequalities involving vectors are understood component-wise: e.g., $(Y_1^2 + Y_2^2)^{1/2} \leq Z$ means that the square root of the sum of the squares of the j-th components of Y_1 and Y_2 do not exceed the j-th component of Z, for each $j \in \{1 \ldots n\}$. The Euclidean vector norm is denoted by $\|\cdot\|$ and the same symbol denotes the induced matrix norm. The vectors $N \in I\!R^n$ and $T = (T_1, T_2) \in \mathcal{A}$ include the discrete normal and tangential Cartesian components of the contact reactions. The linear operators A_{NN}, $A_{TN} = A^t_{NT}$, and A_{TT}, respectively mapping $I\!R^n$ into itself, $I\!R^n$ into $\mathcal{A}$, and $\mathcal{A}$ into itself, are the block components of the condensed compliance matrix of the body, and satisfy the relation:

$$N^tA_{NN}N + N^tA_{NT}T + T^tA_{TN}N + T^tA_{TT}T > 0,$$
$$N \in I\!R^n \setminus \{0\},\ T \in \mathcal{A} \setminus \{(0,0)\}. \tag{13.3}$$

The vectors $D_N \in I\!R^n$ and $D_T \in \mathcal{A}$ are the block components of the condensed displacement. The symmetric positive-definite matrix $\Delta = (\Delta^{11}, \Delta^{12}//\Delta^{21}, \Delta^{22})$ has diagonal blocks and maps $\mathcal{A}$ into itself. Le[illegible] $\Gamma = (\Gamma^{11}, \Gamma^{12}//\Gamma^{21}, \Gamma^{22})$ be the inverse of Δ. The eigenvalues of t[illegible] matrix $(\Gamma^{11}_{jj}, \Gamma^{12}_{jj}//\Gamma^{21}_{jj}, \Gamma^{22}_{jj})$ are the multipliers of the friction coeffic[illegible] $\mu > 0$ along the principal friction directions relevant to the anisot[illegible] friction law involving the j-th components of the contact reactic[illegible] the particular case of a homogeneous friction law, the blocks comprising Δ and Γ are scalar matrices. In the isotropic case, $\Gamma_{jj}^{11} = \Gamma_{jj}^{22} = 1$ and $\Gamma_{jj}^{12} = \Gamma_{jj}^{21} = 0$, for $j \in \{1 \ldots n\}$.

In order to solve the quasi-variational inequality (13.1), the following block-relaxation algorithm is proposed. Let $N_o \leq 0$ be given, the k-th iteration is

- T_k is the solution of the friction problem with given normal contact forces

$$T_k = \arg\min_T \{\frac{1}{2} T^t A_{TT} T + T^t (D_T + A_{TN} N_{k-1}),\ T \in K_{-\mu N_{k-1}}\}; \tag{13.4}$$

- N_k is the solution of the unilateral contact problem with given tangential forces

$$N_k = \arg\min_N \{\frac{1}{2} N^t A_{NN} N + N^t (D_N + A_{NT} T_k),\ N \in I\!R^n_-\}. \tag{13.5}$$

Minimization problems (13.4) and (13.5) have unique solutions, since the involved functionals are strictly convex and the minimization sets are closed convex and not empty.

In a plane problem with homogeneous friction law, the tangential reactions T belong to $I\!R^n \times \{0\}$ and hence the convex K_Z reduces to

$$\overline{K}_Z = \{(Y_1, 0) \in I\!R^n \times \{0\} \colon\ |Y_1| \leq Z\}, \tag{13.6}$$

where $|\cdot|$ operates component-wise. For such a problem, the block-relaxation algorithm was stated and discussed in (Bisegna et al, 2001), and relevant numerical results were discussed in (Bisegna et al, 2002, Bisegna et al). The isotropic three-dimensional case was analyzed in (Bisegna et al, 2002). In this paper, a convergence result for the algorithm in the anisotropic three-dimensional case, summarized by equations (13.4) and (13.5), will be established, with a convergence rate formally identical to the one already established in the isotropic case.

3. Analysis of the algorithm

3.1 Diagonalized formulation

By introducing the following change of variables $\sigma = (A_{NN})^{1/2} N$, $\tau = (A_{TT})^{1/2} T$, the discrete frictional contact problem (13.1) can be written as:

'nd $(\sigma, \tau) \in \mathcal{H} \times \mathcal{K}_{-\mu Q_{\sigma\sigma}}$ such that, for all $(\xi, \theta) \in \mathcal{H} \times \mathcal{K}_{-\mu Q_{\sigma\sigma}}$,

$$(\xi - \sigma)^t (\sigma + C^t \tau + d_\sigma) + (\theta - \tau)^t (C\sigma + \tau + d_\tau) \geq 0 \tag{13.7}$$

where the following positions are understood

$$\begin{array}{ll} Q_\sigma = (A_{NN})^{-1/2}, & Q_\tau = \Delta(A_{TT})^{-1/2}, \\ d_\sigma = (A_{NN})^{-1/2} D_N, & d_\tau = (A_{TT})^{-1/2} D_T, \\ C = (A_{TT})^{-1/2} A_{TN} (A_{NN})^{-1/2}, & \mathcal{H} = \{x \in I\!R^n : \; Q_\sigma x \leq 0\}, \\ \mathcal{K}_z = \{(x_1, x_2) \in \mathcal{A} : \; ((Q_\tau^{11} x_1 + Q_\tau^{12} x_2)^2 + & \\ \qquad (Q_\tau^{21} x_1 + Q_\tau^{22} x_2)^2)^{1/2} \leq z\}, \; z \in I\!R^n_+. & \end{array} \tag{13.8}$$

Both operators Q_σ and Q_τ are invertible. The latter is partitioned in four blocks $(Q_\tau^{11}, Q_\tau^{12} // Q_\tau^{21}, Q_\tau^{22})$, implied by the decomposition $\tau = (\tau_1, \tau_2)$. Analogously, C is partitioned into two blocks $(C^1 // C^2)$, and has the property:

$$\sigma^t \sigma + \sigma^t C^t \tau + \tau^t C \sigma + \tau^t \tau > 0, \; \sigma \in I\!R^n \setminus \{0\}, \tau \in \mathcal{A} \setminus \{(0,0)\}. \tag{13.9}$$

Given $\sigma_o \in \mathcal{H}$, the k-th iteration of the proposed algorithm is written as

$$\tau_k = \arg\min_\tau \{\frac{1}{2} \tau^t \tau + \tau^t (d_\tau + C \sigma_{k-1}), \; \tau \in \mathcal{K}_{-\mu Q_\sigma \sigma_{k-1}}\} \tag{13.10}$$

$$\sigma_k = \arg\min_\sigma \{\frac{1}{2} \sigma^t \sigma + \sigma^t (d_\sigma + C^t \tau_k), \; \sigma \in \mathcal{H}\}. \tag{13.11}$$

Equations (13.10) and (13.11) suggest to introduce the function f, transforming $\mathcal{H}$ into itself, defined by:

$$f \colon \sigma \in \mathcal{H} \to q(-d_\sigma - C^t p(-d_\tau - C\sigma, -\mu Q_\sigma \sigma)), \tag{13.12}$$

where

$$q \colon b \in I\!R^n \to \arg\min_v \{\frac{1}{2} v^t v - v^t b, \; v \in \mathcal{H}\} \tag{13.13}$$

and

$$p \colon (b, c) \in \mathcal{A} \times I\!R^n_+ \to \arg\min_v \{\frac{1}{2} v^t v - v^t b, \; v \in \mathcal{K}_c\}. \tag{13.14}$$

The functions p and q are well defined, since they are projection operators onto nonempty closed convex sets.

For any fixed point $\overline{\sigma} \in \mathcal{H}$ of the transformation f, the couple $(\overline{\sigma}, \overline{\tau}) \in \mathcal{H} \times \mathcal{K}_{-\mu Q_\sigma \overline{\sigma}}$, with $\overline{\tau} = p(-d_\tau - C\overline{\sigma}, -\mu Q_\sigma \overline{\sigma})$, is a solution of the quasi-variational inequality (13.7); conversely, the first component $\overline{\sigma} \in \mathcal{H}$ of a solution $(\overline{\sigma}, \overline{\tau}) \in \mathcal{H} \times \mathcal{K}_{-\mu Q_\sigma \overline{\sigma}}$ of (13.7) is a fixed point of f.

3.2 Preparatory results

Two elementary lemmas, simplifying the exposition of the converger proof, are gathered here and proved in the Appendix, for the sak completeness.

LEMMA 13.1 *Let C be a linear operator from $\mathbb{R}^n$ to $\mathcal{A}$ satisfying equation (13.9). Then $\|C\| < 1$.*

LEMMA 13.2 *Let $Q_\tau = (Q_\tau^{11}, Q_\tau^{12}//Q_\tau^{21}, Q_\tau^{22})$ be an invertible linear operator from $\mathcal{A}$ into itself. The function p defined by equation (13.14) has the property:*

$$\|p(b_1, c_1) - p(b_2, c_2)\| \leq \|b_1 - b_2\| + \|Q_\tau^{-1}\| \; \|c_1 - c_2\|, \qquad (b_1, c_1), (b_2, c_2) \in \mathcal{A} \times \mathbb{R}^n_+. \quad (13.15)$$

3.3 Well-posedness and convergence result

THEOREM 13.3 *Under the hypotheses stated in section 2, there exists a positive constant $\mathcal{M}$ such that for $0 \leq \mu < \mathcal{M}$ the transformation $f\colon \mathcal{H} \to \mathcal{H}$ defined by equation (13.12) is a contraction. As a consequence, for $0 \leq \mu < \mathcal{M}$ the discrete dual condensed formulation (13.7) of the Signorini-Coulomb contact problem has a unique solution $(\overline{\sigma}, \overline{\tau}) \in \mathcal{H} \times \mathcal{K}_{-\mu Q_\sigma \overline{\sigma}}$ and the proposed algorithm converges to this solution for any initial vector $\sigma_o \in \mathcal{H}$. Moreover, a constant $0 \leq \beta < 1$ exists such that the following error estimates hold*

$$\|\sigma_k - \overline{\sigma}\| \leq \frac{\|\sigma_o - \sigma_1\|\beta^k}{1-\beta}, \quad \|\tau_k - \overline{\tau}\| \leq \frac{\|\sigma_o - \sigma_1\|\beta^k}{(1-\beta)\|C\|}, \quad k \in N. \quad (13.16)$$

The constants $\mathcal{M}$ and β are explicitly given in equations (13.19) and (13.18), respectively.

Proof. For any $\tilde{\sigma}, \hat{\sigma} \in \mathcal{H}$, by using the projection theorem (Ciarlet, 1982) and Lemma 13.2, the following estimate is obtained:

$$\|f(\tilde{\sigma}) - f(\hat{\sigma})\| \leq \|C^t(p(-d_\tau - C\tilde{\sigma}, -\mu Q_\sigma \tilde{\sigma}) - p(-d_\tau - C\hat{\sigma}, -\mu Q_\sigma \hat{\sigma}))\| \leq$$
$$\|C\|(\| - C\tilde{\sigma} + C\hat{\sigma}\| + \|Q_\tau^{-1}\| \; \| - \mu Q_\sigma \tilde{\sigma} + \mu Q_\sigma \hat{\sigma}\|) \leq$$
$$\|C\|(\|C\| + \mu\|Q_\sigma\| \; \|Q_\tau^{-1}\|)\|\tilde{\sigma} - \hat{\sigma}\|. \quad (13.17)$$

It follows that f is Lipschitz continuous with Lipschitz coefficient

$$\beta = \|C\|(\|C\| + \mu\|Q_\sigma\| \; \|Q_\tau^{-1}\|). \quad (13.18)$$

From equation (13.9) and Lemma 13.1, it follows that $\|C\| < 1$. Therefore, setting

$$\mathcal{M} = \frac{1 - \|C\|^2}{\|C\| \; \|Q_\sigma\| \; \|Q_\tau^{-1}\|} > 0, \quad (13.19)$$

every $0 \leq \mu < \mathcal{M}$ it turns out that $0 \leq \beta < 1$. Hence, the assert vs from the Banach-Caccioppoli contraction principle, by observing that equations (13.10) and (13.11) can be summarized by the equation $\sigma_k = f(\sigma_{k-1})$. The error estimate for τ_k easily follows from Lemma 13.2.

It is worth observing that the constant $\mathcal{M}$ depends on quantities relevant to the body and sharing a significant mechanical interest: they are $\|C\|$, which is a measure of the energetic coupling between σ and τ, and $\|Q_\sigma\| \; \|Q_\tau^{-1}\|$, related to the direct complementary energies associated to σ and τ.

3.4 Sub-optimality result

A large number of numerical investigations of case studies (Raous et al, 1988) coming out from structural engineering applications showed the effectiveness of the proposed algorithm, even for a friction coefficient μ largely exceeding the established limit value $\mathcal{M}$. On the other hand, the convergence result obtained in the previous section is sub-optimal, since in very special cases the algorithm does not converge for $\mu > \mathcal{M}$, as it is shown by the following counterexample (Tosone, 2003).

THEOREM 13.4 *Let the hypotheses stated in section 2 be satisfied. Moreover, it is assumed that*

a) Q_σ *and* Q_τ *are diagonal matrices;*

b) $C = (C^1//C^2)$ *is such that* C^1 *is diagonal and negative definite;*

c) the diagonal matrix $-\mu(C^1)^t(Q_\tau^{11})^{-1}Q_\sigma$ *has all its diagonal elements not smaller than one;*

d) $d_\sigma \in I\!R^n_+$;

e) $d_\tau = (d_\tau^1//d_\tau^2)$ *is such that* $-d_\tau^2 + C^2 d_\sigma = 0$ *and* $-d_\tau^1 + C^1 d_\sigma > \mu(Q_\tau^{11})^{-1}Q_\sigma d_\sigma$.

Under these assumptions, the algorithm defined in section 2 does not converge when $\sigma_o = 0$.

Proof. Under the assumption *a)*,

$$\begin{aligned} &\mathcal{H} = I\!R^n_- \\ &\mathcal{K}_z = \{(x_1, x_2) \in \mathcal{A}\colon \; ((Q_\tau^{11}x_1)^2 + (Q_\tau^{22}x_2)^2)^{1/2} \le z\}, \; z \in I\!R^n_+, \end{aligned} \tag{13.20}$$

where Q_τ^{11} and Q_τ^{22} are diagonal matrices. Accordingly, for $\sigma_o = 0$, it turns out that

$$\sigma_1 = f(\sigma_o) = q(-d_\sigma - C^t p(-d_\tau, 0)) = q(-d_\sigma) = -d_\sigma, \tag{13.21}$$

since $-d_\sigma \in I\!R^n_-$ by hypothesis *d)*.

Now it remains to compute

$$\sigma_2 = f(\sigma_1) = q(-d_\sigma - C^t p(-d_\tau + Cd_\sigma, \mu Q_\sigma d_\sigma)). \tag{13.22}$$

To this end, it is necessary to compute $p(-d_\tau + Cd_\sigma, \mu Q_\sigma d_\sigma)$, i.e., to project $-d_\tau + Cd_\sigma$ onto the convex $\mathcal{K}_{\mu Q_\sigma d_\sigma}$. This projection is denoted by $\tilde{\tau} = (\tilde{\tau}^1 // \tilde{\tau}^2)$. Moreover, the vector $-d_\tau + Cd_\sigma$ is partitioned into the two components $(\tilde{d}^1 // \tilde{d}^2) = (-d_\tau^1 + C^1 d_\sigma // - d_\tau^2 + C^2 d_\sigma)$. Due to the diagonal structure of Q_τ^{11} and Q_τ^{22}, it is easily seen that $(\tilde{\tau}_i^1, \tilde{\tau}_i^2) \in \mathbb{R}^2$ is the projection of $(\tilde{d}_i^1, \tilde{d}_i^2)$ over the ellipse

$$(x_{1,i}, x_{2,i}) \in \mathbb{R}^2 \colon \ ((Q_{\tau,ii}^{11} x_{1,i})^2 + (Q_{\tau,ii}^{22} x_{2,i})^2)^{1/2} \leq \mu Q_{\sigma,ii} d_{\sigma,i}. \tag{13.23}$$

On the other hand, by assumption *e)*, it follows that

$$\tilde{d}_i^1 > \mu (Q_{\tau,ii}^{11})^{-1} Q_{\sigma,ii} d_{\sigma,i} \quad \text{and} \quad \tilde{d}_i^2 = 0, \tag{13.24}$$

so that $(\tilde{d}_i^1, \tilde{d}_i^2) \in \mathbb{R}^2$ lies on the positive $x_{1,i}$ axis, outside the ellipse (13.23). Thus,

$$\tilde{\tau}_i^1 = \mu (Q_{\tau,ii}^{11})^{-1} Q_{\sigma,ii} d_{\sigma,i} \quad \text{and} \quad \tilde{\tau}_i^2 = 0, \tag{13.25}$$

and

$$p(-d_\tau + Cd_\sigma, \mu Q_\sigma d_\sigma) = (\mu (Q_\tau^{11})^{-1} Q_\sigma d_\sigma, 0). \tag{13.26}$$

As a consequence, by (13.22) it turns out that

$$\sigma_2 = q(-d_\sigma - \mu (C^1)^t (Q_\tau^{11})^{-1} Q_\sigma d_\sigma). \tag{13.27}$$

Since, by hypotheses *c)* and *d)*,

$$-d_\sigma - \mu (C^1)^t (Q_\tau^{11})^{-1} Q_\sigma d_\sigma \geq 0, \tag{13.28}$$

it follows that $\sigma_2 = 0$. As a consequence, the sequence $\{\sigma_n\}$ turns out to be $\{0, -d_\sigma, 0, -d_\sigma, \ldots\}$ and hence it is not regular.

In order to satisfy the hypothesis *c)*, according to assumption *b)*, it suffices to take

$$\mu \geq \max_{i \in \{1 \ldots n\}} \frac{Q_{\tau,ii}^{11}}{-C_{ii}^1 Q_{\sigma,ii}}. \tag{13.29}$$

It appears that the right-hand side of equation (13.29) is greater than or equal to $1/(\|C\| \ \|Q_\sigma\| \ \|Q_\tau^{-1}\|)$, which, in turn, is greater than $\mathcal{M}$.

Acknowledgments

The financial supports of CNR and MIUR are gratefully acknowledged. This research was developed within the framework of Lagrange Laboratory, an European research group between CNRS, CNR, University of Rome "Tor Vergata", University of Montpellier II, ENPC and LCPC.

Appendix

Proof of Lemma 13.1. Choosing $\tau = -C\sigma$ in equation (13.9), it follows that $\sigma^t\sigma - \sigma^t C^t C\sigma > 0$ for $\sigma \in \boldsymbol{R}^n \setminus \{0\}$, or, equivalently, $\|C\sigma\| < 1$, for $\sigma \in \boldsymbol{R}^n : \|\sigma\| = 1$. The compactness of the unit sphere of $\boldsymbol{R}^n$ then yields $\|C\| = \sup_{\|\sigma\|=1} \|C\sigma\| = \max_{\|\sigma\|=1} \|C\sigma\| < 1$.

Proof of Lemma 13.2. Let $\boldsymbol{N}$ be the set of natural numbers. For any $s \in \boldsymbol{N}$, $s \geq 3$, the following function is considered:

$$p_s\colon (b,c) \in \mathcal{A} \times \boldsymbol{R}^n_+ \rightarrow \arg\min_v\{\frac{1}{2}v^t v - v^t b,\ v \in \mathcal{K}^s_c\}, \tag{13.A.1}$$

where, for $\alpha_i = 2(i-1)\pi/s$,

$$\mathcal{K}^s_c = \{(x_1,x_2) \in \mathcal{A}\colon\ (Q^{11}_\tau x_1 + Q^{12}_\tau x_2)\cos\alpha_i + (Q^{21}_\tau x_1 + Q^{22}_\tau x_2)\sin\alpha_i \leq c, i = 1\ldots s\}. \tag{13.A.2}$$

Clearly, p_s projects b onto the closed convex nonempty polyhedron $\mathcal{K}^s_c$ which approximates $\mathcal{K}_c$ from outside when s tends to infinity.

As a matter of fact, it is easily seen that, for every fixed $(b,c) \in \mathcal{A} \times \boldsymbol{R}^n_+$, the sequence $\{p_s(b,c)\}_{s\in\boldsymbol{N}}$ converges to $p(b,c)$. To this end, it is first observed that $\{p_s(b,c)\}_{s\in\boldsymbol{N}}$ is bounded, since

$$\mathcal{K}_c \subseteq \mathcal{K}^s_c \subseteq \mathcal{K}_{c/\cos(\pi/s)} \subseteq \mathcal{K}_{2c}. \tag{13.A.3}$$

Let $\{p_{s_m}(b,c)\}_{m\in\boldsymbol{N}}$ any converging subsequence, whose limit is denoted by $\hat{p}$. By equation (13.A.3) it follows that

$$\hat{p} \in \mathcal{K}_c. \tag{13.A.4}$$

On the other hand, since

$$\|p_{s_m}(b,c) - b\| \leq \|v - b\|,\ v \in \mathcal{K}^{s_m}_c, \tag{13.A.5}$$

it follows that

$$\|\hat{p} - b\| \leq \|v - b\|,\ v \in \mathcal{K}_c, \tag{13.A.6}$$

which, together with (13.A.4), implies that $\hat{p} = p(b,c)$. Hence, $p_s(b,c)$ converges to $p(b,c)$.

As a consequence, the assert of Lemma 13.2 results from the following estimate, which holds for every (b_1,c_1) and (b_2,c_2) belonging to $\mathcal{A} \times \boldsymbol{R}^n_+$:

$$\|p_s(b_1,c_1) - p_s(b_2,c_2)\| \leq \|b_1 - b_2\| + (\cos(\pi/s))^{-1}\|Q^{-1}_\tau\|\ \|c_1 - c_2\|, \tag{13.A.7}$$

which, in turn, follows from the triangle inequality

$$\|p_s(b_1,c_1) - p_s(b_2,c_2)\| \leq \|p_s(b_1,c_1) - p_s(b_2,c_1)\| + \|p_s(b_2,c_1) - p_s(b_2,c_2)\|. \tag{13.A.8}$$

In fact, the estimate of the first term on the right-hand side of equation (13.A.8) by $\|b_1 - b_2\|$ is a classical result (Ciarlet, 1982), whereas the estimate of the second term by $(\cos(\pi/s))^{-1}\|\boldsymbol{Q}^{-1}_\tau\|\ \|c_1 - c_2\|$ is supplied by Lemma A.1.

LEMMA A.1 *Let $Q_\tau = (Q^{11}_\tau, Q^{12}_\tau // Q^{21}_\tau, Q^{22}_\tau)$ be an invertible linear operator fr[...] $\mathcal{A}$ into itself. The function p_s defined by equation (13.A.1) has the property:*

$$\|p_s(b,c_1) - p_s(b,c_2)\| \leq (\cos(\pi/s))^{-1}\|Q^{-1}_\tau\|\ \|c_1 - c_2\|,\ c_1,c_2 \in \boldsymbol{R}^n_+,\ b \in \mathcal{A}. \quad (1[...]$$

Proof of Lemma A.1. Let $b \in \mathcal{A}$ be fixed. The projection $p_s(b,c)$ of b onto $\mathcal{K}_c^s$ activates (i.e., satisfies with the equality sign) a certain subset, depending on $c \in \boldsymbol{R}_+^n$, of the inequality constraints entering the definition of $\mathcal{K}_c^s$. The idea of the proof is to quotient the set $\boldsymbol{R}_+^n$ according to the equivalence relation stating that two c's are equivalent if they activate the same subset of inequality constraints. Given $c \in \boldsymbol{R}_+^n$, let $p_s(b,c) = (x_1, x_2)$, let c_j, A_j and B_j denote the j-th components of c, $Q_\tau^{11}x_1 + Q_\tau^{12}x_2$ and $Q_\tau^{21}x_1 + Q_\tau^{22}x_2$, respectively, and let

- J_o be the set of the indices $j \in [1 \dots n]$ such that $c_j = 0$;
- J_e be the set of the indices $j \in [1 \dots n]$ such that $A_j \cos\alpha_i + B_j \sin\alpha_i = c_j$ for exactly one index i, say $i(j)$; the function $j \in J_e \mapsto i(j) \in [1 \dots s]$ is denoted by i_e;
- J_v be the set of the indices $j \in [1 \dots n]$ such that $A_j \cos\alpha_i + B_j \sin\alpha_i = c_j$ for exactly two consecutive indices i's, say $i(j)$ and $i(j)+1$ (in particular, $i(j)+1$ is intended to be 1, if $i(j) = s$); the function $j \in J_v \mapsto i(j) \in [1 \dots s]$ is denoted by i_v;

It is readily verified that the quintet $(J_o; J_e, i_e; J_v, i_v)$ completely characterizes the set of active constraints, since $A_j \cos\alpha_i + B_j \sin\alpha_i$ may result equal to c_j for none, one, or two consecutive indices i's, unless $c_j = 0$. Two c's are defined to be equivalent if the corresponding quintets are the same. Accordingly, $\boldsymbol{R}_+^n$ is partitioned into $(2+2s)^n$ disjoint equivalence classes, the typical one being denoted by $\mathcal{R}_{J_o;J_e,i_e;J_v,i_v}$.

Let an equivalence class $\mathcal{R}_{J_o;J_e,i_e;J_v,i_v}$ be fixed. The estimate (13.A.9) holds true, provided that $c_1, c_2 \in \mathcal{R}_{J_o;J_e,i_e;J_v,i_v}$, as a consequence of Lemma A.2.

In order to prove the estimate (13.A.9) in the general case, let $c_1, c_2 \in \boldsymbol{R}_+^n$ be given. The segment joining c_1 and c_2, which, of course, belongs to $\boldsymbol{R}_+^n$, is decomposed into the union of a finite number of segments, each having the interior in some equivalence class $\mathcal{R}_{J_o;J_e,i_e;J_v,i_v}$. Provided that $c \mapsto p_s(b,c)$ is continuous, the assert of Lemma A.1 follows by writing the estimate (13.A.9) for the extremes of each of these segments, and then summing up the resulting inequalities.

It remains to prove that the function $c \mapsto p_s(b,c)$ is continuous. To this end, let $\{c_k\}_{k \in \boldsymbol{N}}$ be any sequence in $\boldsymbol{R}_+^n$ converging to some $\hat{c}$ belonging to some equivalence class $\mathcal{R}_{J_o;J_e,i_e;J_v,i_v}$. The assert follows by proving that $p_s(b,c_k)$ converges to $p_s(b,\hat{c})$. Since only a finite number of equivalence classes exist, up to considering a finite number of subsequences, it may be assumed, without loss of generality, that all the c_k's belong to the same equivalence class. Then, $p_s(b,c_k)$ is a Cauchy sequence, according to Lemma A.2, and hence it converges to some $\hat{p} \in \mathcal{A}$. By construction, $p_s(b,c_k) \in \mathcal{K}_{c_k}^s$: hence, taking the limit for $k \to \infty$, it results that:

$$\hat{p} \in \mathcal{K}_{\hat{c}}^s. \tag{13.A.10}$$

On the other hand, for any sufficiently small $\varepsilon > 0$ there exists a natural number ν such that, for $k > \nu$, $\Pi_{\overline{J}_o} c_k > \Pi_{\overline{J}_o}(\hat{c} - \varepsilon)$, where $\overline{J}_o = [1 \dots n] \setminus J_o$ and the minus operator acts component-wise. Then, setting

$$\hat{c}_\varepsilon = \Pi_{\overline{J}_o}^t \Pi_{\overline{J}_o}(\hat{c} - \varepsilon), \tag{13.A.11}$$

it follows that

$$k > \nu \Rightarrow \|p_s(b,c_k) - b\| \le \|v - b\|, \; v \in \mathcal{K}_{\hat{c}_\varepsilon}^s. \tag{13.A.12}$$

king the limit for $k \to \infty$ in equation (13.A.12) and exploiting the arbitrariness of d the continuity of the norm, it is obtained that

$$\|\hat{p} - b\| \le \|v - b\|, \; v \in \mathcal{K}_{\hat{c}}^s, \tag{13.A.13}$$

which, together with (13.A.10), implies that $\hat{p} = p_s(b, \hat{c})$ and hence the continuity of $c \mapsto p_s(b, c)$ follows.

LEMMA A.2 *Let $Q_\tau = (Q_\tau^{11}, Q_\tau^{12} // Q_\tau^{21}, Q_\tau^{22})$ be an invertible linear operator from $\mathcal{A}$ into itself. Let $c_1, c_2 \in \boldsymbol{R}_+^n$ belong to the same equivalence class $\mathcal{R}_{J_o;J_e,i_e;J_v,i_v}$ (defined in the proof of Lemma A.1). Then, the function p_s defined by equation (13.A.1) has the property:*

$$\|p_s(b, c_1) - p_s(b, c_2)\| \leq (\cos(\pi/s))^{-1} \|Q_\tau^{-1}\| \, \|c_1 - c_2\|, \; b \in \mathcal{A}. \tag{13.A.14}$$

Proof of Lemma A.2. Since c_1 and c_2 belong to the same equivalence class $\mathcal{R}_{J_o;J_e,i_e;J_v,i_v}$, the projection operations involved in the computation of $p_s(b, c_1)$ and $p_s(b, c_2)$ activate the same constraints. In order to concisely describe these constraints, some definitions are introduced. Let $J \subseteq [1 \ldots n]$ and n_J denote its cardinality. Let Π_J be the projection operator from $\boldsymbol{R}^n$ onto $\boldsymbol{R}^{n_J}$, associating to any $v \in \boldsymbol{R}^n$ the vector $\Pi_J v \in \boldsymbol{R}^{n_J}$ made of the components of v whose indices belong to J. Let $\Pi_\emptyset$ be the projection operator on the zero-dimensional vector space. Then, the active inequality constraints and the constraint manifold they define can be represented as follows:

$$\mathcal{L} = \{x \in \mathcal{A}\colon \; VPQ_\tau x = D\Pi c\}, \tag{13.A.15}$$

where

$$\Pi = \begin{pmatrix} \Pi_{J_o} \\ \Pi_{J_e} \\ \Pi_{J_v} \end{pmatrix}, \; P = \begin{pmatrix} \Pi & 0 \\ 0 & \Pi \end{pmatrix}, \; V = \begin{pmatrix} I & 0 & 0 & 0 & 0 & 0 \\ 0 & 0 & 0 & I & 0 & 0 \\ 0 & C_e & 0 & 0 & S_e & 0 \\ 0 & 0 & C_v & 0 & 0 & S_v \\ 0 & 0 & C_v' & 0 & 0 & S_v' \end{pmatrix}, \; D = \begin{pmatrix} 0 & 0 & 0 \\ 0 & 0 & 0 \\ 0 & I & 0 \\ 0 & 0 & I \\ 0 & 0 & I \end{pmatrix}. \tag{13.A.16}$$

Here C_e and S_e are diagonal matrices whose nonzero entries are respectively $\cos \alpha_i$ and $\sin \alpha_i$, for i spanning the image of i_e; analogously, C_v and S_v [respectively, C_v' and S_v'] are diagonal matrices whose nonzero entries are $\cos \alpha_i$ and $\sin \alpha_i$ [respectively, $\cos \alpha_{i+1}$ and $\sin \alpha_{i+1}$], for i spanning the image of i_v; moreover, I denotes appropriate identity operators (e.g., I at position $(1, 1)$ of V is $\Pi_{J_o} \Pi_{J_o}^t$, whereas I at position $(3, 2)$ of D is $\Pi_{J_e} \Pi_{J_e}^t$); analogously, 0 denotes appropriate zero operators.

It is observed that $p_s(b, c)$, $c \in \mathcal{R}_{J_o;J_e,i_e;J_v,i_v}$, can be regarded as the projection of b onto the linear manifold $\mathcal{L}$. Moreover, setting $\delta = \sin(2\pi/s)$ and

$$W = \begin{pmatrix} I & 0 & 0 & 0 & 0 \\ 0 & 0 & C_e & 0 & 0 \\ 0 & 0 & 0 & S_v'/\delta & -S_v/\delta \\ 0 & I & 0 & 0 & 0 \\ 0 & 0 & S_e & 0 & 0 \\ 0 & 0 & 0 & -C_v'/\delta & C_v/\delta \end{pmatrix}, \tag{13.A.17}$$

it is readily verified that $Q_\tau^{-1} P^t W$ is a right inverse of VPQ_τ. Then, according to Lemma A.3, for every c_1, c_2 belonging to $\mathcal{R}_{J_o;J_e,i_e;J_v,i_v}$ and every b belonging to $\mathcal{A}$, it results that:

$$\|p_s(b, c_1) - p_s(b, c_2)\| \leq \|Q_\tau^{-1}\| \, \|P^t\| \, \|WD\| \, \|\Pi\| \, \|c_1 - c_2\|. \tag{13.A.18}$$

As a consequence, the estimate (13.A.14) follows by noting that P^t and Π are projection operators, and

$$WD = \begin{pmatrix} 0 & 0 & 0 \\ 0 & C_e & 0 \\ 0 & 0 & (S'_v - S_v)/\delta \\ 0 & 0 & 0 \\ 0 & S_e & 0 \\ 0 & 0 & -(C'_v - C_v)/\delta \end{pmatrix}, \tag{13.A.19}$$

has an Euclidean norm not greater than $(\cos(\pi/s))^{-1}$.

LEMMA A.3 *Let F be a linear operator from $\boldsymbol{R}^k$ onto $\boldsymbol{R}^m$, $b \in \boldsymbol{R}^k$ and $d \in \boldsymbol{R}^m$. Then the orthogonal projection of b onto the linear manifold $\{x \in \boldsymbol{R}^k\colon Fx = d\}$, defined by*

$$\overline{x} = \arg\min_x \{\frac{1}{2}x^t x - x^t b,\ x \in \boldsymbol{R}^k,\ Fx = d\} \tag{13.A.20}$$

is given by

$$\overline{x} = (I - \mathcal{P})b + \mathcal{P}Md \tag{13.A.21}$$

where I is the identity of $\boldsymbol{R}^k$, $\mathcal{P}$ is the orthogonal projection operator onto the image of F^t and M is any linear operator from $\boldsymbol{R}^m$ to $\boldsymbol{R}^k$ such that FM is the identity of $\boldsymbol{R}^m$ (i.e., M is a right inverse of F).

Proof of Lemma A.3. It is well known (e.g., (Ciarlet, 1982)) that

$$\overline{x} = b + F^t\lambda, \tag{13.A.22}$$

where $\lambda \in \boldsymbol{R}^m$ is the Lagrange multiplier of the constraint $Fx = d$, and satisfies the equation:

$$Fb + FF^t\lambda = d. \tag{13.A.23}$$

Since F is onto, FF^t is invertible, and equation (13.A.23) provides a unique λ, which, after substitution into equation (13.A.22), yields equation (13.A.21). Indeed, $\mathcal{P} = F^t(FF^t)^{-1}F$ is the orthogonal projection operator onto $\Im F^t$, since it transforms $\ker F$ into the null vector, and the orthogonal to $\ker F$ (i.e., $\Im F^t$) into itself.

References

Alart P., Curnier A., A mixed formulation for frictional contact problems prone to Newton like solution methods, Computer Methods in Applied Mechanics and Engineering 92 (1991) 353–375.

Bisegna P., Lebon F., Maceri F., D-PANA: a convergent block-relaxation solution method for the discretized dual formulation of the Signorini–Coulomb contact problem, C. R. Acad. Sci. Paris 333 (2001) 1053–1058.

Bisegna P., Lebon F., Maceri F., The frictional contact of a piezoelectric body with a rigid support, in: Martins J., Monteiro Marques M. (Eds.), Contact Mechanics, Kluwer, Dordrecht, 2002.

Bisegna P., Lebon F., Maceri F., Relaxation procedures for solving Signorini-Coulomb contact problems, Computers & Structures (to appear).

Bisegna P., Maceri F., Tocchetti, F., An algorithmic approach to the discretized three-dimensional Signorini-Coulomb contact problem in dual formulation, in: Baniotopoulos, C.C. (Ed.), Nonsmooth/Nonconvex Mechanics with Applications in Engineering, Ziti Editions, Tessaloniki, 2002.

Chabrand P., Dubois F., Raous M., Various numerical methods for solving unilateral contact problems with friction, Mathematical and Computer Modelling 28 (1998) 97–108.

Ciarlet P. G., Introduction à l'analyse numérique matricielle et à l'optimisation, Masson, Paris, 1982.

Cryer C. W., The solution of a quadratic programming problem using systematic overrelaxation, SIAM Journal of Control 9 (1971) 385–392.

Dostàl Z., Vondrak V., Duality based solution of contact problems with Coulomb friction, Archives of Mechanics 49 (1997) 453–460.

Duvaut G., Lions J. L., Inequalities in mechanics and physics, Springer Verlag, Berlin, 1976.

Giambanco, G., Mróz, Z., The interphase model for the analysis of joints in rock masses and masonry structures, Meccanica 36 (2001) 111–130.

Glowinski R., Lions J. L., Trémolières R., Analyse numérique des inéquations variationnelles, Dunod, Paris, 1975.

Haslinger J., Approximation of the Signorini problem with friction obeying Coulomb law, Mathematical Methods in the Applied Science 5 (1983) 422–437.

Haslinger J., Hlavàček I., Nečas J., Numerical methods for unilateral problems in solid mechanics, in: Ciarlet P. G., Lions J. L. (Eds.), Handbook of numerical analysis vol. IV, Elsevier, Amsterdam, 1996.

Haslinger J., Panagiotopoulos P. D., The reciprocal variational approach of the Signorini problem with friction. Approximation results, Proceedings Royal Society Edinburgh 98 (1984) 365–383.

Hjiaj, M., de Saxcé, G., Mróz, Z., A variational inequality-based formulation of the frictional contact law with a non-associated sliding rule, European Journal of Mechanics A/Solids 21 (2002) 49–59.

Jean M., The non smooth contact dynamics method, Computer Methods in Applied Mechanics and Engineering 177 (1999) 235–257.

Kalker J. J., On the contact problem in elastostatics, in: Del Piero G., Maceri F. (Eds.), Unilateral problems in structural analysis, Springer Verlag, Berlin, 1983.

Kikuchi N., Oden J. T., Contact problems in elasticity: a study of variational inequalities and finite element methods, SIAM, Philadelphia, 1988.

Klarbring A., Quadratic programs in frictionless contact problems, International Journal of Engineering Science 24 (1986) 1207–1217.

Klarbring A., Bjorkman G., Solution of large displacement contact problems with friction using Newton's method for generalized equations, Internat. J. Numer. Methods Engrg. 34 (1992), 249–269.

Lebon F., Raous M., Friction modelling of a bolted junction under internal pressure loading, Computers & Structures 43 (1992) 925–933.

Maceri F., Bisegna P., The unilateral frictionless contact of a piezoelectric body with a rigid support, Mathematical and Computer Modelling 28 (1998) 19–28.

Moreau J. J., Unilateral contact and dry friction in finite freedom dynamics, in: Moreau J. J., Panagiotopoulos P. D. (Eds.), Nonsmooth mechanics and applications, Springer Verlag, Berlin, 1988.

Mróz, Z., Contact friction models and stability problems, in: Martins J.A.C., Raous M. (Eds.), Friction and instabilities, CISM Courses and Lectures 457, Springer Verlag, Berlin, 2002.

Panagiotopoulos P. D., A nonlinear programming approach to the unilateral contact and friction boundary value problem in the theory of elasticity, Ing.-Arch. 44 (1975) 421–432.

Panagiotopoulos P. D., Inequality problems in mechanics and applications. Convex and nonconvex energy functions, Birkhauser, Boston, 1985.

Panagiotopoulos P. D., Haslinger J., On the dual reciprocal variational approach to the Signorini-Fichera problem. Convex and nonconvex generalization, ZAMM Journal of Applied Mathematics and Mechanics 72 (1992) 497–506.

Panagiotopoulos P. D., Hemivariational inequalities. Applications in mechanics and engineering, Springer Verlag, Berlin, 1993.

Raous M., Chabrand P., Lebon F., Numerical methods for frictional contact problem and applications, Journal of Theoretical and Applied Mechanics 7 (1988) 111–128.

Simo J., Laursen A., An augmented lagrangian treatment of contact problems involving friction, Computers & Structures 42 (1992) 97–116.

Telega J. J., Topics on unilateral contact problems of elasticity and inelasticity, in: Moreau J. J., Panagiotopoulos P. D. (Eds.), Nonsmooth mechanics and applications, Springer Verlag, Berlin, 1988.

Telega J. J., Quasi-static Signorini's contact problem with friction and duality, Internat. Ser. Numer. Math. 101 (1991) 199–214.

Tosone C., Private communication, 2003.

Wriggers P., Finite elements algorithms for contact problems, Archives of Computational Methods in Engineering 2 (1995) 1–49.

Zmitrowicz, A., An equation of anisotropic friction with sliding path curvature effects, International Journal of Solids and Structures 36 (1999) 2825–2848.

III

FLUID MECHANICS

Chapter 14

A BRIEF HISTORY OF DROP FORMATION

Jens Eggers
School of Mathematics, University of Bristol, University Walk, Bristol BS8 1TW, United Kingdom
jens.eggers@bris.ac.uk

Abstract Surface-tension-related phenomena have fascinated researchers for a long time, and the mathematical description pioneered by Young and Laplace opened the door to their systematic study. The time scale on which surface-tension-driven motion takes place is usually quite short, making experimental investigation quite demanding. Accordingly, most theoretical and experimental work has focused on static phenomena, and in particular the measurement of surface tension, by physicists like Eötvös, Lenard, and Bohr. Here we will review some of the work that has eventually lead to a closer scrutiny of time-dependent flows, highly non-linear in nature. Often this motion is self-similar in nature, such that it can in fact be mapped onto a pseudo-stationary problem, amenable to mathematical analysis.

Keywords: drop formation, hydrodynamics, non-linear fluid mechanics, stroboscopic method

Introduction

Flows involving free surfaces lend themselves to observation, and thus have been scrutinized for hundreds of years. The earliest theoretical work was concerned almost exclusively with the equilibrium shapes of fluid bodies, and with the stability of the motion around those shapes. Experimentalists, always being confronted with physical reality, were much less able to ignore the strongly non-linear nature of hydrodynamics. Thus many of the non-linear phenomena, that are the focus of attention today, had already been reported 170 years ago. However, with no theory in place to put these observations into perspective, non-linear phenomena took the back seat to other issues, and were soon forgotten.

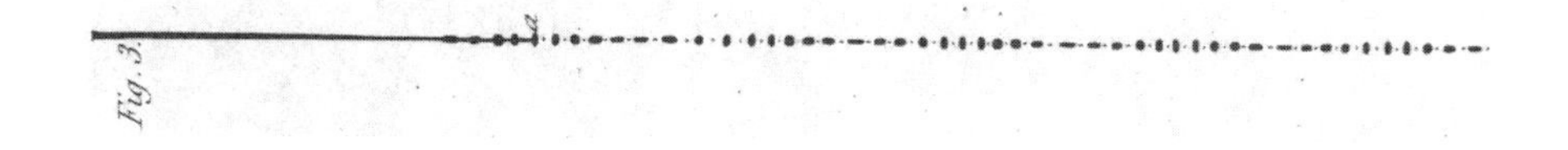

Figure 14.1. A figure from Savart's original paper (Savart, 1833), showing the breakup of a liquid jet 6 mm in diameter. It clearly shows the succession of main and satellite drops as well as drop oscillations.

Here we report on the periodic rediscovery of certain non-linear features of drop formation, by retracing some of the history of experimental observation of surface tension driven flow. Recently there has been some progress on the theoretical side, which relies on the self-similar nature of the dynamics close to pinching.

1. Savart and Plateau

Modern research on drop formation begins with the seminal contribution of (Savart, 1833). He was the first to recognize that the breakup of liquid jets is governed by laws independent of the circumstance under which the jet is produced, and concentrated on the simplest possible case of a circular jet. Without photography at one's disposal, experimental observation of drop breakup is very difficult, since the timescale on which it is taking place is very short.

Yet Savart was able to extract a remarkably accurate and complete picture of the actual breakup process using his naked eye alone. To this end he used a black belt, interrupted by narrow white stripes, which moved in a direction parallel to the jet. This effectively permitted a stroboscopic observation of the jet. To confirm beyond doubt the fact that the jet breaks up into drops and thus becomes discontinuous, Savart moved a "slender object" swiftly across the jet, and found that it stayed dry most of the time. Being an experienced swordsman, he undoubtedly used this weapon for his purpose (Clanet, 2003). Savart's insight into the *dynamics* of breakup is best summarized by Fig.14.1 taken from his paper (Savart, 1833).

To the left one sees the continuous jet as it leaves the nozzle. Perturbations grow on the jet, until it breaks up into drops, at a point labeled "a". Near a an elongated neck has formed between two bulges which later become drops. After breakup, in between two such drops, a much smaller "satellite" drop is always visible. Owing to perturbations received when they were formed, the drops continue to oscillate around a spherical shape. Only the very last moments leading to drop formation are not quite resolved in Fig.14.1.

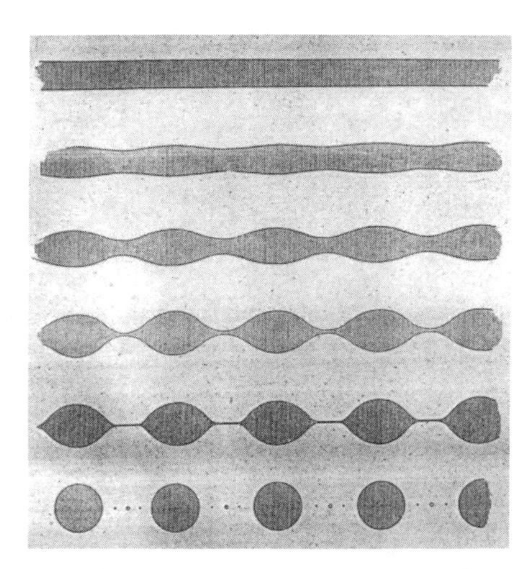

Figure 14.2. Breakup of a liquid column of oil, suspended in a mixture of alcohol and water (Plateau, 1849). First small perturbations grow, leading to the formation of fine threads. The threads each break up leaving three satellites.

From a theoretical point of view, what is missing is the realization that surface tension is the driving force behind drop breakup, the groundwork for the description of which was laid by (Young, 1804) and (Laplace, 1805). Savart however makes reference to mutual attraction between molecules, which make a sphere the preferred shape, around which oscillations take place. The crucial role of surface tension was recognized by (Plateau, 1849), who confined himself mostly to the study of equilibrium shapes. From this insight it follows as well whether a given perturbation imposed on a fluid cylinder will grow or not. Namely, any perturbation that will lead to a reduction of surface area is favored by surface tension, and will thus grow. This makes all sinusoidal perturbations with wavelength longer than 2π unstable. At the same time as Plateau, Hagen published very similar investigations, without quite mastering the mathematics behind them (Hagen, 1849). The ensuing quarrel between the two authors, published as letters to *Annalen der Physik*, is quite reminiscent of similar debates over priority today.

A little earlier Plateau had developed his own experimental technique to study drop breakup (Plateau, 1843), by suspending a liquid bridge in another liquid of the same density in a so-called "Plateau tank", thus eliminating the effects of gravity. Yet this research was focused on predicting whether a particular configuration would be stable or not. However Plateau also included some experimental sketches (cf. Fig.14.2) that offer interesting insight into the nonlinear dynamics of breakup for a

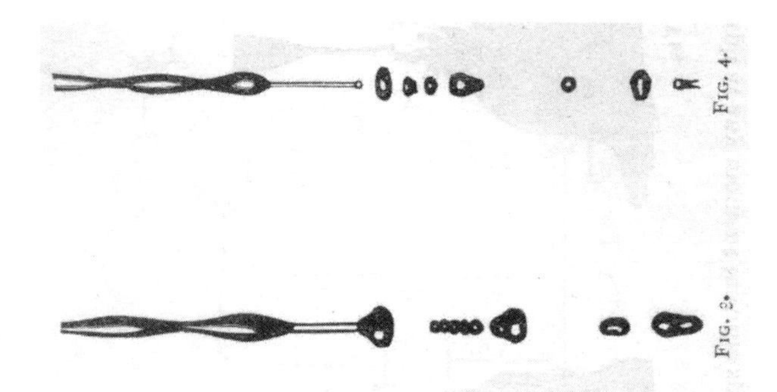

Figure 14.3. Two photographs of water jets taken by (Rayleigh, 1891), using a short-duration electrical spark.

viscous fluid: first a very thin and elongated thread forms, which has its minimum in the middle. However, the observed final state of a satellite drop in the center, with even smaller satellite drops to the right and left indicates that the final stages of breakup are more complicated: the thread apparently broke at 4 different places, instead of in the middle.

Following up on Plateau's insight, (Rayleigh, 1879) added the flow dynamics to the description of the breakup process. At low viscosities, the time scale τ of the motion is set by a balance of inertia and surface tension:

$$\tau = \sqrt{\frac{r^3 \rho}{\gamma}}. \tag{14.1}$$

Here r is the radius of the (water) jet, ρ the density, and γ the surface tension. For the jet shown in Fig.14.1, this amounts to $\tau = 0.02\ s$, a time scale quite difficult to observe with the naked eye. Rayleigh's linear stability calculation of a fluid cylinder only permits to describe the initial growth of instabilities as they initiate near the nozzle. It certainly fails to describe the details of drop breakup leading to, among others, the formation of satellite drops. Linear stability analysis is however quite a good predictor of important quantities like the continuous length of the jet.

2. Photography

Rayleigh was well aware of the intricacies of the last stages of breakup, and published some experimental pictures himself (Rayleigh, 1891). Unfortunately, these pictures were produced by a single short spark, so they only transmit a rough idea of the *dynamics* of the process. However, it is again clear that satellite drops, or entire sequences of them, are pro-

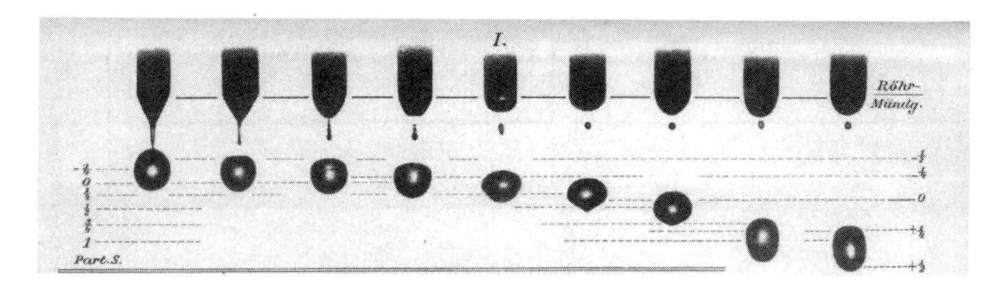

Figure 14.4. A sequence of pictures of a drop of water falling from a pipette (Lenard, 1887). For the first time, the sequence of events leading to satellite formation can be appreciated.

duced by elongated necks between two main drops. Clearly, what is needed for a more complete understanding is a sequence of photographs showing one to evolve into the other.

The second half of the 19th century is an era that saw a great resurgence of the interest in surface tension related phenomena, both from a theoretical and experimental point of view. The driving force was the central role it plays in the quest to understand the cohesive force between fluid particles (Rowlinson, 2002), for example by making precise measurements of the surface tension of a liquid. Many of the most well-known physicists of the day contributed to this research effort, some of whom are known today for their later contributions to other fields (Eötvös, 1886, Quincke, 1877, Lenard, 1887, Bohr, 1909). A particular example is the paper by (Lenard, 1887), who observed the drop oscillations that remain after break-up, already noted by Savart. By measuring their frequency, the value of the surface tension can be deduced.

To record the drop oscillations, Lenard used a stroboscopic method, which enabled him to take an entire sequence with a time resolution that would otherwise be impossible to achieve. As more of an aside, Lenard also records a sequence showing the dynamics close to breakup, leading to the separation of a drop. It shows for the first time the origin of the satellite drop: first the neck breaks close to the main drop, but before it is able to snap back, it also pinches on the side toward the nozzle. The presence of a slender neck is intimately linked to the profile near the pinch point being very asymmetric: on one side it is very steep, fitting well to the shape of the drop. On the other side it is very flat, forcing the neck to be flat and elongated.

However, as noted before, few people took note of the fascinating dynamics close to breakup. From a theoretical point of view, tools were limited to Rayleigh's linear stability analysis, which does not include a

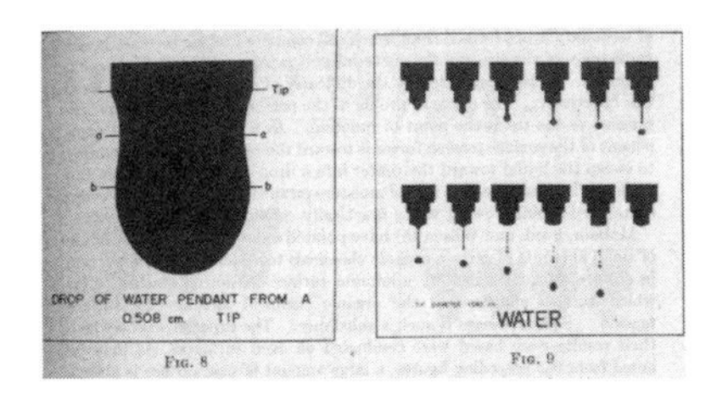

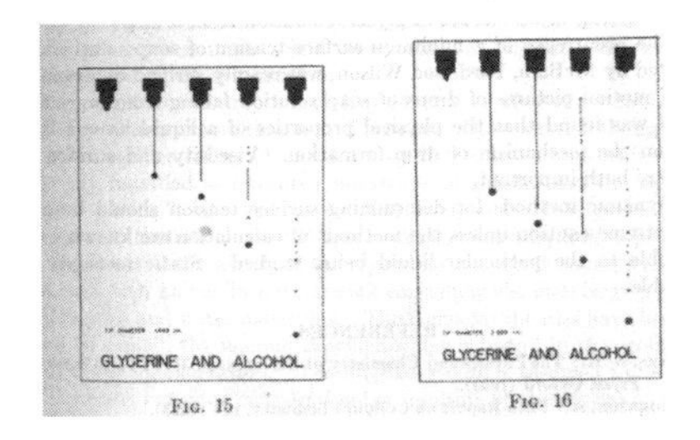

Figure 14.5. A drop of water (left) and a glycerol-alcohol mixture (right) falling from a pipette (Edgerton et al., 1937). The drop of viscous fluid pulls out long necks as it falls.

mechanism for satellite formation. Many years later, the preoccupation was still to find simple methods to measure surface tension, one of them being the "drop weight method" (Harkins and Brown, 1919). The idea of the method is to measure surface tension by measuring the weight of a drop falling from a capillary tubes of defined diameter. Harold Edgerton and his colleagues looked at time sequences of a drop of fluid of different viscosities falling from a faucet (Edgerton et al., 1937), rediscovering some of the features observed originally by Lenard, but adding some new insight.

Fig.14.5 shows a water drop falling from a faucet, forming quite an elongated neck, which then decays into *several* satellite drops. The measured quantity of water thus comes from the main drop as well as from some of the satellite drops; some of the satellite drops are projected upward, and thus do not contribute. The total weight thus depends on a very subtle dynamical balance, that can hardly be a reliable measure of surface tension. In addition, as Fig.14.5 demonstrates, a high viscosity fluid like glycerol forms extremely long threads, that break up into myriads of satellite drops. In particular, the drop weight cannot be a function of surface tension alone, but also depends on viscosity, making the furnishing of appropriate normalization curves unrealistically complicated.

3. Modern times

After Edgerton's paper, the next paper that could report significant progress in illuminating *non-linear* aspects of drop breakup was published in 1990 (Peregrine, 1990). Firstly, it contains a detailed sequence of a drop of water falling from a pipette $D = 5.2mm$ in diameter, renewing efforts to understand the underlying dynamics. Secondly, it was proposed that close to pinch-off the dynamics actually becomes quite

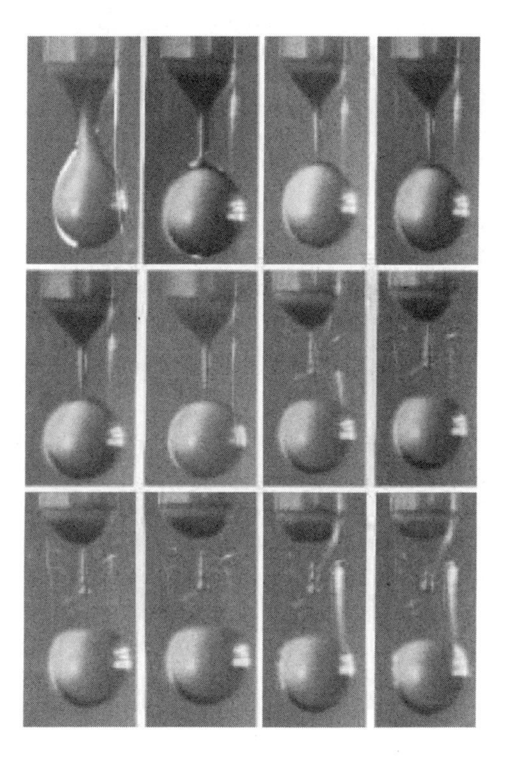

Figure 14.6. A high-resolution sequence showing the bifurcation of a drop of water (Peregrine, 1990).

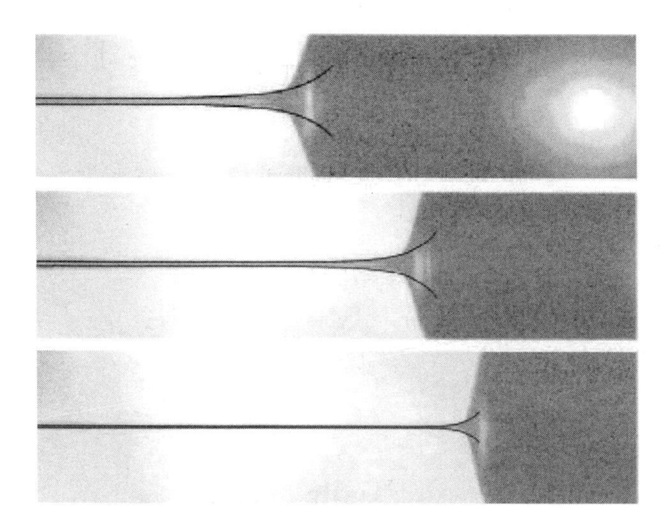

Figure 14.7. A sequence of interface profiles of a jet of glycerol close to the point of breakup (Kowalewski, 1996). The experimental images correspond to $t_0 - t = 350\mu s, 298\mu s$, and $46\mu s$. Corresponding analytical solutions based on self-similarity of the entire profile are superimposed.

simple, since any *external* scale cannot play a role. Namely, if the minimum neck radius h_{min} is the only relevant length scale, and if viscosity does not enter the description, than at a time $t_0 - t$ away from breakup on must have

$$h_{min} \propto \left(\frac{\gamma}{\rho}\right)^{2/3} (t_0 - t)^{2/3} \tag{14.2}$$

for dimensional reasons. At some very small scale, one expects viscosity to become important. The only length scale that can be formed from fluid parameters alone is

$$\ell_\nu = \frac{\nu^2 \rho}{\gamma}. \tag{14.3}$$

Thus the validity of (14.2) is limited to the range $D \gg h_{min} \gg \ell_\nu$ between the external scale and this inner viscous scale.

These simple similarity ideas can in fact be extended to obtain the laws for the entire profile, not just the minimum radius (Eggers, 1993). Namely, one supposes that the profile around the pinch point remains the same throughout, while it is only its radial and axial length scales which change. In accordance with (14.2), these length scales are themselves power laws in the time distance from the singularity. In effect, by making this transformation one has reduced the extremely rapid dynamics close to break-up to a static theory, and simple analytical solutions are possible.

The experimental pictures in Fig.14.7 are again taken using a stroboscopic technique, resulting in a time resolution of about $10\mu s$ (Kowalewski, 1996). Since for each of the pictures the temporal distance away from breakup is known, the form of the profile can be predicted without adjustable parameters. The result of the theory is superimposed on the experimental pictures of a glycerol jet breaking up as black lines. In each picture the drop about to form is seen on the right, a thin thread forms on the left. The neighborhood of the pinch point is described quite well; in particular, theory reproduces the extreme *asymmetry* of the profile. We already singled out this asymmetry as responsible for the formation of satellite drops.

One of the conclusions of this brief overview is that research works in a fashion that is far from straightforward. Times of considerable interest in a subject are separated by relative lulls, and often known results, published in leading journals of the day, had to be rediscovered. However from a broader perspective one observes a development from questions of (linear) stability and the measurement of static quantities, to a focus that is more and more on the (non-linear) dynamics that makes fluid mechanics so fascinating.

Acknowledgments

I have the pleasure to acknowledge very helpful input from Christophe Clanet and David Quéré.

References

F. Savart, Ann. Chim. **53**, 337; plates in vol. 54, (1833).

C. Clanet (2003). I am relying on remarks by Christophe Clanet, a scholar of Savart's life and achievements.

J. Plateau, Acad. Sci. Bruxelles Mém. **XVI**, 3 (1843).

J. Plateau, Acad. Sci. Bruxelles Mém. **XXIII**, 5 (1849).

G. Hagen, *Verhandlungen Preuss. Akad. Wissenschaften*, (Berlin), p. 281 (1849).

T. Young, Philos. Trans. R. Soc. London **95**, 65 (1804).

P. S. de Laplace, *Méchanique Celeste,* Supplement au X Libre (Courier, Paris, 1805)

Lord Rayleigh, Proc. London Math. Soc. **10**, 4 (1879). (appeared in the volume of 1878)

L. Eötvös, Wied. Ann. **27**, 448 (1886).

G.H. Quincke, Wied. Ann. **2**, 145 (1877).

P. Lenard, Ann. Phys. **30**, 209 (1887).
Lord Rayleigh, Nature **44**, 249 (1891).
N. Bohr, Phil. Trans. Roy. Soc. A **209**, 281 (1909).
W.D. Harkins and F. E. Brown, J. Am. Chem. Soc. **41**, 499 (1919).
H.E. Edgerton, E. A. Hauser, and W. B. Tucker, J. Phys. Chem. **41**, 1029 (1937).
D.H. Peregrine, G. Shoker, and A. Symon, J. Fluid Mech. **212**, 25 (1990).
J. Eggers, Phys. Rev. Lett. **71**, 3458 (1993).
T.A. Kowalewski, Fluid Dyn. Res. **17**, 121 (1996).
J.S. Rowlinson, *Cohesion*, Cambridge (2002).

Chapter 15

SEMICLASSICAL APPROACH OF THE "TETRAD MODEL" OF TURBULENCE

Aurore Naso
and Alain Pumir
INLN, CNRS, 1361, route des Lucioles, F-06560, Valbonne, France
Aurore.Naso@inln.cnrs.fr
Alain.Pumir@inln.cnrs.fr

Abstract We consider the 'tetrad model' of turbulence, which describes the coarse grained velocity derivative tensor as a function of scale with the help of a mean field approach. The model is formulated in terms of a set of stochastic differential equations, and the fundamental object of this model is a tetrahedron (tetrad) of Lagrangian particles, whose characteristic size lies in the inertial range. We present the approximate 'semiclassical' method of resolution of the model. Our first numerical results show that the solution correctly reproduces the main features obtained in Direct Numerical Simulations, as well as in experiments (van der Bos et al., 2002).

Keywords: Hydrodynamic turbulence, Lagrangian tetrad, semiclassical approximation

Introduction

Turbulence in fluids is a complex process, which involves subtle hydrodynamic effects (Moffatt, 1981). The existence of coherent vortical structures has a profound impact on turbulent motion. In this sense, the constraint discovered by Prof. Moreau (Moreau, 1961, see also Moffatt, 1969) to whom this article is dedicated, was a major reminder that turbulent flows cannot be understood without properly addressing fluid dynamical issues.

Another fascinating aspect of turbulent flows comes from the existence of a broad range of excited length scales. The successful prediction of a

$k^{-5/3}$ spectrum of velocity fluctuations (Frisch, 1995) and the discovery that simple scaling laws do not adequately describe turbulence has been a major source of inspiration for theoretical work on turbulence. The more so as the study of turbulence has relied for many years on measurements and analysis of a signal (such as a velocity component) at one or two points. Much effort has been devoted to the investigation of the nth order structure function and its scale dependence. However, although these structure functions are very suitable for the study of scaling laws, they do not provide much information on dynamical processes such as the vorticity stretching. Indeed, structure functions take into account only one of the eight independent components of the velocity gradient tensor.

This observation leads to the suggestion to describe the velocity field through the whole velocity gradient tensor, more particularly through its average over a volume Γ whose characteristic scale lies in the inertial range: $M_{ab} = \int_\Gamma d\vec{r}\, \partial_b v_a\,(\vec{r})$ (Chertkov et al., 1999), referred to in the following as the *coarse-grained* velocity gradient tensor. Thus the flow is described with a mean field approach.

The first step for the construction of the model consists in parameterizing these small volumes. As the minimal parametrization of a tridimensional volume is a cluster of four points, we propose to follow the Lagrangian dynamics of a tetrad of Lagrangian particles, what gives access to the velocity finite difference. The phenomenological model introduced in (Chertkov et al., 1999) describes the M-dynamics, as well as the geometrical deformation of the tetrahedron. In order to take into account this geometry, the g-tensor is introduced: $g = (\rho\rho^+)^{-1}$, where ρ is the reduced coordinates tensor of the tetrahedron vertices. In the original paper, an approximate solution was obtained in the *classical* limit. In a path integral formalism, the method consisted in constructing the Green's function by simply following the deterministic trajectory. This amounts to neglect the effect of the noise. Crude as this approximation was, it led to numerical results in semiquantitative agreement with DNS results.

In this short note, we propose to improve the solution by incorporating the effect of noise in a *semiclassical* approach.

Section 1 is devoted to the coarse-grained velocity gradient tensor M and to its connection to the local topology of the flow. In Section 2 we briefly remind the main aspects of the phenomenological stochastic model for M- and g- dynamics. In Section 3 we discuss the path integral representation of the solution, as well as the classical approximation. In Section 4 we discuss the semiclassical approximation. We present numerical results which demonstrate that our analysis captures very well the numerical results, as well as the experimental results.

1. The M-tensor

Before presenting the model, we insist on the advantages of the velocity gradient tensor M for the geometric description of the flow. Indeed, its eigenvalues permit to characterize the topology of the flow at a given scale. In the incompressible case, the three eigenvalues can be completely characterized by the two invariants Q and R of the tensor M:

$$\begin{cases} Q = -\frac{1}{2} Tr\, \mathbf{M}^2 \\ R = -\frac{1}{3} Tr\, \mathbf{M}^3 \end{cases}$$

The local topology of the flow qualitatively depends on the sign of the discriminant $D = 27R^2 + 4Q^3$. Thus, for $D > 0$, the M-tensor has two complex conjugate eigenvalues and a real one, hence the flow is elliptic, with locally swirling streamlines. For $D < 0$, the three eigenvalues of M are real: physically strain dominates and the flow is locally hyperbolic. Hence the PDF of Q and R gives a geometric description of the flow. It is also highly informative to represent the densities of quantities, such as the entropy, the energy flux or the vortex stretching in the (R,Q) plane, as shown in Chertkov et al., 1999.

One of our main goals here is to compute the PDF of the Q, R invariants. We stress that this quantity can be computed in Direct Numerical Simulations of turbulence, and is now accessible in experiments (van der Bos et al., 2002), suggesting challenging comparisons between theory and experiments.

2. The phenomenological model

In order to construct the coarse-grained approximation of the velocity field over a finite size region of fluid comoving with the flow, we consider a set of points. The minimal parametrization of a tridimensional volume is a cluster of four points, hence we propose here to follow the dynamics of four Lagrangian particles. The motion of the tetrad's center of mass results from the overall advection of a parcel of fluid, of no interest for the study of the small scale properties of turbulence. In the following, we consider the reduced coordinates tensor ρ, and in particular, the moment of inertia tensor g, defined by:

$$\mathbf{g} = \left(\rho\rho^{+}\right)^{-1} \tag{15.1}$$

The evolution equation of this tensor is derived from the ρ one. Writing the reduced velocities tensor v as the sum of a coherent component and of a fluctuating one, and approximating the first one by the best linear fit (which defines the coarse-grained velocity gradient tensor M), we obtain the following equation:

$$\frac{d\rho}{dt} = \mathbf{v} = \rho\mathbf{M} + \widetilde{\zeta} \tag{15.2}$$

Equations (15.1) and (15.2) imply:

$$\frac{d\mathbf{g}}{dt} + \mathbf{M}\mathbf{g} + \mathbf{g}\mathbf{M}^{+} = \zeta \tag{15.3}$$

where ζ is a fluctuating term. We assume here that it is a Gaussian noise, white in time.

The M evolution equation is reminiscent of the equation for the velocity gradient $A_{ij} \equiv \partial_i v_j$:

$$\frac{d\mathbf{A}_{ij}}{dt} + \mathbf{A}_{ij}^2 - \frac{1}{3}(Tr\,\mathbf{A}^2)\delta_{ij} = \mathbf{H}_{ij}$$

where H_{ij} is the pressure hessian term, up to viscosity effects.

The case where the right hand side of this equation is set to zero is the *Restricted Euler dynamics*, derived from the Euler equation (Vieillefosse, 1984) with the assumption that the pressure term is isotropic. This approximation leads to a finite time singularity.

In the present model, the fundamental object is the averaged velocity gradient tensor, and the g-tensor is introduced in order to take into account the shape of the tetrad, thus allowing an anisotropic model of pressure. The postulated evolution equation for M is:

$$\frac{d\mathbf{M}}{dt} + (1-\alpha)\left(\mathbf{M}^2 - \mathbf{\Pi}\,Tr\mathbf{M}^2\right) = \eta \tag{15.4}$$

where $\mathbf{\Pi} \equiv \frac{\mathbf{g}}{Tr\,\mathbf{g}}$.

In (15.4), non local pressure effects are modeled as the sum of a coherent component, which is chosen to renormalize the dynamics by an α factor ($0 < \alpha < 1$), and a fluctuating term η. The coupling term between M- and g- dynamics, Π, maintains incompressibility at any time. Its form was chosen to ensure that the pressure term does not contribute to the energy balance, as it should in an homogeneous isotropic flow. The noise term ζ is also assumed to be Gaussian, white in time.

Moreover, we assume that the ζ and η statistics obey the K41 scaling. Hence the expressions of their variances are:

$$\begin{aligned}\langle \zeta_{ab}(\rho;t) . \zeta_{cd}(0;0)\rangle &= \frac{1}{2}(\delta_{ac}\delta_{bd} + \delta_{bc}\delta_{ad}) \\ &\times \delta(t)\left[\frac{a}{\rho^4}\sqrt{Tr\left(\mathbf{MM}^+\right)} + b\frac{\varepsilon^{1/3}}{\rho^{14/3}}\right] \quad (15.5)\end{aligned}$$

$$\langle \eta_{ab}(\rho;t) . \eta_{cd}(0;0)\rangle = \left(\delta_{ac}\delta_{bd} - \frac{1}{3}\delta_{ab}\delta_{cd}\right)\frac{\varepsilon}{\rho^2}\delta(t) \quad (15.6)$$

where ε is the energy dissipation rate introduced in the Kolmogorov theory, and α, a and b are constants of the model.

The equations (15.3), (15.4), (15.5) and (15.6) completely specify the stochastic model.

Details about the derivation can be found in Chertkov et al., 1999, 2001.

3. Resolution of the model - Path integral representation

The system (15.3,15.4,15.5,15.6) defines a well-posed stochastic problem: it can be interpreted as the Langevin equation which governs the evolution of a Brownian particle in the (M, g) space. A Fokker-Planck equation for the Eulerian probability distribution function can thus be readily derived:

$$\partial_t P(M, g, t) = \mathbf{L} P(M, g, t)$$

We try to find the steady state solution of the problem. One also has to impose the proper normalization for the PDF. The system is solved subject to the boundary condition that the velocity field is Gaussian at large (integral) scales. This latter constraint is consistent with many experimental observations. In order to satisfy these three constraints, we write the system:

$$\begin{cases} \mathbf{L}P = 0 \\ \int d\mathbf{M}\, P(\mathbf{M}, \mathbf{g}) = 1 \\ P\left(\mathbf{M}, \mathbf{g} = L^{-2}\, Id\right) \sim \exp\left[-\frac{c\, Tr\left(\mathbf{MM}^+\right)}{(\varepsilon L^{-2})^{2/3}}\right] \end{cases}$$

where c is a constant introduced in the model, and L is the integral scale.

Then the solutions can be expressed in terms of the Green's function of the system. This allows us to express the PDF of M and g as a function of the corresponding PDF of M' and g', at integral scales:

$$P(\mathbf{M},\mathbf{g}) = \int d\mathbf{M}' \int dT \qquad G_{-T}\left(\mathbf{M};\mathbf{g}|\mathbf{M}';\mathbf{g}' = L^{-2}\,Id\right) \times P\left(\mathbf{M}',\mathbf{g}' = L^{-2}\,Id\right) \qquad (15.7)$$

The Green's function can be written with the help of the following path integral representation, in which all the trajectories connecting (M, g) to (M', g') must be considered:

$$G_{-T}\left(\mathbf{M};\mathbf{g}|\mathbf{M}';\mathbf{g}'\right) = \int_{M(-T)=M'}^{M(0)=M} [\mathcal{D}m''] \times \int_{g(-T)=g'}^{g(0)=g} [\mathcal{D}g''] \quad \exp\left[-S\left(\mathbf{m}'';\mathbf{g}''\right)\right] \qquad (15.8)$$

where S is the usual action for each trajectory. The equations (15.7) and (15.8) lead to the following expression for the PDF:

$$P(\mathbf{M},\mathbf{g}) = \int \quad d\mathbf{m}' \quad \int dT \int_{m(-T)=m'}^{m(0)=m} [\mathcal{D}m''] \times \int_{g(-T)=g'}^{g(0)=g} [\mathcal{D}g''] \quad \exp - \left[S\left(\mathbf{m}'';\mathbf{g}''\right) + Tr\left(\mathbf{m}'.\mathbf{m}'^{+}\right)\right] \qquad (15.9)$$

Phenomenologically, one starts at a point (M', g'), at the integral scale, and integrates the system in time until a given scale in the inertial range. In principle, it is necessary to take into account all the trajectories, which could be possible numerically using a Monte-Carlo approach. However, due to the large number of degrees of freedom (large dimension of the phase space), this is a daunting task. We thus resort to a simplifying approximation.

If one ignores for a moment the noise term, the dynamics becomes deterministic. The unique trajectory governed by the deterministic part of the Lagrangian dynamics is the only considered in the calculation. Its action is equal to zero, which leads to a very simple form of the PDF expression. Indeed, the Green's function is equal to the identity, and it is enough to integrate the deterministic equations backward in time. The results of this so called *classical* approximation are presented in (Chertkov et al., 1999). In this paper, the authors compare results of the classical approximation and of direct numerical simulations for the PDF of both invariants Q and R, as well as the densities of several dynamically significant quantities. A semiquantitative agreement with the DNS as well as with some experimental measurements (van der Bos et al., 2002) was found.

4. Semiclassical approximation

4.1 Theoretical approach

To go further, one uses the so called *semiclassical* approximation which consists in considering small, but non-zero, noises. In that case, one can consider only the trajectory for which the action is minimal. At fixed initial and final conditions, this trajectory can be calculated by integrating the Euler-Lagrange equations. However, by fixing Q, R and a scale r_0 in the inertial range, there remain seventeen free parameters to define the initial conditions M, $\dot{M}$, g, and $\dot{g}$. Rigorously, it would be necessary to take into account all their possible values, which would be very costly in computer time. Instead, we use the saddle approximation. Thereafter, we will keep only the initial conditions which contribute the most to the PDF, that is those for which $\exp - \left[S\left(M; g; \dot{M}; \dot{g}\right) + Tr\left(MM^{+}\right)\right]$ is maximal. This maximization over all possible initial conditions is realized numerically by using the *amebsa* algorithm (Press et al., 1992), a mixture of the *amoeba* algorithm and of the simulated annealing method.

4.2 Results

In Figure 15.1 are represented the isoprobability contours of the invariants Q and R, for two different scales. The values of Q (R) have been scaled by $\langle Tr(s^2)\rangle$ ($\langle Tr(s^2)\rangle^{3/2}$). These contours are qualitatively much closer to the DNS pictures (Chertkov et al., 1999) than the isoprobability contours obtained in the classical approximation of the model. In particular, the low probability "gulf" along the $R > 0$ axis is very significantly smoothed. This is due to our better treatment of the fluctuations (ζ, $\eta \neq 0$). Interestingly, the isoprobability contours do not change much when r decreases from $r = L/2$ to $r = L/4$. This quasi self-similar behavior is consistent with the trend observed in DNS.

It is also interesting to investigate physical quantities such as strain variance $Tr(s^2)$, where $s = \frac{M+M^{+}}{2}$. As a first step, we have represented the average of this quantity $\langle Tr(s^2)\rangle$ as a function of the normalized scale r/L, L being the integral scale. The variance of strain is consistent with the (Kolmogorov) scaling law $\langle Tr(s^2)\rangle \propto r^{-4/3}$ (Figure 15.2), which results from $\langle(\Delta u)^p\rangle \propto r^{p/3}$.

Conclusion

We have presented the semi-classical approximation, which allows us to solve the stochastic equations governing the 'tetrad model' in the

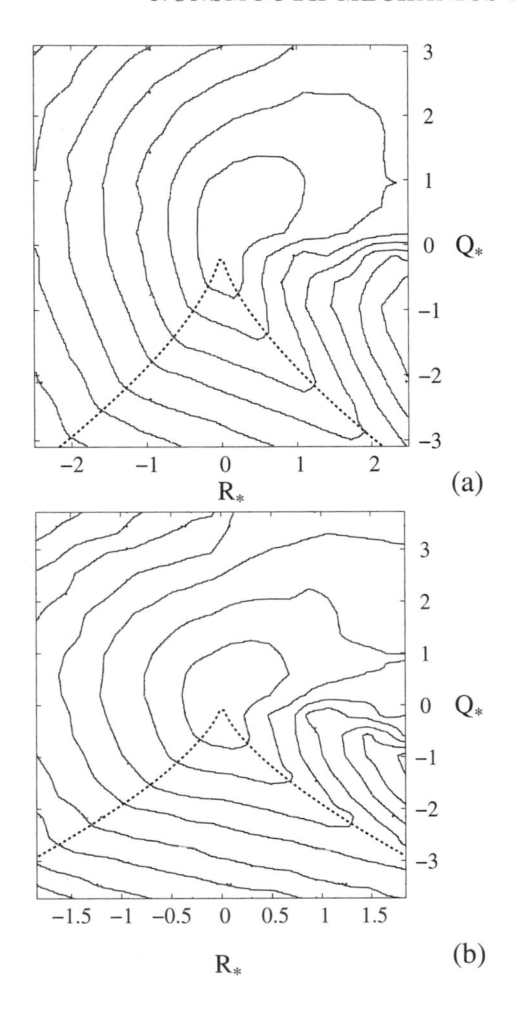

Figure 15.1. PDF of Q_*, R_*, invariants normalized by the variance of strain $\langle Tr(s^2)\rangle$: $Q_* = Q/\langle Tr(s^2)\rangle^{3/2}$, $R_* = R/\langle Tr(s^2)\rangle$, calculated in the semiclassical resolution of the model, at: (a) $r = L/2$ and (b) $r = L/4$. The isoprobability contours are logarithmically spaced, separated by factors of $\sqrt{10}$. The dashed line is the separatrix: $27R^2 + 4Q^3 = 0$.

limit where the noise term is weak. We have obtained numerical solutions which show much better agreement with the DNS results than the 'classical' approximation.

The stochastic problem, formulated by Eq.(15.5,15.6) involves several free parameters. The precise behavior of the model is expected to depend on the chosen values of these parameters. A systematic parametric study is under way.

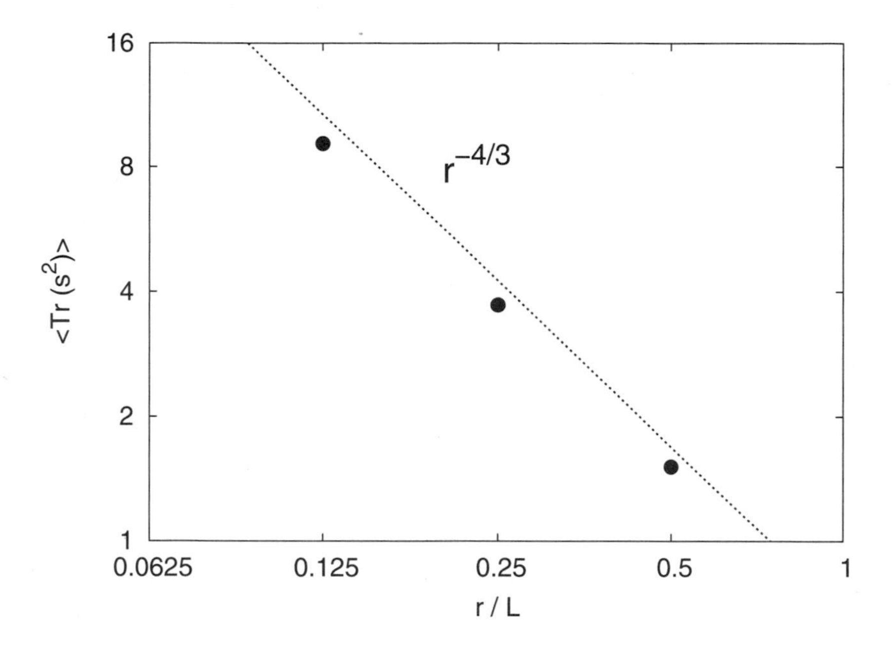

Figure 15.2. Variance of strain $\langle Tr(s^2) \rangle$ as a function of scale normalized by L, in log-log scale (black dots). The dashed straight line has a slope of $-4/3$.

It is a pleasure to acknowledge many discussions with M. Chertkov and B. Shraiman.

References

Chertkov, M., Pumir, A. and Shraiman, B.I. (1999). "Lagrangian tetrad dynamics and the phenomenology of turbulence", Phys. Fluids **11**, 2394–2410.

Chertkov, M., Pumir, A. and Shraiman, B.I. (2001). "Statistical geometry and lagrangian dynamics of turbulence", in 'Intermittency in turbulent flows', edited by C. Vassilicos, Cambridge University Press, 243–261.

Frisch, U. (1995). "Turbulence: The legacy of AN Kolmogorov", Cambridge University Press, Cambridge.

Moffatt, H.K. (1969). "The degree of knottedness of tangled vortex lines", J. Fluid Mech., **35**, 117–129.

Moffatt, H.K. (1981). "Some developments in the theory of turbulence", J. Fluid Mech., **106**, 27–47.

Moreau, J.J. (1961). "Constantes d'un îlot tourbillonaire d'un fluide parfait barotrope", CRAS, **252**, 2810–2812.

Press, W.H., Teukolsky, S.A., Vetterling, W.T. and Flannery, B.P. (1992). "Numerical Recipes in C", Cambridge University Press, 451–455.

Van der Bos, F., Tao, B., Meneveau, C. and Katz, J. (2002). "Effects of small-scale turbulent motions on the filtered velocity gradient tensor as deduced from holographic particle image velocimetry measurements", Phys. Fluids **14**, 2456–2474.

Vieillefosse, P. (1984). "Internal motion of a small element of a fluid in inviscid flow", Physica **125A**, 150–162.

IV

MULTIBODY DYNAMICS: NUMERICAL ASPECTS

Chapter 16

THE GEOMETRY OF NEWTON'S CRADLE*

Christoph Glocker
and Ueli Aeberhard
IMES - Center of Mechanics, ETH Zentrum, Tannenstrasse 3, CH-8092 Zürich, Switzerland
glocker@imes.mavt.ethz.ch

Abstract A geometric approach to the multi-impact problem of perfect collisions in finite freedom dynamics is presented. The set of all admissible post-impact velocities at an impact is derived from kinematic, kinetic and energetic compatibility conditions. Impact events are classified with respect to their topology, intensity and dissipation behavior. A quantity to characterize the influence of wave effects on the impacts, and thus their deviation from the mechanical impact theory is introduced. The method is applied to Newton's cradle with three balls.

Keywords: Newton's cradle, unilateral constraint, impact, collision, rigid body

1. Introduction

The paper is devoted to a geometric study of perfect collisions in dynamics with the aim to find invariant characteristics of impacts. In section 2 the concept of a breathing Riemannian manifold is chosen to embed impulsive behavior in the same mathematical setting as classical mechanics. External impulsive excitation and time-dependent inequality constraints are introduced, allowing the mechanical system to impact against perfect boundaries. This general approach is narrowed down to the scleronomic case without excitations in section 3, in which the geometric picture of the impact in the tangent space is developed. A crucial

*Parts of this research were conducted within the European project SICONOS IST-2001-37172 and supported by the Swiss Federal Office for Education and Science (BBW)

point in the entire analysis plays the impact with maximal dissipation as introduced by J.J. Moreau, which is used as the reference for developing classification concepts. This approach is illustrated by a discussion of Newton's cradle with three balls in section 4. To anticipate the conclusion, the present findings have already been extended to the rheonomic case without and with external impulsive excitation, but still demand further investigations.

2. Collision Model

In order to set up the kinematical and kinetical framework of our problem, we consider an f-dimensional breathing Riemannian configuration manifold $\mathcal{M}(t)$ with $t \to \mathcal{M}(t)$ "smooth". We denote by $\mathbf{q}$ the f-tuple of local coordinates in use, and, for every fixed time t, by $\mathbf{M}(\mathbf{q}, t)$ the representation with respect to $\mathbf{q}$ of the (kinetic) metric associated with $\mathcal{M}$. The tangent cone to a set $\mathcal{D}$ at a point $\mathbf{q} \in \mathcal{D}$ is denoted by $\mathcal{T}_\mathcal{D}(\mathbf{q})$ as defined in (Aubin and Ekeland, 1984). For $\mathcal{D} = \mathcal{M}$, the tangent cone simplifies to the classical tangent space with dual $\mathcal{T}^\star_\mathcal{M}(\mathbf{q}) = \mathbf{M}\,\mathcal{T}_\mathcal{M}(\mathbf{q})$. Inner products on $\mathcal{T}_\mathcal{M}(\mathbf{q})$ are denoted by a dot, $\mathbf{a} \cdot \mathbf{b} := \mathbf{a}^T\mathbf{M}\mathbf{b}$ for $\mathbf{a}, \mathbf{b} \in \mathcal{T}_\mathcal{M}(\mathbf{q})$, and the associated norm is written as $\|\mathbf{a}\| := \sqrt{\mathbf{a} \cdot \mathbf{a}}$.

We will first formulate the kinematics of the collision problem. We assume n functions g^i of class C^1 defined by

$$g^i : \quad \mathcal{M} \times \mathbb{R} \to \mathbb{R}, \quad (\mathbf{q}, t) \to g^i(\mathbf{q}, t), \tag{16.1}$$

which will be recognized as the gaps between bodies in a collision model of a finite dimensional mechanical system. By this definition, there exist n contact velocity mappings

$$\gamma^i : \quad \mathcal{T}_\mathcal{M}(\mathbf{q}) \to \mathbb{R}, \quad \mathbf{u} \to \gamma^i(\mathbf{u}) := \nabla g^i \cdot \mathbf{u} + g^i_{,t} \tag{16.2}$$

with gradient $\nabla g^i := \mathbf{M}^{-1}\,\partial g^i/\partial\mathbf{q}$ and $g^i_{,t} := \partial g^i/\partial t$. The mappings (16.2) are linearly affine in $\mathbf{u}$, thus

$$\gamma^i(\mathbf{a} + \mathbf{b}) = \gamma^i(\mathbf{a}) + \nabla g^i \cdot \mathbf{b}. \tag{16.3}$$

Consider now a curve on $\mathcal{M}$ which is absolutely continuous in time,

$$\mathbf{q} : \quad \mathbb{R} \to \mathcal{M}(t), \quad t \to \mathbf{q}(t). \tag{16.4}$$

Due to absolute continuity of $\mathbf{q}(t)$, the associated tangents $\mathbf{u} = \dot{\mathbf{q}}(t)$ exist almost everywhere and admit a left limit $\mathbf{u}_-$ and a right limit $\mathbf{u}_+$ at every time t. In the same way behave the gaps (16.1) as functions of time, $g^i(\mathbf{q}(t), t)$, provided that the mapping (16.1) preserves absolute continuity. Thus, γ^i is related to $\dot{g}^i$ thanks to (16.1) and (16.2) by

$$\dot{g}^i(\mathbf{q}(t), t) = \gamma^i(\mathbf{u}) \tag{16.5}$$

whenever $\mathbf{u} = \dot{\mathbf{q}}(t)$ exists. We further set

$$\gamma^i(\mathbf{u}_-) =: \gamma^i_-, \quad \gamma^i(\mathbf{u}_+) =: \gamma^i_+. \tag{16.6}$$

For a discontinuity point of $\dot{\mathbf{q}}(t)$ being caused by a collision, the values of $\gamma^i_\pm$ provide the pre- and post-impact contact velocities. Using the gap functions (16.1), we introduce inequality constraints in the form $g^i(\mathbf{q}, t) \geq 0$, which turn $\mathcal{M}$ into a manifold with boundary, characterized by the set of admissible displacements $\mathcal{C}$,

$$\mathcal{C}(t) := \{\mathbf{q} \mid g^i(\mathbf{q}, t) \geq 0, \ \forall i = 1, \ldots, n\}. \tag{16.7}$$

Furthermore, let $\mathcal{H}$ denote the set of active constraints, i.e.

$$\mathcal{H}(\mathbf{q}) := \{i \mid g^i(\mathbf{q}, t) = 0\}. \tag{16.8}$$

The tangent cone which characterizes the set $\mathcal{C}$ in some neighborhood of a point $\mathbf{q} \in \mathcal{C}$, is then obtained by

$$\mathcal{T}_\mathcal{C}(\mathbf{q}) := \{\mathbf{u} \in \mathcal{T}_\mathcal{M}(\mathbf{q}) \mid \nabla g^i \cdot \mathbf{u} \geq 0, \ \forall i \in \mathcal{H}\}. \tag{16.9}$$

Note that this definition also includes the case that $\mathbf{q}$ is not a boundary point of $\mathcal{C}$, for which $\mathcal{H} = \emptyset$ and the tangent cone becomes identical to the tangent space as a consequence. In general, one has $\mathcal{T}_\mathcal{C}(\mathbf{q}) \subset \mathcal{T}_\mathcal{M}(\mathbf{q})$. We further introduce the two polyhedral convex sets

$$\mathcal{P}^+_\mathcal{C}(\mathbf{q}) := \{\mathbf{u} \in \mathcal{T}_\mathcal{M}(\mathbf{q}) \mid \gamma^i(\mathbf{u}) \geq 0, \ \forall i \in \mathcal{H}\}, \tag{16.10}$$

$$\mathcal{P}^-_\mathcal{C}(\mathbf{q}) := \{\mathbf{u} \in \mathcal{T}_\mathcal{M}(\mathbf{q}) \mid \gamma^i(\mathbf{u}) \leq 0, \ \forall i \in \mathcal{H}\}. \tag{16.11}$$

Their definition is motivated by the fact that a necessary condition for a closed gap $g^i(\mathbf{q}(t), t) = 0$ not to leave the admissible region $g^i \geq 0$ in some adjoint time interval, is, that it has approached this value according to (16.5) and (16.6) by a left velocity $\gamma^i_- \leq 0$, and it has to proceed in time by a right velocity $\gamma^i_+ \geq 0$, respectively. Thus, $\mathcal{P}^\pm_\mathcal{C}(\mathbf{q})$ describe the set of admissible pre- and post-impact velocities $\mathbf{u}^\pm$.

In the spirit of the virtual work approach in classical dynamics, kinetics is treated in the framework of duality as an equality of measures on the cotangent space $\mathcal{T}^\star_\mathcal{M}(\mathbf{q})$. We denote by $\delta\mathbf{q}$ the virtual displacements of the system, by $\mathrm{d}\mathbf{u}$ the measure accelerations related to the velocities $\mathbf{u}(t)$, see e.g. (Moreau, 1988a, Moreau, 1988b, Ballard, 2000), and by $\mathrm{d}\mathbf{F}$ the measure of external forces. The virtual action $\delta\mathrm{d}W$ has to vanish for arbitrary virtual displacements, when the system is to be in the dynamic equilibrium,

$$\forall \delta\mathbf{q} \in \mathcal{T}_\mathcal{M}(\mathbf{q}) : \quad \delta\mathrm{d}W := \delta\mathbf{q}^\mathsf{T}(\mathbf{M}\,\mathrm{d}\mathbf{u} - \mathrm{d}\mathbf{F}) = 0. \tag{16.12}$$

The force measure $\mathrm{d}\mathbf{F}$ can be split into the reactions $\mathrm{d}\mathbf{R}$ due to the inequality constraints (16.7) and the remaining external forces $\mathrm{d}\mathbf{P}$,

$$\mathrm{d}\mathbf{F} = \mathrm{d}\mathbf{R} + \mathrm{d}\mathbf{P}. \tag{16.13}$$

Following (Ballard, 2000, Glocker, 2001) we define a set of inequality constraints to be perfect if and only if the reactions satisfy

$$-\delta\mathbf{q}^{\mathsf{T}}\mathrm{d}\mathbf{R} \leq 0 \quad \forall\delta\mathbf{q} \in \mathcal{T}_{\mathcal{C}}(\mathbf{q}). \tag{16.14}$$

In this case, the reactions are called constraint forces. The set of possible reactions constitutes by (16.14) a cone in the cotangent space of $\mathcal{M}$,

$$\mathcal{N}_{\mathcal{C}}(\mathbf{q}) := \{\boldsymbol{\xi} \mid \delta\mathbf{q}^{\mathsf{T}}\boldsymbol{\xi} \leq 0, \ \forall\delta\mathbf{q} \in \mathcal{T}_{\mathcal{C}}(\mathbf{q})\}, \tag{16.15}$$

which is the polar cone of $\mathcal{T}_{\mathcal{C}}(\mathbf{q})$, and which enables us to express the constraint forces by the inclusion $-\mathrm{d}\mathbf{R} \in \mathcal{N}_{\mathcal{C}}(\mathbf{q})$. By combining this representation with (16.12)–(16.14), the dynamics is then stated as a measure inclusion on the cotangent space,

$$\mathbf{M}\,\mathrm{d}\mathbf{u} - \mathrm{d}\mathbf{P} \in -\mathcal{N}_{\mathcal{C}}(\mathbf{q}). \tag{16.16}$$

In order to allow for a geometric interpretation of this inclusion together with the kinematical restrictions (16.7)–(16.11), we map (16.16) to the tangent space via the natural isomorphism available by the metric on $\mathcal{M}$. We set $\mathrm{d}\boldsymbol{\pi} := \mathbf{M}^{-1}\mathrm{d}\mathbf{P}$ and $\mathcal{T}_{\mathcal{C}}^{\perp}(\mathbf{q}) := \mathbf{M}^{-1}\mathcal{N}_{\mathcal{C}}(\mathbf{q})$ to arrive with

$$\mathrm{d}\mathbf{u} - \mathrm{d}\boldsymbol{\pi} \in -\mathcal{T}_{\mathcal{C}}^{\perp}(\mathbf{q}). \tag{16.17}$$

Finally, this inclusion is integrated over a singleton $\{t\}$ in time to obtain the impact equations of the system,

$$\mathbf{u}_{+} - \mathbf{u}_{-} - \boldsymbol{\pi} \in -\mathcal{T}_{\mathcal{C}}^{\perp}(\mathbf{q}). \tag{16.18}$$

Here, $\boldsymbol{\pi}$ represents any impulsive forces mapped to the tangent space that are externally applied on the system, $-\mathcal{T}_{\mathcal{C}}^{\perp}(\mathbf{q})$ contains the impulsive reactions due to the unilateral perfect constraints, and $\mathbf{u}_{+} - \mathbf{u}_{-}$ is the instantaneous change in velocities with which the system dynamically reacts. Finally, it can be seen from (16.9) and (16.15) that $-\mathcal{T}_{\mathcal{C}}^{\perp}(\mathbf{q})$ is positively generated by the gradients of g^i,

$$\mathcal{T}_{\mathcal{C}}^{\perp}(\mathbf{q}) = \{-\nabla g^i \varLambda_i \mid \varLambda_i \geq 0, \ \forall i \in \mathcal{H}\} \tag{16.19}$$

with Lagrangian multipliers $\varLambda_i \geq 0$ that act as compressive impulsive scalar forces in the associated contacts $i \in \mathcal{H}$. As subsets of the tangent space, we call $\mathcal{T}_{\mathcal{C}}(\mathbf{q})$ and $\mathcal{T}_{\mathcal{C}}^{\perp}(\mathbf{q})$ a pair of cones that are orthogonal to each other in the kinetic metric.

3. Impacts without Excitation

In this section we give a geometric characterization of impacts due to collisions and confine ourselves to the special case of no external excitation. This assumption affects both the kinematic and kinetic excitation. Kinematic excitation acts on the system via the inhomogeneity terms $g^i_{,t}$ in (16.2). If such an excitation is excluded for a certain impact configuration ($g^i_{,t} = 0 \; \forall i \in \mathcal{H}$), then the sets of admissible pre- and post-impact velocities (16.10), (16.11) are directly related to the tangent cone (16.9) by $\mathcal{P}^+_{\mathcal{C}}(\mathbf{q}) = -\mathcal{P}^-_{\mathcal{C}}(\mathbf{q}) = \mathcal{T}_{\mathcal{C}}(\mathbf{q})$. Thus, pre-impact velocities $\mathbf{u}_- \in -\mathcal{T}_{\mathcal{C}}(\mathbf{q})$ have to be assumed. We understand by kinematic compatibility that the collision ends up with feasible post-impact velocities

$$\mathbf{u}_+ \in \mathcal{T}_{\mathcal{C}}(\mathbf{q}). \tag{16.20}$$

Kinetic excitation is excluded from the problem by setting the associated external impulsive forces $\boldsymbol{\pi}$ in (16.18) equal to zero. This yields

$$\mathbf{u}_+ \in \mathbf{u}_- - \mathcal{T}^{\perp}_{\mathcal{C}}(\mathbf{q}), \tag{16.21}$$

which we call kinetic compatibility. A third and independent property that we demand is energetic consistency. We do not allow for any active behavior of the contacts as, for example, observed in pin ball machines, but assume a dissipative character only. This means that the kinetic energy T of the system must not increase during the collision, $T_+ - T_- \leq 0$. From $T = \frac{1}{2}\|\mathbf{u}\|^2$ for a scleronomic system, it follows that the dissipative character of the impact is assured if $\|\mathbf{u}_+\| \leq \|\mathbf{u}_-\|$. This condition is expressed by the inclusion

$$\mathbf{u}_+ \in \mathcal{B}_{\|\mathbf{u}_-\|}(\mathbf{q}), \tag{16.22}$$

where $\mathcal{B}_{\|\mathbf{u}_-\|}(\mathbf{q})$ denotes the energy ball with radius $\|\mathbf{u}_-\|$ and midpoint placed at the origin. Thus, kinetic, kinematic and energetic consistency restricts the set of admissible post-impact velocities by (16.20)–(16.22) to

$$\mathcal{S} := (\mathbf{u}_- - \mathcal{T}^{\perp}_{\mathcal{C}}(\mathbf{q})) \cap \mathcal{T}_{\mathcal{C}}(\mathbf{q}) \cap \mathcal{B}_{\|\mathbf{u}_-\|}(\mathbf{q}). \tag{16.23}$$

This convex set is shown together with the orthogonal pair of cones $\mathcal{T}_{\mathcal{C}}(\mathbf{q})$, $\mathcal{T}^{\perp}_{\mathcal{C}}(\mathbf{q})$ and the energy ball $\mathcal{B}_{\|\mathbf{u}_-\|}(\mathbf{q})$ in figure 16.1. Note the special case that $\mathbf{q}$ is not a boundary point of $\mathcal{C}$, i.e. $\mathbf{q} \in \operatorname{int} \mathcal{C}$. For this situation, $\mathcal{T}^{\perp}_{\mathcal{C}}(\mathbf{q}) = 0$, and $\mathcal{S}$ reduces to a single element $\mathcal{S} = \{\mathbf{u}_-\}$. Thus, it holds that $\mathbf{u}^+ = \mathbf{u}^-$ which means that there is no velocity jump.

The post-impact velocities $\mathbf{u}_+$ are restricted by equation (16.23) to the set $\mathcal{S}$, but are still not uniquely defined. In order to pick a particular element out of this set, one needs an impact law in the sense of an

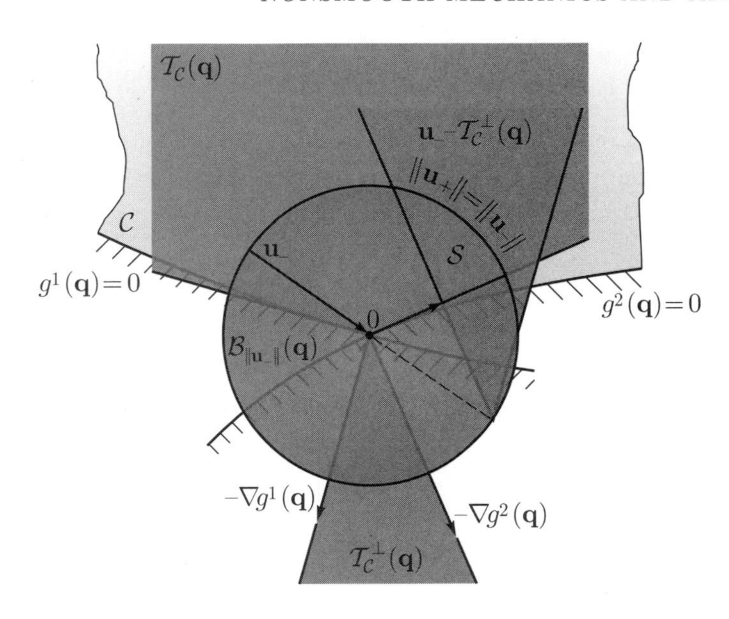

Figure 16.1. Geometry of multi-contact collisions without excitation.

additional, independent equation. One possible approach is suggested by J.J. Moreau in (Moreau, 1988a). It requires the (unique) orthogonal decomposition of the pre-impact velocity $\mathbf{u}_-$ into two parts $\mathbf{z}$ and $\mathbf{z}^\perp$ with respect to the orthogonal pair of cones $\mathcal{T}_\mathcal{C}(\mathbf{q})$ and $\mathcal{T}_\mathcal{C}^\perp(\mathbf{q})$ respectively, and can be stated as follows: For $\mathbf{u}_- = \mathbf{z} + \mathbf{z}^\perp$ find $\mathbf{u}_+ = \mathbf{z} - \varepsilon\mathbf{z}^\perp$ such that

$$\mathcal{T}_\mathcal{C}(\mathbf{q}) \ni \mathbf{z} \perp \mathbf{z}^\perp \in \mathcal{T}_\mathcal{C}^\perp(\mathbf{q}). \tag{16.24}$$

The tangential component $\mathbf{z}$ in this collision law is left unchanged by the impact, whereas the normal component $\mathbf{z}^\perp$ is inverted and multiplied by a dissipation coefficient ε to give $-\varepsilon\mathbf{z}^\perp$, see figure 16.2. From this figure, one recognizes also that $\mathbf{z}$ and $\mathbf{z}^\perp$ are the nearest points to $\mathbf{u}_-$ in the sets $\mathcal{T}_\mathcal{C}(\mathbf{q})$ and $\mathcal{T}_\mathcal{C}^\perp(\mathbf{q})$, respectively, which can be expressed in terms of proximations as $\mathbf{z} = \mathrm{prox}_{\mathcal{T}_\mathcal{C}(\mathbf{q})}(\mathbf{u}_-)$ and $\mathbf{z}^\perp = \mathrm{prox}_{\mathcal{T}_\mathcal{C}^\perp(\mathbf{q})}(\mathbf{u}_-)$. For $\varepsilon = 0$, the impact law yields $\mathbf{u}_+ = \mathbf{z}$, which corresponds to the unique post-impact velocity of minimal length under the restrictions (16.23). We call the associated point in $\mathcal{S}$ the point of maximal dissipation. Thus, a uniquely determined half-line passing this point and emanating from the tip $\mathbf{u}_-$ of the translated cone $-\mathcal{T}_\mathcal{C}^\perp(\mathbf{q})$ is accessible through Moreau's impact law by values $\varepsilon \geq -1$. It is easily seen that $\varepsilon = 0$ addresses the point of maximal dissipation, whereas conservation of kinetic energy is assured for $\varepsilon = 1$. Values of ε in between correspond to different levels of dissipation, defining a family of concentric spheres in the tangent space. As discussed in (Moreau, 1988a, Glocker, 2001a), Moreau's half-line does

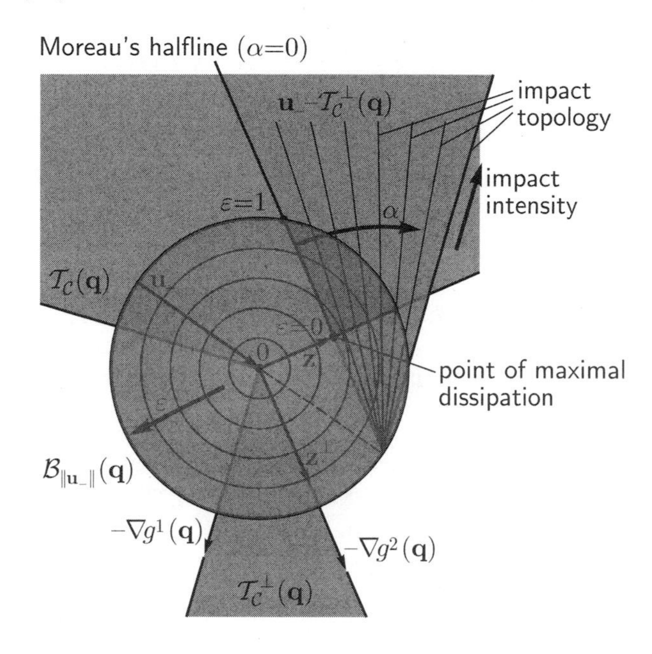

Figure 16.2. Geometric characterization of impacts without excitation.

not comprise events of collisions that benefit from impulse transfer between non-neighboring bodies via traveling waves. This behavior is also excluded in classical text books on single impacts by saying that wave effects can be neglected. The framework achieved by this assumption is usually called the *mechanical impact theory.* However, a precise mathematical formulation of this term has never been given. We therefore suggest to *define* the mechanical impact theory as to consist precisely of those impact events that can be accessed through Moreau's half-line.

Obviously, not every post-impact velocity in $\mathcal{S}$ can be reached by Moreau's law. In view of a future parameterization of $\mathcal{S}$ we want to give already here a characterization of general collisions based on geometric properties. We define the *impact topology* as the entity connected to the different half-lines of $\mathcal{T}_C^\perp(\mathbf{q})$. In other words, all post-impact velocities pointing for some given $\Lambda_i > 0$ to the same half-line $\mathbf{u}_- + \mathbb{R}_0^+ \Lambda_i \nabla g^i$ in $\mathbf{u}_- - \mathcal{T}_C^\perp(\mathbf{q})$ are said to have the same impact topology. This classification corresponds to a constant impact impulse ratio, as called by other authors. We further propose to take the angle α between a certain half-line of $\mathcal{T}_C^\perp(\mathbf{q})$ and Moreau's half-line as a measure of deviation from the mechanical impact theory, which also quantifies the wave effects at the collision. We further take as *impact intensity* the distance of the post-impact velocity $\mathbf{u}_+$ from the tip of the translated cone $\mathbf{u}_- - \mathcal{T}_C^\perp(\mathbf{q})$, which is nothing else than $\|\mathbf{u}_+ - \mathbf{u}_-\|$. As soon as the post-impact ve-

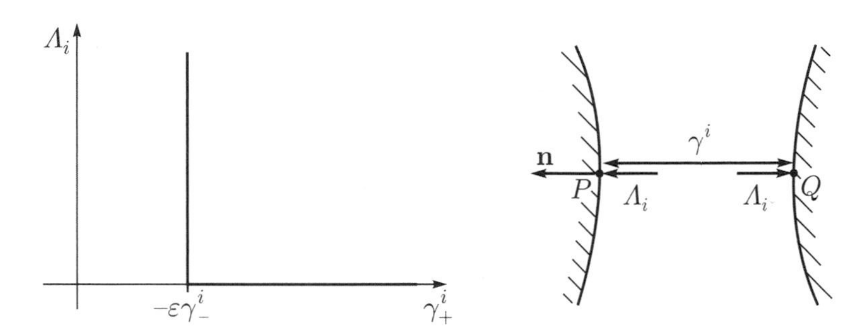

Figure 16.3. Representation of Moreau's impact law in contact coordinates.

locity $\mathbf{u}_+$ has been located in the set $\mathbf{u}_- - \mathcal{T}_C^\perp(\mathbf{q})$, it can in addition be characterized by the dissipation coefficient ε, associated with the sphere on which it is placed.

In (Glocker, 2002) a representation of Moreau's impact law in local contact coordinates has been derived, showing that

$$0 \leq \Lambda_i \perp (\gamma^i_+ + \varepsilon\, \gamma^i_-) \geq 0 \tag{16.25}$$

for each contact i in the active set $\mathcal{H}$. The complementarity conditions in (16.25) express that each contact that takes an impulsive force $\Lambda_i > 0$ has to fulfill the classical Newtonian impact law $\gamma^i_+ = -\varepsilon\, \gamma^i_-$. However, if the contact point is to be regarded as not to participate in the impact ($\Lambda_i = 0$), then the post-impact relative velocity is allowed for having values $\gamma^i_+ > -\varepsilon\gamma^i_-$. This impact behavior is said to have a *global dissipation index*, because the *same* restitution coefficient ε is taken in (16.25) for each individual contact. In order to extend this concept of local representation to general impacts in $\mathcal{S}$, a matrix of impact coefficients ε^i_j has to be introduced in the inequality impact law (16.25) as proposed in (Frémond, 1995),

$$0 \leq \Lambda_i \perp (\gamma^i_+ + \varepsilon^i_j\, \gamma^j_-) \geq 0. \tag{16.26}$$

The matrix of impact coefficients allows to access impact topologies different from Moreau's half-line, such as spatially separated chain-like contacts which directly interact with each other. Restrictions on the coefficients ε^i_j such that $\mathbf{u}_+ \in \mathcal{S}$ is guaranteed have still to be worked out.

4. Newton's Cradle

As an example, we investigate the impact configuration in Newton's cradle with three balls of equal masses $m_1 = m_2 = m_3 =: m$. The

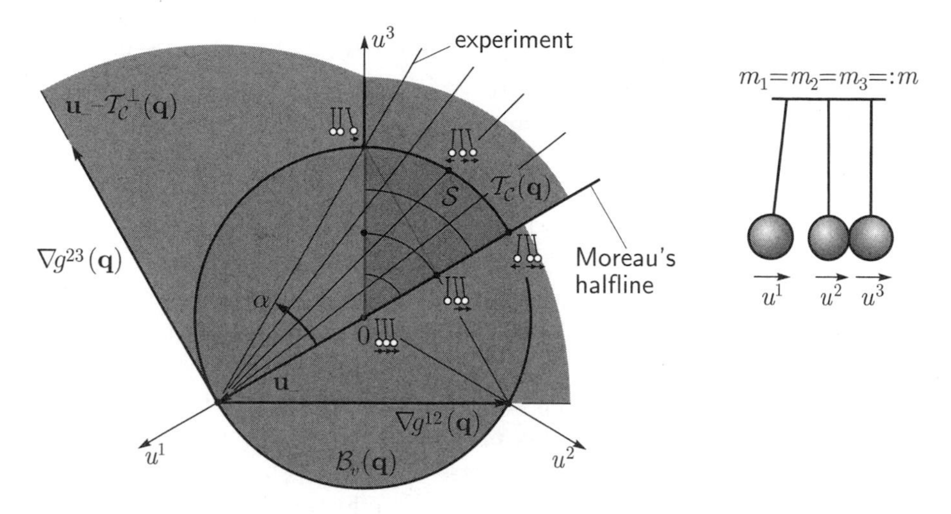

Figure 16.4. Impact geometry of Newton's cradle with three balls.

horizontal absolute displacements of the three masses are denoted by q^1, q^2, q^3, and the associated velocities by u^1, u^2, u^3, respectively. The inequality contact constraints are expressed by the two gap functions $g^{12} = q^2 - q^1$ and $g^{23} = q^3 - q^2$. For the pre-impact state we assume that balls two and three are at rest ($u^2_- = u^3_- = 0$) by touching each other ($g^{23} = 0$), and that ball one hits ball two ($g^{12} = 0$) with a velocity $u^1_- = v > 0$. We therefore have a pre-impact velocity state $\mathbf{u}_- = (v, 0, 0)^\mathsf{T}$ with both contacts contained in the set of active constraints $\mathcal{H}$ from (16.8). The impact equations (16.18) for this configuration are

$$\begin{pmatrix} u^1_+ - u^1_- \\ u^2_+ - u^2_- \\ u^3_+ - u^3_- \end{pmatrix} = \frac{1}{m}\begin{pmatrix} -1 \\ 1 \\ 0 \end{pmatrix} \Lambda_{12} + \frac{1}{m}\begin{pmatrix} 0 \\ -1 \\ 1 \end{pmatrix} \Lambda_{23}, \tag{16.27}$$

where Λ_{12} and Λ_{23} denote the impulsive forces between balls 1-2 and 2-3, respectively. From this equation one also identifies the gradients of the gap functions as $\nabla g^{12} = \frac{1}{m}(-1, 1, 0)$ and $\nabla g^{23} = \frac{1}{m}(0, -1, 1)$, which define by (16.19) the edges of $\mathcal{T}_\mathcal{C}^\perp(\mathbf{q})$. As shown in figure 16.4, kinetic compatibility (16.21) yields a subset $\mathbf{u}_- - \mathcal{T}_\mathcal{C}^\perp(\mathbf{q})$ in the deviatoric plane of the u^1-u^2-u^3 trihedral, which is the plane that intersects the three axes by equal values v. The set $\mathcal{S}$ of admissible post-impact velocities $\mathbf{u}^+$ is obtained according to (16.23) by subsequent intersection with the tangent cone $\mathcal{T}_\mathcal{C}(\mathbf{q})$ and the energy ball $\mathcal{B}_v(\mathbf{q})$.

The point of maximal dissipation in $\mathcal{S}$ as the point in $\mathbf{u}_- - \mathcal{T}_\mathcal{C}^\perp(\mathbf{q})$ with minimal distance from the origin, is recognized as the post-impact motion, for which the three balls move together as one rigid body with a common velocity $u_+^{1,2,3} = \frac{1}{3}v$. This point defines Moreau's half-line, which emanates from the tip of $\mathbf{u}_- - \mathcal{T}_C^{\perp}(\mathbf{q})$ and carries the topology of impacts with global dissipation index. On its intersection with the energetic sphere that assures conservation of kinetic energy, balls two and three remain in contact after the impact and move with a common velocity of $u_+^{2,3} = \frac{2}{3}v$ to the right, whereas ball one bounces back to the left with a velocity of $u_+^1 = -\frac{2}{3}v$. In reality, balls one and two stand nearly idle ($u_+^{12} = 0$) after the impact, and ball three leaves with an approximate velocity $u_+^3 = v$ to the right. This behavior is on the same energetic sphere, but takes place on the topological half-line that deviates by a maximal possible angle ($\alpha = 30°$) from Moreau's impact law. Note that only one point of this half-line belongs to $\mathcal{S}$. As a result, Newton's cradle can be seen as a device that shows experimentally a collision behavior that maximizes wave effects under all imaginable experiments with three arbitrary bodies of equal mass.

References

Aubin, J., and Ekeland, I. (1984). *Applied Nonlinear Analysis.* New York: John Wiley & Sons.

Ballard, P. (2000). "The Dynamics of Discrete Mechanical Systems with Perfect Unilateral Constraints," *Arch. Rational Mech. Anal.* **154** 3, pp. 199–274.

Frémond, M. (1995). "Rigid Bodies Collisions," *Physics Letters A* **204**, pp. 33–41.

Glocker, Ch. (2001). *Set-Valued Force Laws: Dynamics of Non-Smooth Systems.* Lecture Notes in Applied Mechanics 1. Berlin, Heidelberg: Springer.

Glocker, Ch. (2001a). "On Frictionless Impact Models in Rigid-Body Systems," *Phil. Trans. Royal. Soc. Lond.* A 359, pp. 2385–2404.

Glocker, Ch. (2002). "The Geometry of Newtonian Impacts with Global Dissipation Index for Moving Sets," *Proc. of the Int. Conf. on Nonsmooth/Nonconvex Mechanics* (ed. Baniotopoulos, C.C.), pp. 283 - 290. Thessaloniki: Editions Ziti.

Moreau, J.J. (1988a). "Unilateral Contact and Dry Friction in Finite Freedom Dynamics," in *Non-Smooth Mechanics and Applications, CISM Courses and Lectures Vol. 302* (eds. Moreau, J.J. and Panagiotopoulos, P.D.), pp. 1–82. Wien: Springer.

Moreau, J.J. (1988b). "Bounded Variation in Time," in *Topics in Nonsmooth Mechanics* (eds. Moreau, J.J., Panagiotopoulos, P.D., and Strang, G.), pp. 1–74. Basel: Birkhäuser Verlag.

Chapter 17

USING NONSMOOTH ANALYSIS FOR NUMERICAL SIMULATION OF CONTACT MECHANICS

Pierre Alart
David Dureisseix
and Mathieu Renouf
LMGC, University Montpellier 2, CNRS UMR5508, CC048, Place Eugène Bataillon, F-34095 Montpellier Cedex 5, France
alart@lmgc.univ-montp2.fr
dureisseix@lmgc.univ-montp2.fr
renouf@lmgc.univ-montp2.fr

Abstract Two different approaches for simulation of contact mechanics problems are investigated. First, the equivalence of the Large Time INcrement (LATIN) method for unilateral contact problems, in terms of an augmented Lagrangian approach dedicated to nonsmooth problems, is proved (with the non trivial introduction of two constraints: an equality constraint and an inequality-type constraint). Second, the Finite Element Tearing and Interconnecting method for Contact (FETI-C) is interpreted as a projected gradient with projection, acting on the dual problem. An extension to frictional contact is presented in the context of granular media.

1. Introduction

In nonsmooth analysis and mechanics, Jean Jacques Moreau proclaims to be firstly a mechanician and a mathematician who develops only what is strictly necessary for mechanical modeling. He was not initially a numerician even if his last works on granular material require efficient computer resources and algorithms. His mathematical tools proved to be very useful in physical / mechanical modeling (friction, plasticity...) and in applied mathematics related to the solution of partial differential equations and in optimization theory. The theory of duality in Convex Analysis plays an important role in the mathematical modeling of the aforementioned fields and in the design of efficient solution schemes based, for instance, on augmented Lagrangian functions. But a lot of numerical developments in intensive computational mechanics, when limited to linear large scale problems, did not call for sophisticated notions of Convex Analysis. This research field led to specific concepts already difficult to handle in a linear context (preconditioning, domain decomposition methods...).

The motivation of this work is the re-engineering of algorithms arising from 'intuitive' approaches of computational mechanics, using convex and nonsmooth analysis. This could be a guide for algorithm extensions and development of new approaches.

1.1 Model problem

To simplify the presentation, let us consider the case of the frictionless Signorini problem depicted in Figure 1.1: only one elastic body Ω pressed against a rigid plate, under small perturbations, with a quasi-static evolution process. The potential contact area Γ will be called an interface. $\underline{n}$ designates the outward unitary normal vector on the boundary $\partial\Omega$.

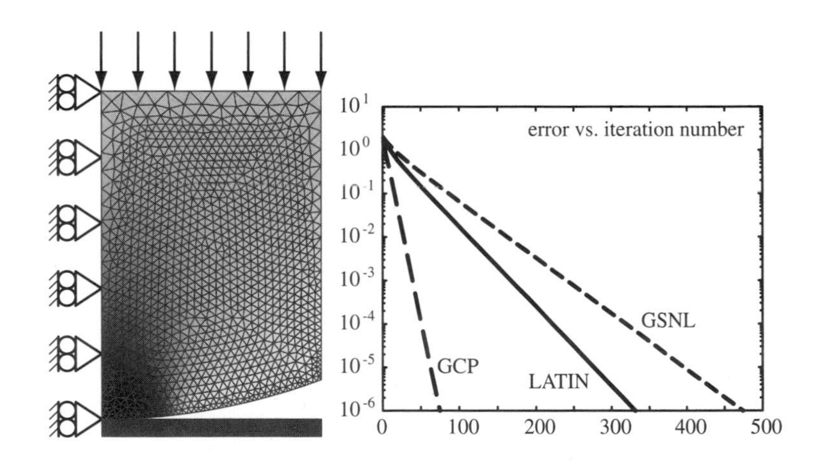

Figure 17.1. Left: vertical stress field solution on the model problem (cylindrical punch); right: convergence results for different iterative algorithms

If $\underline{F}$ is the force applied to the body by the plate, and if $\underline{V}$ is the displacement of the body on the interface Γ, the interface behavior is: (i) no friction, i.e. $(\mathbf{1} - \underline{nn^T})\underline{F} = 0$, and (ii) complementarity conditions (Karush-Khun-Tucker):

- no penetration: $z = \underline{n}^T \underline{V} \geq 0$, i.e. $z \in \mathcal{C}_1$ (the positive cone);
- no adhesion: $w = \underline{n}^T \underline{F} \geq 0$, i.e. $w \in \mathcal{C}_1$;
- no dissipation: $wz = 0$.

The problem consists in finding a displacement field $\underline{U}(M)$, and a stress field $\boldsymbol{\sigma}(M)$, such that:

- $\underline{U}$ is sufficiently regular (the strain field is: $\boldsymbol{\varepsilon} = (\text{grad}\, \underline{U})_{\text{sym}}$), and equals some prescribed values on a first part of the boundary: $\underline{U}_{|\partial_1 \Omega} = \underline{U}_d$, and on the interface: $\underline{n}^T \underline{U}_{|\Gamma} = z$; a couple $(\underline{U}, z)$ satisfying these conditions is said kinematically admissible (KA);
- the stress field balances the external body force $\underline{f}_d$ and the prescribed forces on a second part of the boundary: $\underline{\text{div}}\, \boldsymbol{\sigma} + \underline{f}_d = 0$ in Ω, $\boldsymbol{\sigma} \underline{n}_{|\partial_2 \Omega} = \underline{F}_d$, and on the interface: $\boldsymbol{\sigma} \underline{n}_{|\Gamma} = w \underline{n}$; a couple $(\boldsymbol{\sigma}, w)$ satisfying these conditions is said statically admissible (SA);
- the constitutive relation of the body (linear elasticity with $\boldsymbol{D}$ as Hooke tensor): $\boldsymbol{\sigma} = \boldsymbol{D}\boldsymbol{\varepsilon}$, and of the interface (see above).

There are several ways to formulate this problem: minimization with inequality-like constraints, variational inequalities, linear complementarity problems, differential inclusions... The formulations of the next Sections can be derived from the following: the problem is to find $(\underline{U}, z, \boldsymbol{\sigma}, w)$ minimizing the so-called error in constitutive relation η for contact problems (Coorevits et al., 1999):

$$\eta^2(\underline{U}, z, \boldsymbol{\sigma}, w) = \frac{1}{2} \int_\Omega \text{Tr}[(\boldsymbol{\sigma} - \boldsymbol{D}\boldsymbol{\varepsilon}(\underline{U}))\boldsymbol{D}^{-1}(\boldsymbol{\sigma} - \boldsymbol{D}\boldsymbol{\varepsilon}(\underline{U}))] d\Omega + \int_\Gamma wz d\Gamma$$

on the set of admissible fields $(\underline{U}, z)$ KA, $(\boldsymbol{\sigma}, w)$ SA, and $z \in \mathcal{C}_1$, $w \in \mathcal{C}_1$ (on this set, η^2 is obviously positive, and is zero if and only if the constitutive relations are satisfied).

Using Stoke's theorem, for admissible fields, the expression of η^2 can be split into two uncoupled terms: $\eta^2 = \eta_1^2(\underline{U}, z) + \eta_2^2(\boldsymbol{\sigma}, w)$ with:

$$\eta_1^2(\underline{U}, z) = \frac{1}{2} \int_\Omega \text{Tr}[\boldsymbol{\varepsilon}(\underline{U}) \boldsymbol{D} \boldsymbol{\varepsilon}(\underline{U})] d\Omega - \int_\Omega \underline{f}_d \cdot \underline{U} d\Omega - \int_{\partial_2 \Omega} \underline{F}_d \cdot \underline{U} dS$$

$$\eta_2^2(\boldsymbol{\sigma}, w) = \frac{1}{2} \int_\Omega \text{Tr}[\boldsymbol{\sigma} \boldsymbol{D}^{-1} \boldsymbol{\sigma}] d\Omega - \int_{\partial_1 \Omega} \underline{U}_d \cdot \boldsymbol{\sigma} \underline{n} dS$$

The minimization of the first part only leads to the primal formulation of the problem, while the minimization of the second one leads to the dual (in the Legendre-Fenchel way) formulation. Mixed formulations can be obtained by dualizing each of them (in the Arrow-Hurwicz way) using a Lagrange multiplier (De Saxce, 1989).

1.2 Discretized formulations

The discretized version of the primal formulation, once finite elements are used, is:

$$\min_{\boldsymbol{Bu}\geq 0} \frac{1}{2}\boldsymbol{u}^T\boldsymbol{K}\boldsymbol{u} - \boldsymbol{u}^T\boldsymbol{f}$$

where $\boldsymbol{u}$ is the set of finite element degrees of freedom (dof), $\boldsymbol{K}$ is the stiffness matrix, and $\boldsymbol{f}$ are the prescribed generalized forces. $\boldsymbol{B}$ is a Boolean operator that selects the dof on the interface; $\boldsymbol{z} = \boldsymbol{Bu}$. $\boldsymbol{B}$ can also change to the local basis on normal direction; for the sake of simplicity, we assume in all of the following that no such rotations are required. Therefore, the previous problem is easily condensed on the contact dof $\boldsymbol{z}$ and leads to

$$\min_{\boldsymbol{z}\geq 0} \phi(\boldsymbol{z}) \quad \text{with} \quad \phi(\boldsymbol{z}) = \frac{1}{2}\boldsymbol{z}^T\boldsymbol{M}\boldsymbol{z} + \boldsymbol{z}^T\boldsymbol{q} \tag{17.1}$$

where $\boldsymbol{M}$ is the condensed stiffness matrix, and $\boldsymbol{q}$ the opposite of the condensed generalized forces.

Using the indicator function $\psi_{\mathcal{C}_1}$ of the discretized cone $\mathcal{C}_1$, this is equivalent to:

$$\min_{\boldsymbol{z}} \phi(\boldsymbol{z}) + \psi_{\mathcal{C}_1}(\boldsymbol{z}) \quad \Leftrightarrow \quad 0 \in \nabla\phi(\boldsymbol{z}) + \partial\psi_{\mathcal{C}_1}(\boldsymbol{z}) \tag{17.2}$$

The problem can therefore be re-stated as to find $\boldsymbol{a} \in \partial\psi_{\mathcal{C}_1}(\boldsymbol{z})$ such that $\nabla\phi(\boldsymbol{z}) + \boldsymbol{a} = 0$. $\boldsymbol{a}$ must satisfy: $\psi^\star_{\mathcal{C}_1}(\boldsymbol{a}) + \psi_{\mathcal{C}_1}(\boldsymbol{z}) - \boldsymbol{a}^T\boldsymbol{z} \leq 0$, where $\psi^\star_{\mathcal{C}_1}$ is the Legendre-Fenchel transform of $\psi_{\mathcal{C}_1}$. As the Legendre-Fenchel inequality should be satisfied, $\psi^\star_{\mathcal{C}_1}(\boldsymbol{a}) + \psi_{\mathcal{C}_1}(\boldsymbol{z}) - \boldsymbol{a}^T\boldsymbol{z} \geq 0$, the previous inequality reduces to an equality. This equality is equivalent to: $\boldsymbol{a} \leq 0$, $\boldsymbol{z} \geq 0$ and $\boldsymbol{a}^T\boldsymbol{z} = 0$. Using $\boldsymbol{w} = -\boldsymbol{a}$, the problem is then to find $(\boldsymbol{z}, \boldsymbol{w})$ such that

$$\boldsymbol{M}\boldsymbol{z} - \boldsymbol{w} = -\boldsymbol{q} \quad \text{with} \quad 0 \leq \boldsymbol{z} \perp \boldsymbol{w} \geq 0 \tag{17.3}$$

This is a classical linear complementarity problem (LCP), (Cottle et al., 1992). The unknown is the couple $(\boldsymbol{z}, \boldsymbol{w})$.

A dual formulation can be built from (17.2): with the indicator function $\psi_{\mathcal{C}_1}(\boldsymbol{z}) = \sup_{\boldsymbol{w}\geq 0}(-\boldsymbol{z})^T\boldsymbol{w}$, it is equivalent to: $\inf_{\boldsymbol{z}}[\sup_{\boldsymbol{w}\geq 0}\phi(\boldsymbol{z}) - \boldsymbol{z}^T\boldsymbol{w}]$. With sensible assumptions (convexity, semicontinuity) this is again equivalent to: $\sup_{\boldsymbol{w}\geq 0}[\inf_{\boldsymbol{z}}\phi(\boldsymbol{z}) - \boldsymbol{z}^T\boldsymbol{w}]$. The inner minimization problem leads to $\boldsymbol{z} = \boldsymbol{M}^{-1}(-\boldsymbol{q} + \boldsymbol{w})$. Substituting this last value in the previous problem leads to the dual problem:

$$\sup_{\boldsymbol{w}\geq 0} -\frac{1}{2}\boldsymbol{w}^T\boldsymbol{M}^{-1}\boldsymbol{w} + \boldsymbol{w}^T\boldsymbol{M}^{-1}\boldsymbol{q} \tag{17.4}$$

2. The LATIN method for contact problems

The LATIN method (Ladevèze, 1999) was originally designed for time-dependent problems. However, it has been applied to different situations, and among them domain decomposition, contact problems (Champaney et al., 1995)... We are interested herein with the version dealing with contact between substructures, for which the method can be viewed as a mixed domain decomposition method, with nonlinear behavior of interfaces. We will focus on frictionless contact. The principles of the LATIN method will be briefly recalled, especially the splitting it introduces for the problem unknowns and constraints, since the augmented Lagrangian approaches are often compared to operator splitting methods (Fortin and Glowinski, 1982, Glowinski and Le Tallec, 1989).

2.1 The LATIN algorithm

To simplify the presentation, we will derive all the presentation for the discretized problem (17.3), though the method is directly applicable to the continuum media problem. In each case, the unknowns are both the displacement $\boldsymbol{z}$ and the forces $\boldsymbol{w}$. The LATIN method is based on three mechanical principles:

- splitting the difficulties: the linear equations (possibly global) are collected into the set $\boldsymbol{A}_d$, while the local equations (possibly non linear) are collected into the second set $\boldsymbol{\Gamma}$. The solution is the intersection of these sets.

  For the problem (17.3), $(\boldsymbol{w}, \boldsymbol{z})$ belongs to $\boldsymbol{A}_d$ iff $\boldsymbol{M}\boldsymbol{z} - \boldsymbol{w} = -\boldsymbol{q}$, and $(\hat{\boldsymbol{w}}, \hat{\boldsymbol{z}})$ belongs to $\boldsymbol{\Gamma}$ iff $0 \leq \hat{\boldsymbol{z}} \perp \hat{\boldsymbol{w}} \geq 0$;

- building a 2-stage iterative algorithm: the solution is searched with the construction of two series of approximations belonging alternatively to $\boldsymbol{A}_d$ and $\boldsymbol{\Gamma}$.

  At iteration n, the local stage consists in finding $(\hat{\boldsymbol{w}}, \hat{\boldsymbol{z}}) \in \boldsymbol{\Gamma}$ with a first search direction $(\hat{\boldsymbol{w}} - \boldsymbol{w}_n, \hat{\boldsymbol{z}} - \boldsymbol{z}_n) \in \boldsymbol{E}^+$. $(\boldsymbol{w}_n, \boldsymbol{z}_n)$ is known at this stage.

  The linear stage consists in finding $(\boldsymbol{w}_{n+1}, \boldsymbol{z}_{n+1}) \in \boldsymbol{A}_d$ with a second search direction $(\boldsymbol{w}_{n+1} - \hat{\boldsymbol{w}}, \boldsymbol{z}_{n+1} - \hat{\boldsymbol{z}}) \in \boldsymbol{E}^-$, while $(\hat{\boldsymbol{w}}, \hat{\boldsymbol{z}})$ is known from the previous stage;

- using an *ad hoc* representation of the unknowns. This last principle is suited to time-dependent problems, and won't be used herein.

The choice for the search direction is classically to use a parameter k homogeneous to an interface stiffness. For the local stage, the search direction is: $(\hat{\boldsymbol{w}} - \boldsymbol{w}_n) - k(\hat{\boldsymbol{z}} - \boldsymbol{z}_n) = 0$. The local stage is a non-linear and local updating problem, but its solution $(\hat{\boldsymbol{z}}, \hat{\boldsymbol{w}})$ is explicit:

$$\hat{\boldsymbol{z}} = -k^{-1}\langle \boldsymbol{w}_n - k\boldsymbol{z}_n \rangle_-$$
$$\hat{\boldsymbol{w}} = \langle \boldsymbol{w}_n - k\boldsymbol{z}_n \rangle_+$$

$\langle \bullet \rangle_-$ and $\langle \bullet \rangle_+$ designate respectively the negative part and the positive part of their argument.

For the linear stage, the unknowns are $(\boldsymbol{z}_{n+1}, \hat{\boldsymbol{w}})$ and the conjugate search direction is used: $(\boldsymbol{w}_{n+1} - \hat{\boldsymbol{w}}) + k(\boldsymbol{z}_{n+1} - \hat{\boldsymbol{z}}) = 0$. The linear stage is a global and linear updating problem: $\boldsymbol{z}_{n+1}$ is obtained with

$$(\boldsymbol{M} + k\mathbf{1})\boldsymbol{z}_{n+1} = (\hat{\boldsymbol{w}} + k\hat{\boldsymbol{z}}) - \boldsymbol{q} \quad \text{and} \quad \boldsymbol{w}_{n+1} = (\hat{\boldsymbol{w}} + k\hat{\boldsymbol{z}}) - k\boldsymbol{z}_{n+1}$$

Here, k is a parameter of the method. The value of k has no influence on the solution at convergence, but it modifies the convergence rate. Numerical simulations show that there is an optimal value, which is related to the interface operator obtained when condensing the problem on the interface. For the present model problem, this operator is $\boldsymbol{M}$, and k can be diagonal values of $\boldsymbol{M}$, or an average of these.

Figure 1.1 reports convergence results (with an energy-norm error) for different iterative algorithms (CPG: conjugate gradient with projection (Renouf and Alart, 2004), NLGS: non-linear Gauss-Seidel (Moreau, 1999, Jean, 1999)). For this test case, the convergence rate comparison is similar to the one that can be obtained in linear case, for which conjugate gradient has usually a higher convergence rate than LATIN, itself higher than for Gauss-Seidel.

2.2 Interpretation as a nonsmooth augmented Lagrangian approach

For perfect interfaces, the LATIN algorithm has been proved to be equivalent to the augmented Lagrangian interpretation (Glowinski and Le Tallec, 1990) of the approach proposed in (Lions, 1990), and more precisely to the so-called 'ALG3' algorithm in (Fortin and Glowinski, 1982), see (Champaney et al., 1995). We proposed herein to prove the equivalence to an augmented Lagrangian algorithm for the case of contact without friction.

Beginning with the minimization problem (17.1), the first step is to split the unknown $\boldsymbol{z}$, using an intermediate unknown $\hat{\boldsymbol{z}}$, an to introduce an equality constraint $\boldsymbol{z} = \hat{\boldsymbol{z}}$, as well as an inequality-like on $\hat{\boldsymbol{z}}$:

$$\min_{\boldsymbol{z}=\hat{\boldsymbol{z}},\ \hat{\boldsymbol{z}}\geq 0} \phi(\boldsymbol{z}) \tag{17.5}$$

The second step is to use two multipliers $\boldsymbol{w}_c$ and $\hat{\boldsymbol{w}}_c$ and convex analysis methodology for dualizing the previous constraints (Rockafellar, 1976). The augmented Lagrangian is obtained with a classical dualization and regularization of the equality constraint, while the Moreau-Yosida technique is applied to the inequality-like constraint, using a positive parameter k (Yosida, 1964, Moreau, 1965):

$$\begin{aligned}\mathcal{L}(\boldsymbol{z}, \hat{\boldsymbol{z}}; \boldsymbol{w}_c, \hat{\boldsymbol{w}}_c) = \phi(\boldsymbol{z}) + \\ -\frac{1}{2}\boldsymbol{w}_c^T k^{-1} \boldsymbol{w}_c + \frac{1}{2}[\boldsymbol{w}_c + k(\boldsymbol{z} - \hat{\boldsymbol{z}})]^T k^{-1} [\boldsymbol{w}_c + k(\boldsymbol{z} - \hat{\boldsymbol{z}})] + \\ -\frac{1}{2}\hat{\boldsymbol{w}}_c^T k^{-1} \hat{\boldsymbol{w}}_c^T + \frac{1}{2}\langle \hat{\boldsymbol{w}}_c + k\hat{\boldsymbol{z}}\rangle_-^T k^{-1} \langle \hat{\boldsymbol{w}}_c + k\hat{\boldsymbol{z}}\rangle_- \end{aligned} \quad (17.6)$$

The stationary conditions for this augmented Lagrangian are

$$\begin{aligned}\frac{\partial \mathcal{L}}{\partial \boldsymbol{z}} &= 0 = \boldsymbol{M}\boldsymbol{z} + \boldsymbol{q} + \boldsymbol{w}_c + k(\boldsymbol{z} - \hat{\boldsymbol{z}}) \\ \frac{\partial \mathcal{L}}{\partial \boldsymbol{w}_c} &= 0 = -k^{-1}\left\{\boldsymbol{w}_c - [\boldsymbol{w}_c + k(\boldsymbol{z} - \hat{\boldsymbol{z}})]\right\} \\ \frac{\partial \mathcal{L}}{\partial \hat{\boldsymbol{z}}} &= 0 = -\boldsymbol{w}_c - k(\boldsymbol{z} - \hat{\boldsymbol{z}}) + \langle \hat{\boldsymbol{w}}_c + k\hat{\boldsymbol{z}}\rangle_- \\ \frac{\partial \mathcal{L}}{\partial \hat{\boldsymbol{w}}_c} &= 0 = -k^{-1}\left\{\hat{\boldsymbol{w}}_c - \langle \hat{\boldsymbol{w}}_c + k\hat{\boldsymbol{z}}\rangle_-\right\}\end{aligned}$$

One can then remark that for the solution, $\boldsymbol{z} = \hat{\boldsymbol{z}}$ and $\boldsymbol{w}_c = \hat{\boldsymbol{w}}_c = -\boldsymbol{w}$. Designing an iterative Uzawa-like algorithm consists in expressing the successive stationary conditions with the previous obtained values of the unknowns. With the previous remark, the latest values of $\boldsymbol{z}$ or $\hat{\boldsymbol{z}}$ can be substituted one to the other, and similarly for $\boldsymbol{w}_c$ and $\hat{\boldsymbol{w}}_c$. The successive steps are the following:

- step 1: assuming that the values $\hat{\boldsymbol{w}}_c^{(n)}$ and $\hat{\boldsymbol{z}}^{(n)}$ are the latest ones, the stationary with respect to $\boldsymbol{z}$ gives the next value of $\boldsymbol{z}^{(n+1)}$: $\boldsymbol{M}\boldsymbol{z}^{(n+1)} + \boldsymbol{q} + k(\boldsymbol{z}^{(n+1)} - \hat{\boldsymbol{z}}^{(n)}) + \hat{\boldsymbol{w}}_c^{(n)} = 0$
- step 2: with this last updated value $\boldsymbol{z}^{(n+1)}$, the stationary with respect to $\boldsymbol{w}_c$ gives the next value of $\boldsymbol{w}_c^{(n+1)}$: $\boldsymbol{w}_c^{(n+1)} - [\hat{\boldsymbol{w}}_c^{(n)} + k(\boldsymbol{z}^{(n+1)} - \hat{\boldsymbol{z}}^{(n)})] = 0$
- step 3: with these last values $\boldsymbol{z}^{(n+1)}$ and $\boldsymbol{w}_c^{(n+1)}$, the stationary with respect to both $\hat{\boldsymbol{z}}$ and $\hat{\boldsymbol{w}}_c$ allows to update their values: first, $-\boldsymbol{w}_c^{(n+1)} - k(\boldsymbol{z}^{(n+1)} - \hat{\boldsymbol{z}}^{(n+1)}) + \langle \boldsymbol{w}_c^{(n+1)} + k\boldsymbol{z}^{(n+1)}\rangle_- = 0$

that leads to

$$\begin{aligned}\hat{\boldsymbol{z}}^{(n+1)} &= k^{-1}\left\{(\boldsymbol{w}_c^{(n+1)} + k\boldsymbol{z}^{(n+1)}) - \langle \boldsymbol{w}_c^{(n+1)} + k\boldsymbol{z}^{(n+1)}\rangle_-\right\} \\ &= k^{-1}\langle \boldsymbol{w}_c^{(n+1)} + k\boldsymbol{z}^{(n+1)}\rangle_+\end{aligned}$$

and second, $\hat{\boldsymbol{w}}_c^{(n+1)} - \langle \boldsymbol{w}_c^{(n+1)} + k\boldsymbol{z}^{(n+1)}\rangle_- = 0$

The steps 1 and 2 are the linear stage of the LATIN, while the step 3 is the local stage.

Though other permutations may lead to other augmented Lagrangian versions for nonsmooth problems, the proposed one recovers the LATIN method that appears to be efficient for the simulation of large scale assemblies of deformable parts (Champaney et al., 1997).

3. Gradient methods for large scale contact problems

We focuss herein on two kinds of large scale problems. Such large scale problems require sophisticated tools arising from the optimization field, traditionally relying on convex and nonsmooth analysis (Bonnans et al., 2003). The first one is related to finite element simulation of assemblies with a dual domain decomposition method. This method is well suited to problems with a large number of degrees of freedom and a few contact conditions between subdomains. The second one deals with simulation of granular materials and structures: the contact conditions are numerous in this case, between rigid bodies in a large collection.

3.1 Structural frictionless contact problems

The FETI-C method has been built as an extension of the Finite Element Tearing and Interconnecting (FETI) method (Farhat and Roux, 1991) for the case of contact without friction between subdomains (Dureisseix and Farhat, 2001). The numerical and parallel scalability (i.e. the ability to converge with a number of iterations depending weakly on the number of subdomains) of the FETI method have been examplified, thanks to the presence of a 'coarse' problem acting on the rigid body displacement of the floating subdomains (subdomains without prescribed displacement). Nevertheless, for simplification, no floating subdomain will be considered here, and only one subdomain is assumed to have an interface with a rigid plate (as for the model problem of Figure 1.1).

The extension to interfaces that exhibit an unilateral contact behavior is performed with the following remarks:

- the procedure should act on admissible reaction forces: $\boldsymbol{w} \geq 0$;

- the classical residual $\boldsymbol{z}$, which is the displacement jump across the interfaces, should be replaced by the contact residual $\tilde{\boldsymbol{z}} = \langle \boldsymbol{z} \rangle_-$ to monitor the convergence;
- the 'gluing' condition should act on the *active* contact interface (defined as the part Γ_a of Γ where $\boldsymbol{w} > 0$), using an adapted Boolean operator: $\boldsymbol{B_w r} = \begin{cases} \boldsymbol{r} & \text{on } \Gamma_a \\ 0 & \text{on } \Gamma \backslash \Gamma_a \end{cases}$

With $\boldsymbol{B}_{\boldsymbol{w}}^T$ as the transpose of the matrix $\boldsymbol{B_w}$ for a *given* $\boldsymbol{w}$, one gets the optimization problem (non quadratic due to the dependence of $\boldsymbol{B_w}$ on $\boldsymbol{w}$):

$$\min_{\boldsymbol{w} \geq 0} \frac{1}{2} \boldsymbol{w}^T (\boldsymbol{B}_{\boldsymbol{w}}^T \boldsymbol{M}^{-1} \boldsymbol{B_w}) \boldsymbol{w} - \boldsymbol{w}^T (\boldsymbol{B}_{\boldsymbol{w}}^T \boldsymbol{M}^{-1} \boldsymbol{q}) \tag{17.7}$$

The straightforward extension of FETI approach consists in applying formally a conjugate gradient algorithm to the problem (17.7) as if this problem is quadratic, driving $\tilde{\boldsymbol{z}}$ to zero, and reusing classical preconditioners. It appears to be scalable for the case where there are floating subdomains, for which an additional projection is used (Dureisseix and Farhat, 2001). Of course, the conjugating property is not satisfied for all of the previous search directions because the $\boldsymbol{B_w}$ operator projects the iterate, and the $\boldsymbol{B}_{\boldsymbol{w}}^T$ operator projects the gradient.

Therefore, this algorithm can be interpreted as a projected conjugate gradient with projection, applied to the classical quadratic dual problem as expressed in (17.4). Work is in progress on adapting FETI-DP version (Farhat et al., 2001) to contact and friction. Such an adaptation would take advantage of the aforementioned interpretation, and of the developments presented in the next Section.

3.2 Granular frictional contact problems

Independently, P. Alart and M. Renouf developed a conjugate gradient type algorithm for granular media characterized with a large number of bodies and consequently of contacts (Renouf and Alart, 2003).

A granular material is by essence a discrete medium, where each 'subdomain' is a rigid body. The problem is therefore stated from the point of view of interactions between these bodies. As an example, Figure 3.2 illustrates the simulation of a sandbox experiment, useful in geomechanics for soil layer formation. About 40 000 grains in interaction (involving 88 000 contact conditions) are modeled.

Each interaction, as an 'interface', deals with the relative velocity between bodies, $\boldsymbol{z} = \boldsymbol{Bv}$ (where $\boldsymbol{v}$ stores the velocities of the bodies), as well as the contact impulsion $\boldsymbol{w}$. This is of course a time-dependent

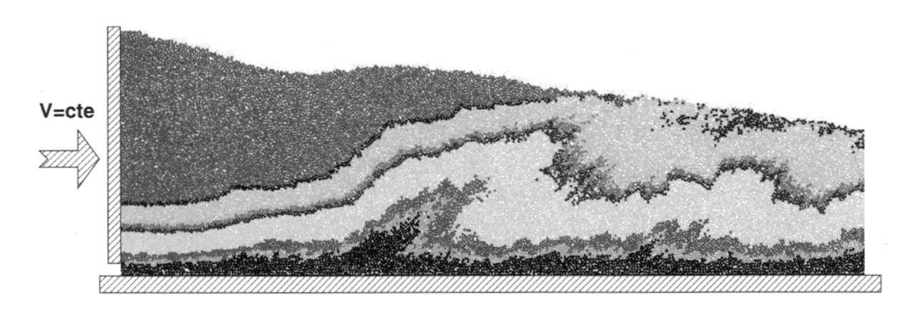

Figure 17.2. Sandbox simulation (40 000 grains)

problem, and for a given time step, the dynamic evolution of the bodies is $\boldsymbol{M}\boldsymbol{v} = \boldsymbol{B}^T\boldsymbol{w} - \boldsymbol{q}$ where $\boldsymbol{q}$ stores external impulsions and contributions of relative velocities at the end of the previous time step. $\boldsymbol{M}$ is in this case the block-diagonal mass matrix of all the bodies.

For the case of frictionless interaction, the constitutive relations are simply $0 \leq \boldsymbol{z} \perp \boldsymbol{w} \geq 0$. Expressing $\boldsymbol{v}$ from the dynamic evolution, one gets $\boldsymbol{z} = \boldsymbol{W}\boldsymbol{w} - \widetilde{\boldsymbol{q}}$, with $\boldsymbol{W} = \boldsymbol{B}\boldsymbol{M}^{-1}\boldsymbol{B}^T$ and $\widetilde{\boldsymbol{q}} = \boldsymbol{B}\boldsymbol{M}^{-1}\boldsymbol{q}$. The resulting problem is therefore similar to the previous ones.

For the case of frictional contact, normal and tangential components at each node have to be separated, and we will use lowerscript n and t to designate them. For simplicity, we will consider only bidimensional analysis. The constraints on impulsion $\boldsymbol{w}$ must belong to the Coulomb's friction cone, i.e. $\boldsymbol{w} \in \mathcal{C}(\mu \boldsymbol{w}_n)$, with μ as the Coulomb's friction coefficient, $\boldsymbol{w}_n$ the normal component of the impulsion, and

$$\mathcal{C}(\mu \boldsymbol{w}_n) = \prod_{\alpha=1}^{n_c} \mathbb{R}^+ \times [-\mu \boldsymbol{w}_{\alpha,n}, \mu \boldsymbol{w}_{\alpha,n}]$$

where n_c is the number of potential contact nodes. As for the frictionless contact, the matrix $\boldsymbol{W} = \boldsymbol{B}\boldsymbol{M}^{-1}\boldsymbol{B}^T$ is semi definite positive and therefore can be singular. The resulting quasi-optimization problem is:

$$\boldsymbol{w} \in \underset{\boldsymbol{x} \in \mathcal{C}(\mu \boldsymbol{w}_n)}{\operatorname{argmin}} \frac{1}{2}\boldsymbol{x}^T \boldsymbol{W} \boldsymbol{x} - \boldsymbol{x}^T \widetilde{\boldsymbol{q}} \tag{17.8}$$

The solving procedure relies on a fixed point method for the Tresca threshold $\mu \boldsymbol{w}_n$ and on an inner conjugate projected gradient. The proposed approach is a diagonalized version of this procedure (with a one shot single loop). The projection operators are replaced by specific corrections which are not exactly projections, but may be derived from a mathematical analyses in such a way that the converged solution satisfies the relation (17.8), (Renouf and Alart, 2004).

Figure 17.2 is related to the first time step of the compaction under gravity of a box filled with 33 000 grains. The convergence of 3 different algorithms (CPG: conjugate projected gradient, NLGS: non-linear Gauss-Seidel, and PCPG: preconditioned conjugate projected gradient) is compared for increasing values of Coulomb's friction coefficient μ. For all of them, the convergence requires more iterations when friction increases. The conjugate gradient methods are more efficient than the Gauss-Seidel one, but the gain decreases as the friction increases.

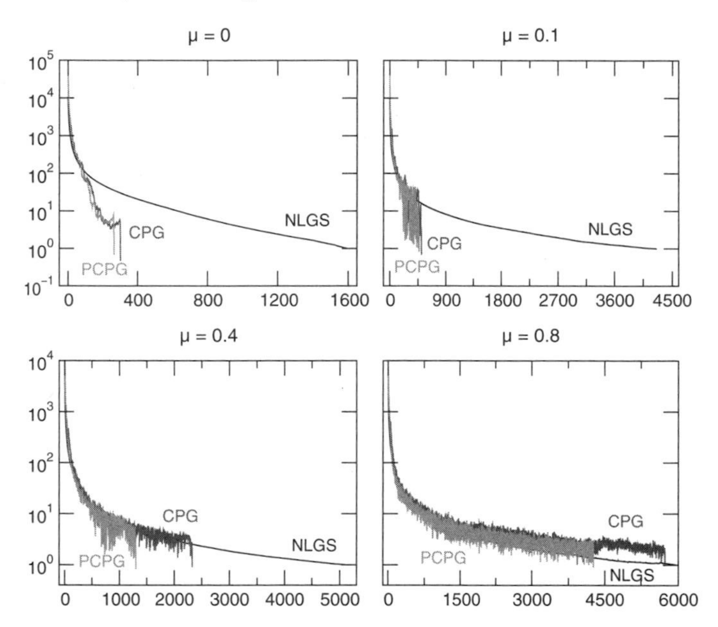

Figure 17.3. Influence of friction coefficient on convergence

4. Structural versus granular

Different scientific communities usually develop different dedicated tools that can sometimes be re-unified. Such a gap bridging is done in Section 2.2 where the LATIN algorithm for frictionless contact is reinterpreted as a Lagrangian approach for nonsmooth problems. Convergence results in this case can be transferred to Lagrangian approches as convergence is proved (Ladevèze, 1999) for $0 < k < \infty$. The construction of the equivalence can also lead to other versions of the method.

Moreover, between structural analysis and granular simulation communities, the approach used for frictional contact and gradient-based simulations of a large number of grains can guide the development of extensions to frictional case for dual domain decomposition methods. The primal domain decomposition method already got benefits from convex and nonsmooth analysis (Alart et al., 2004).

References

Alart, P., Barboteu, M., and Renouf, M. (2004). Parallel computational strategies for multicontact problems: Applications to cellular and granular media. *Int. J. for Multiscale Comput. Engng.* to appear.

Bonnans, J.-F., Gilbert, J.-C., Lemaréchal, C., and Sugastizàbal, C. (2003). *Numerical Optimization – Theoretical and Practical Aspect.* Springer Verlag.

Champaney, L., Cognard, J.-Y., Dureisseix, D., and Ladevèze, P. (1995). Une approche modulaire pour l'analyse d'assemblages de structures tridimensionnelles. Application au contact avec frottement. In *Proceedings of StruCoMe 95*, pages 295–306, Paris.

Champaney, L., Cognard, J.-Y., Dureisseix, D., and Ladevèze, P. (1997). Large scale applications on parallel computers of a mixed domain decomposition method. *Comput. Mech.*, (19):253–263.

Coorevits, P., Hild, P., and Pelle, J.-P. (1999). Contrôle des calculs éléments finis pour les problèmes de contact unilatéral. *Revue Européenne des Éléments Finis*, 8(1):7–29.

Cottle, R. W., Pang, J.-S., and Stone, R. E. (1992). *The Linear Complementarity Problem.* Academic Press, Boston.

De Saxce, G. (1989). Sur quelques problèmes de mécanique des solides considérés comme matériaux à potentiel convexe. volume 118 of *Publications de la Faculté des Sciences appliquées.* Université de Liège.

Dureisseix, D. and Farhat, C. (2001). A numerically scalable domain decomposition method for the solution of frictionless contact problems. *Int. J. for Numer. Meth. Engng.*, 50:2643–2666.

Farhat, C., Lesoinne, M., Le Tallec, P., Pierson, K., and Rixen, D. (2001). FETI-DP: a dual-primal unified FETI method - part I: a faster alternative to the two-level FETI method. *Int. J. for Numer. Meth. Engng.*, 50(7):1523–1544.

Farhat, C. and Roux, F.-X. (1991). A method of finite element tearing and interconnecting and its parallel solution algorithm. *Int. J. for Numer. Meth. Engng.*, 32:1205–1227.

Fortin, M. and Glowinski, R. (1982). Méthodes de lagrangien augmenté. In Lions, P.-L., editor, *Méthodes Mathématiques de l'informatique*, volume 9.

Glowinski, R. and Le Tallec, P. (1989). Augmented lagrangian and operator-splitting methods in nonlinear mechanics. volume 9 of *SIAM Studies in Appl. Math.* SIAM.

Glowinski, R. and Le Tallec, P. (1990). Augmented lagrangian interpretation of the nonoverlapping Schwarz alternating method. In Chan, T. F., Glowinski, R., Périaux, J., and Widlund, O. B., editors, *Third Int. Symp. on Domain Decomposition Methods for Part. Differ. Equations*, pages 224–231. SIAM.

Jean, M. (1999). The non-smooth contact dynamics method. *Comput. Meth. Appl. Mech. Engng.*, 117(3-4):235–257.

Ladevèze, P. (1999). *Nonlinear Computational Structural Mechanics — New Approaches and Non-Incremental Methods of Calculation.* Springer Verlag.

Lions, P.-L. (1990). On the Schwarz alternating method III: a variant for nonoverlapping subdomains. In Chan, T. F., Glowinski, R., Périaux, J., and Widlund, O. B., editors, *Third Int. Symp. on Domain Decomposition Methods for Partial Differential Equ.*, pages 202–223. SIAM.

Moreau, J. J. (1965). Proximité et dualité dans un espace hilbertien. *Bulletin de la Société Mathématique de France*, 93:273–299.

Moreau, J. J. (1999). Numerical aspects of the sweeping process. *Comput. Meth. Appl. Mech. Engng.*, 117(3-4):329–349.

Renouf, M. and Alart, P. (2003). Un nouvel algorithme de quasi-optimisation pour la résolution des problèmes multicontacts et application aux milieux granulaires. In *Proc. of the 16e Congrès Français de Mécanique*, Nice.

Renouf, M. and Alart, P. (2004). Conjugate gradient type algorithms for frictional multicontact problems: applications to granular materials. *Comput. Meth. Appl. Mech. Engng.* submitted.

Rockafellar, R. T. (1976). Augmented lagrangians and applications of the proximal point algorithm in convex programming. *Math. of Operations Research*, 1(2):97–116.

Yosida, K. (1964). *Functional analysis.* Springer Verlag, Berlin.

Chapter 18

NUMERICAL SIMULATION OF A MULTIBODY GAS

Michel Jean

Laboratoire de Mécanique et d'Acoustique, LMMA-CNRS, EGIM, 31 chemin Joseph Aiguier, 13402 Marseille, Cedex 20, France

mjean@imtumn.imt-mrs.fr

Abstract Cet article traite du comportement d'une collection de disques rigides, enfermés dans un réservoir de forme circulaire, en collision avec une loi de choc parfaitement élastique (et frottement nul). Des considérations sur les propriétés de similitude des solutions des équations de la dynamique avec choc et plusieurs résultats de simulation numérique obtenus avec la méthode *NonSmooth Contact Dynamics* permettent de conduire une discussion sur la notion de gaz parfait.

This paper deals with the behavior of a collection of rigid disks enclosed in a circular reservoir, colliding with a perfect elastic shock law (frictionless). Considerations concerning similarity properties of the solutions of the dynamical equations of colliding bodies together with numerical results using the *NonSmooth Contact Dynamics* method, allow for a discussion about the notion of perfect gas.

1. Introduction

Is it possible to come to the idea of perfect gas, a basic concept in classical thermodynamics, through multibody behavior introducing as reduced as possible statistical considerations? The following numerical experiments are fine but cheap experiments, in the sense that the number of grains used in the collection is rather small (2406 grains), and the duration of evolution is not really long enough to ensure the existence of a steady regime. There would be so far no inconvenience in increasing the number of grains and the duration of computation. Nevertheless, some encouraging prospects emerge. Similar simulations could be made using molecular dynamics or event driven methods with quite a large number of grains. Actually, besides the question of perfect gas, is raised also the question of presumably other kinds of perfect gas, for instance rotating dipoles. The software *LMGC90* is well equipped to deal with such collections, but they will not be considered in this paper.

The *Contact Dynamics* method was originated by J.J. Moreau around 1984, see (Moreau, 1988), (Moreau, 1994), (Moreau, 1999). It was then extended to deal with more general applications, such as finite element modelling, to which we refer as *NonSmooth Contact Dynamics* method, (Jean and Moreau, 1991), (Jean and Moreau, 1992), (Jean, 1995), (Jean, 1999). For details about *NSCD* devoted to granular materials, see (Cambou and Jean, 2001). For details about the software *LMGC90* see F. Dubois, M. Jean, http://www.lmgc.univ-montp2.fr.

2. The mechanical and mathematical framework

Since similarity properties of the solutions of the dynamical problem with perfect elastic shocks will be pointed out, here is a brief recall of the governing equations.

$$\begin{cases} M d\dot{q} = r d\theta \,, \\ U^{\alpha} = H^{*\alpha}(q)\,\dot{q} \;,\; r = \sum\limits_{\alpha} r^{\alpha} = \sum\limits_{\alpha} H^{\alpha}(q)\,R^{\alpha} \,, \\ \dot{g}^{+\alpha} = U_N^{+\alpha} \,, \\ \text{if a shock occurs at time } \tau,\; U_N^{\alpha\,+}(\tau) = -U_N^{\alpha\,-}(\tau) \,, \\ \text{initial conditions and boundary conditions.} \end{cases} \tag{18.1}$$

where q is the Lagrange variable (the gravity centers coordinates and the rotation vector of the rigid bodies), r is the generalized forces (impulses) representative of the local reaction forces (impulses) for the variable q; it is a density of measure with respect to a real positive measure $d\theta$ (in some circumstances, see assumption below, a sum of Dirac measures at times where shocks occur). The mass matrix is noted M. The relative velocity between contactors for the candidate to contact labelled α is U^{α}, R^{α} is the reaction force (impulse) between contactors, g^{α} is the gap. The relative velocity and the local reaction have normal components (subscript $_N$) expressed in the local frame at the candidate contacting point (the tangential frame vector lies in the common tangent plane to both contactors). Right values, superscript $(^+)$, represent values just after the instant of a possible shock, and left values, superscript $(^-)$, are values just before the instant of a possible shock. The linear mappings $H^{\alpha}(q)$ and its transpose $H^{*\alpha}(q)$ allow to map the local reaction into its generalized force (impulse) representative, and the time derivative of the Lagrange variable to the relative velocity. The first equation in 18.1 is the equation of dynamics written in the sense of measures. The second set of equations is the kinematic relations. The third equation expresses that the right derivative of the gap function is the right relative velocity. The fourth equation stands for perfect elastic shock law. *The gravity* is **assumed** *to be zero, and no external loads are applied on the grains.*

3. The numerical experimental setting

Figure 18.1. The packed sample of disks.

A reference sample of two-dimensional grains is composed of $N = 2406$ rigid disks, packed between two circles, see figures 18.1, 18.2. This sample is polydisperse [1] Such a distribution of radii allows for a compact packing avoiding the formation of crystal like clusters. We assume that the disks are actually cylinders of unit length, so that the "volume" of a disk is nothing but the area of the disk. All disks, labelled $a = 1, N$, have the same specific mass [2] generating a distribution of masses $m_a, a = 1, N$

[1] 42 grains of radius $0.75\,10^{-2}m$, 311 grains of radius $0.65\,10^{-2}m$, 845 grains c $0.55\,10^{-2}m$, 845 grains of radius $0.45\,10^{-2}m$, 311 grains of radius $0.35\,10^{-2}m$, 42 radius $0.25\,10^{-2}m$, and 12 grains of radius $0.85\,10^{-2}m$ (what are you doing here po Furthermore, 8 extra grains of radius $0.25\,10^{-2}m$ are used as projectiles to gene kinetic energy.

[2] The specific mass is $0.58\,10^{3}Kg/m^{3}$.

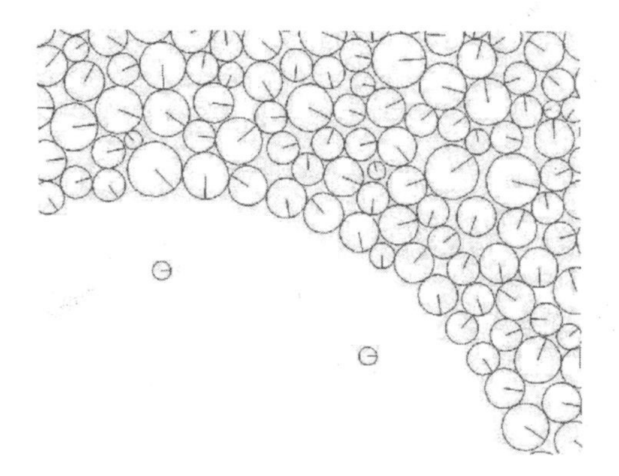

Figure 18.2. A snapshot of a packed sample of disks; two projectiles can be seen.

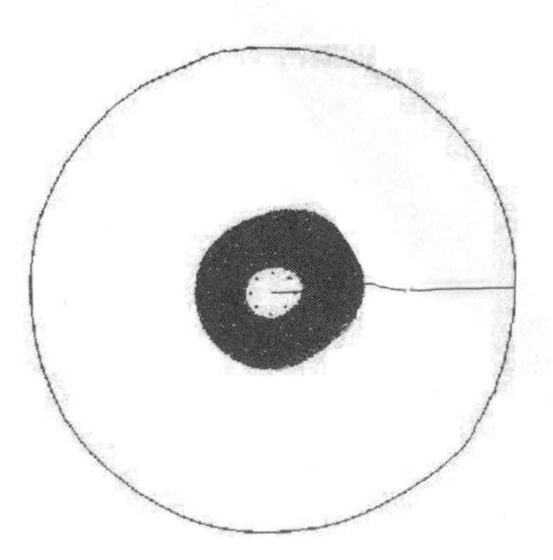

Figure 18.3. Initial state, $v/v_0 = 10$.

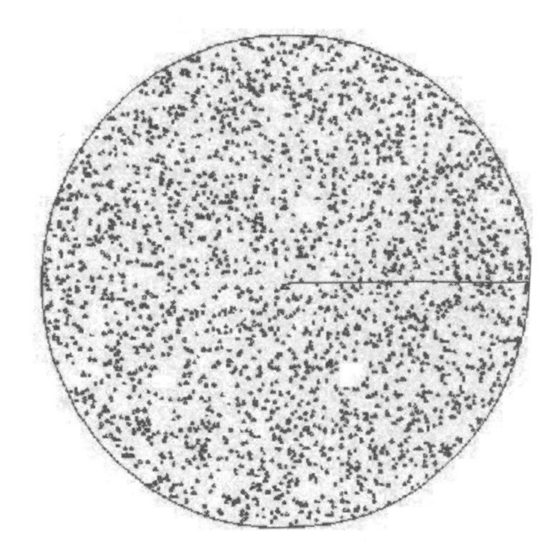

Figure 18.4. State at time 150 s, $v/v_0 = 10$.

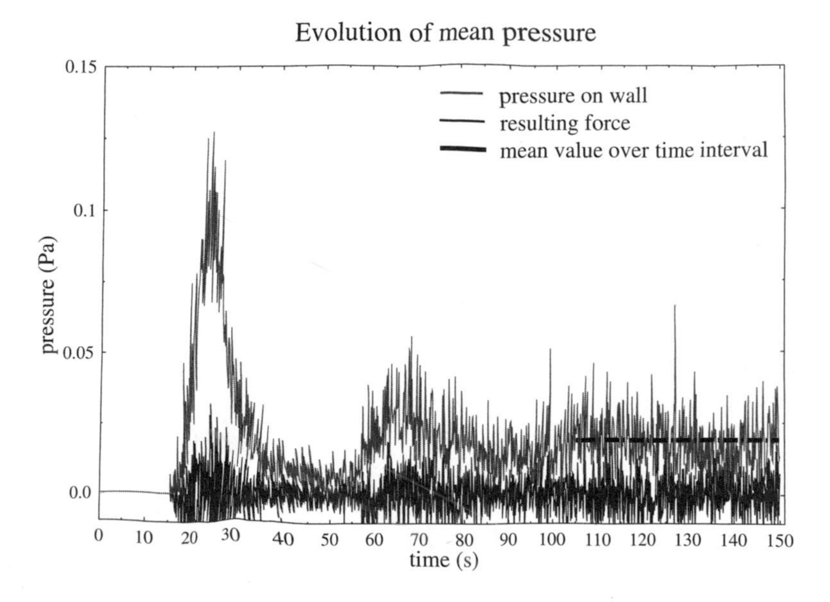

Figure 18.5. Mean pressures and forces (average impulses over sub intervals 0.125 s) against time; pressure on the reservoir boundary [pressure on wall]; mean resulting force on the reservoir boundary [resulting force]; an average of the pressure on the reservoir boundary between times 105s, 150s, is shown [mean value over time interval]; granular temperature $\mathcal{T} = 0.1893\, 10^{-4}\, J/mole$, $v/v_0 = 10$.

according to the distribution of the radii. The disks collide accordir a perfect elastic frictionless shock law.

Replica of the reference packed sample of grains are enclosed successively in different circular reservoirs with different radii; an example is shown in figure 18.2. The volume of the reservoir is equal to the area of the reservoir times a unit length. The reservoirs are labelled after the ratio v/v_0 where v is the volume of the reservoir and v_0 is some reference volume [3]. As for the initial conditions, the packed sample is at rest while the 8 enclosed projectile grains are given initial velocities directed outward. The initial kinetic energy (the kinetic energy of projectiles), is the same for all simulations [4]. When the projectiles reach the packed sample, simultaneous shocks occur within the sample and a kinematic wave propagates. After some time, grains reach the boundary and a reflexive wave propagates backwards. The motion of grains degenerates more and more tending to a random motion, though some reminiscence of the wave regime remains still visible, see figure 18.5, where mean pressures and forces (i.e. average impulses over time intervals of $0.125s$) are plotted as a function of time, namely the mean pressure on the reservoir boundary (pressure on wall) and the mean force resultant on the reservoir boundary. The mean value of this force is vanishing. In this figure, the volume of the reservoir is 10 times the reference volume v_0. For a more precise definition of the mean pressure and of the temperature, see subsection below 4.1. Figure 18.2 shows the initial state, and figure 18.4 shows the state at time 150 s.

4. Inducing some macroscopic behavior law

Is it possible to come out with a behavior law between the pressure acting on the cylindrical (circular) boundary of the reservoir and the volume (cross section area) of the reservoir, in a steady state regime? The *mean pressure* is defined in the following way. Let $t \rightarrow R^\alpha(t)$ be the reaction density of impulse exerted by some grain in the neighborhood of the reservoir boundary (expressed in the local frame); $R^\alpha(t)$ is zero unless a shock occurs. The mean pressure for a time interval $[t_1, t_2]$ is defined as the integral of the function R^α_N, the normal component of the

[3]The sum of the volumes of the grains belonging to the sample is about $0.1955\, m^3$, the smaller circular reservoir able to contain the whole packed collection of grains has a volume about $0.25\, m^3$ with a radius $0.28\, m$; the reference volume is chosen as $0.2827\, m^3$. In fact, the volume of a reservoir with radius $0.3\, m$, is just large enough to enclose the sample.

[4]The projectiles initial outward velocity is $1\, m/s$ in the simulations $v/v_0 = 1.5, 2, 2.5, 3, 6$; initial outward velocity $0.75\, m/s$ is applied to 4 projectiles and $1.1198957\, m/s$ to the er 4 projectiles in the simulations $v/v_0 = 10, 18$, such initial conditions generating less propagation; the final state of experiment $v/v_0 = 18$ at time $200\, s$, is used as initial ions for the simulation $v/v_0 = 34$. The time step is $0.5\, 10^{-3}$s in the simulations 1.5, 2, 2.5, 3, 6, and $1\, 10^{-3}$ in other simulations

density of impulse reaction, for the Dirac measure, with summation over all grains colliding the wall during the interval $[t_1, t_2]$. This is has then to be divided by $t_2 - t_1$ and the boundary circumference of the reservoir. Similarly, the mean resultant force on the reservoir wall is the integral of the function r, the reaction forces exerted on each disk at its potential contacts (the components of which are expressed in the general frame) [5].

4.1 An attempt to define a concept of temperature

For a given sample, i.e. a given set of disks with its distribution of radii and masses $\{m_i\}$, the following property is true: Let $X(t)$ be the vector of the coordinates of disks centers at time t, and $r(t)$ the generalized impulse vector;

> If, $X : t \mapsto X(t)$, is a motion solution of the multibody collision problem with initial conditions, $X(0) = X_0$, $\dot{X}(0) = \dot{X}_0$,
> together with the generalized impulse, $r : t \mapsto r(t)$,
> then, $X^\lambda : t \mapsto X^\lambda(t) = X(\lambda t)$, is also a solution of the problem with initial conditions, $X^\lambda(0) = X_0$, $\dot{X}^\lambda(0) = \lambda \dot{X}_0$,
> together with the generalized impulse, $r^\lambda : t \mapsto r^\lambda(t) = \lambda r(t)$.

Observe that the trajectories for motions X and X^λ are the same, but the kinematics are different. In time intervals where no collision occurs, i.e. during a free flight, a grain is moving with a constant velocity vector, while the velocity vector of the motion X^λ is λ times the velocity of the motion X during the same corresponding free flight. One observes also that at some time τ where a shock occurs in the configuration $X(\tau)$, a homologue shock occurs at time τ/λ in the same configuration $X^\lambda(\tau/\lambda) = X(\tau)$, and the velocity vectors before and after the impact time for the X^λ motion are respectively λ times the velocity vectors before and after the impact time for the X motion, so that the reaction impulse for the X^λ motion is λ times the reaction impulse for the X motion. The above result is true thanks to peculiar circumstances inherent to this experiment; the gravity is **assumed** *to be zero, and no external loads are exerted.*

[5] In fact the reservoir wall is a special contactor, a circle with a supplementary de freedom, a variable radius (set here to a constant value). The dual variable, the ger reaction for this supplementary degree of freedom, at time t, is the sum of the im tween all grains and the reservoir wall. Then a time-averaged value is recorded ov time intervals. In these simulations, we have $t_2 - t_1 = 0.125\ s$.

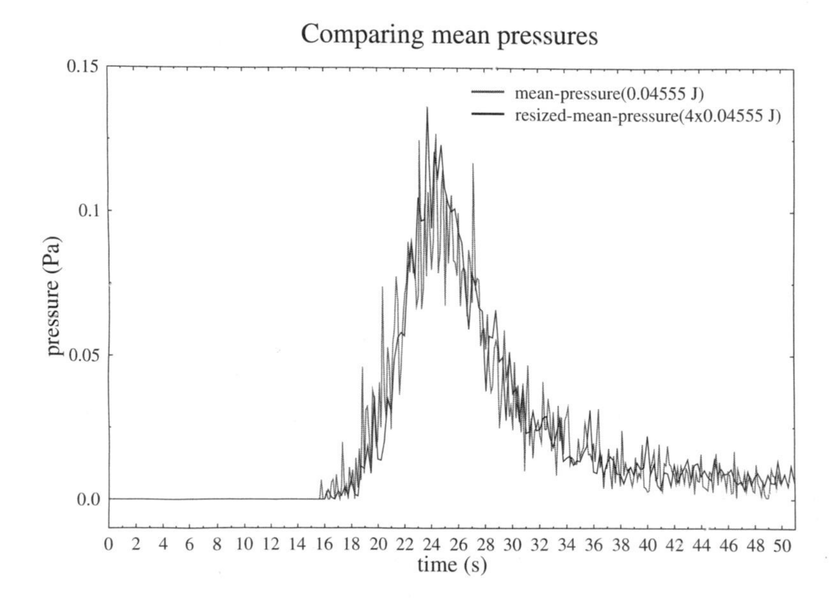

Figure 18.6. Comparing mean pressures for two values of the kinetic energy: mean pressure(0.04555 J); rescaled-mean-pressure($4 \times 0.04555\ J$)=1/4 mean pressure($4 \times 0.04555\ J$), together with rescaled-time=2× time. $v/v_0 = 10$.

The number of grains being large enough, it is **assumed** that *when a steady state is reached, the shocks on the reservoir boundary are occuring at random but "uniformly time distributed".* [6] Since the reaction impulses are multiplied by λ and since the number of shocks on the reservoir wall occuring in a given time interval for the solution X^λ, is λ times the number of shocks for the solution X, it appears that the pressure varies as λ^2. This point can be checked in figure 18.6 [7].

Our numerical simulation shows that, during a run, (i.e. computing a numerical approximation of the solution X with some initial condition $X(0), \dot{X}(0)$) the total kinetic energy of the collection, $\frac{1}{2}\sum_i^N m_i\dot{X}_i(t).\dot{X}_i(t)$ is conserved, as expected. This numerical accuracy is a consequence of the fact that the implicit NSCD algorithm, is conservative in these circumstances [8]. Any $X(t_0), \dot{X}(t_0)$, at some time t_0 may be taken as initial value at time t_0 for the motion X, so that the

distribution can be studied numerically.

numerical experiment, the initial velocities of the numerical experiment shown in 5 have been multiplied by 2

an event driven method would also preserve the kinetic energy. In explicit-scheme ding some damping to ensure stability, the kinetic energy will degrade.

λ scaling property holds. It may thus be inferred that the rele\ rameter is the kinetic energy, which will be noted $\frac{1}{2}\sum_i^N m_i\dot{X}_i.\dot{X}_i$. parameter was already suggested in 1978 by S. Ogawa (Ogawa, 19. and called *granular temperature* with extensive use in kinetic theori of granular gases. This parameter is not a new variable, though in the granular case it emerges only through multibody dynamics considerations, with a touch of randomness and equidistribution properties.

Actually, more analysis is necessary before ending up with a relevant parameter. Indeed, let us now assume that the specific mass of the grains is increased by a factor λ. The following proposition is true:

> If, $X : t \mapsto X(t)$, is a motion solution of the multibody collision problem with initial conditions, $X_0\,, \dot{X}_0$, for the distribution of masses $\{m_i\}$, together with the generalized impulse, $r : t \mapsto r(t)$,
>
> then, $X : t \mapsto X(t)$, is also a solution of the problem with initial conditions, $X_0\,, \dot{X}_0$, for the distribution of masses $\{\lambda m_i\}$, together with the generalized impulse, $r^\lambda : t \mapsto r^\lambda(t) = \lambda r(t)$.

Although for a given multibody sample the distribution of masses is unchanged, it is worthwile to notice that, when the kinetic energy of the sample is increased by a factor λ^2 by increasing $\dot{X}(0)$ by a factor λ or the mass by a factor λ^2, the pressure is increased by a factor λ^2.

When a steady state regime is reached, it is expected that any subcollection of the sample is governed by the same law as the one governing the whole sample; a sub-reservoir containing a sub-collection of the sample should bear the same pressure as the reservoir. If so, **assuming** *that the kinetic energy is equally distributed in all parts of the reservoir, and that the grains masses are also equally distributed,* [9] the proper parameter appears to be the average kinetic energy by grain,

$$\mathcal{T} = \frac{\frac{1}{2}\sum_i^N m_i\dot{X}_i.\dot{X}_i}{N}.$$

The above expression will be called hereafter *granular temperature.* One has to notice that this definition is different, up to the number of grains, from definitions used in the literature, the whole kinetic energy, or the kinetic energy per unit of volume. Also, it should be noticed that the granular temperature is a concept different from *Kelvin temperature.* The unit of granular temperature is $\mathcal{T}$ is $J/mole$, so that $pv/N\mathcal{T}$ is dimensionless, wheseas that of the Kelvin temperature T is dimensionless,

[9]These distributions can be numerically studied.

the unit of pv/NT is $J/mole\,K$. In the case of a perfect gas, $\mathcal{T} = 8.3144\, J/mole\,K$.

.2 Trying a law

$$pv = f(v)N\mathcal{T}$$

The ratio $pv/N\mathcal{T}$, with $\mathcal{T} = 0.1893\,10^{-4}\ J/mole$ is plotted as a function

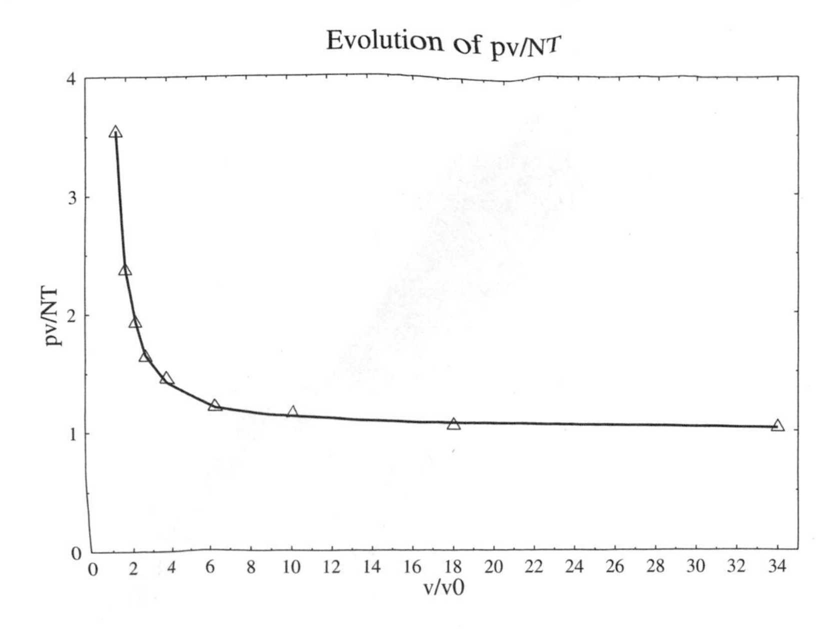

Figure 18.7. Evolution of $pv/N\mathcal{T}$, $\mathcal{T} = 0.1893\,10^{-4}\ J/mole$

of v/v_0 in figure 4.2. Though a steady regime is not really reached, a plausible steady mean pressure is evaluated as a mean value over a period of reminiscence of the wave regime, an interval of time from 40 to 70 s, as shown in figure 18.5. One observes that the ratio is decreasing as the collection of disks gets more and more dilute, and this ratio tends to 1.

5. Comments and conclusion

Alhough the existence of a steady state of pressure is not proved, and though properties of time equidistribution of shocks and of spatial homogeneity may not be satisfactorily realized with such a small number of grains, the graph plotted in figure 4.2 suggests the existence of a perfect gas behavior, with a physical constant 1. Deeper investigations are still necessary to ensure the relevance of the multibody perfect gas behavior. Estimations may be obtained (a suggestion from J.J. Moreau) [10] from the so called viriel equation,

$$x.m\frac{du}{dt} = x.r \quad .$$

One obtains the following relation, when a steady regime is reached,

$$N\mathcal{T} = (p - \Delta p)v + [xr]_{bulk} + [mxu] \, .$$

The term $[mxu]$ (an average kinetic momentum) is vanishing at steady state, and both terms Δp, (a boundary pressure grain size corrective term) and $[xr]_{bulk}$, (an average internal reaction momentum term) are vanishing when the ratio $\frac{l_{mean}}{L}$ tends to zero (l_{mean} is the grain mean radius, L is the radius of the reservoir). The perfect gas law,

$$pv = N\mathcal{T} \, ,$$

is thus relevant for a dilute gas, as stated in statistical thermodynamics.

Acknowledgments

A l'occasion de ce colloque, je tiens à dire à quel point côtoyer Jean Jacques Moreau est pour moi (pour nous tous), scientifiquement enrichissant, stimulant, dérangeant, étonnant et rassurant, tout cela à la fois.

References

J.J. Moreau. *Unilateral contact and dry friction in finite freedom dynamics*, volume 302 of *International Centre for Mechanical Sciences, Courses and Lectures.* pp.1–82, J.J. Moreau, P.D. Panagiotopoulos, Springer, Vienna, 1988.

J.J. Moreau. *Some numerical methods in multybody dynamics: application to granular materials*, Eur. J. Mech. , A/Solids, Vol. 13, n.4 - suppl., pp. 93–114, 1994.

J.J. Moreau. *Some basics of unilateral dynamics*, in F. Pfeiffer & C. Glocker (ed), Unilateral Multibody Dynamics, Kluwer, Dordrect, 1999.

M. Jean, J.J. Moreau. Dynamics of elastic or rigid bodies with frictional contact and numerical methods. In R. Blanc, P. Suquet, M. Raous, (ed), *Publications du LMA*, pp. 9–29, 1991.

[10] J.J. Moreau has also performed numerical "adiabatic compression" experiments of such collections of grains (private communication).

M. Jean, J.J. Moreau. Unilaterality and dry friction in the dynamics of rigid bodies collections. In A. Curnier, (ed), *Proc. of Contact Mech. Int. Symp.*, pp. 31–48, 1992.

M. Jean. *Frictional contact in rigid or deformable bodies: numerical simulation of geomaterials*, pp. 463–486, A.P.S. Salvadurai J.M. Boulon, Elsevier Science Publisher, Amsterdam, 1995.

M. Jean. The NonSmooth Contact Dynamics Method, *in Computer Methods in Applied Mechanics and Engineering,* special issue on computational modeling of contact and friction, J.A.C. Martins, A. Klarbring (ed), 177, pp. 235 – 257, 1999.

B. Cambou, M. Jean. *Micro Mécanique des Matériaux Granulaires.* Hermes, Paris, 2001.

S. Ogawa. *Multitemperature theory of granular materials* Proc. of *US-Jpn Semin. Contin. Mech. and Stat. App. Mech. Granular Mat. Tokyo (1978)* . 208-217.

Chapter 19

GRANULAR MEDIA AND BALLASTED RAILWAY TRACKS MILIEUX GRANULAIRES ET VOIES BALLASTÉES

Catherine Cholet
Gilles Saussine
Pierre-Etienne Gautier
and Louis-Marie Cléon

SNCF - Direction de la Recherche et de la Technologie, 45 rue de Londres, 75379 Paris Cedex 08, France

catherine.cholet@sncf.fr

gilles.saussine@sncf.fr

pierre-etienne.gautier@sncf.fr

louis-marie.cleon@sncf.fr

Abstract

Since the beginnings of the railway track, ballasted track is widely employed because of its mechanical properties and its flexibility for construction and maintenance but also for their capacity to fulfil the mechanical requirements fixed by the transport of heavy loads at a long distance. The study of the track deterioration is one of the principle assignments in railway engineering, specially for SNCF to preserve a good quality of track and to ensure the comfort and the safety of circulation.

In spite of the experience on the railway track structure, the lifespan of a railway remains difficult to estimate and the mechanisms that lead to its deterioration remain insufficiently known. The answers come primarily from experiments or an empirical approach. A SNCF research project is to study the behaviour of the ballast so as to integrate this understanding in a strategy of control of the costs for the design and the maintenance, in particular the renewal of the railway.

In this context, it appears necessary to improve the understanding of the fundamental mechanisms that lead in the medium-to-long term to the appearance of the track defects. The simulation methods resulting from research on the granular medium yielded significant results in

Civil Engineering which can be applied to railway problems and make it possible to study the phenomena related to the microstructure of the ballast. The interest of these methods was largely improved by work of Professor J.J. Moreau, in particular on the avalanches and old monuments (Moreau, 2000).

The work performed at SNCF in collaboration with LMGC concerns the development of a two-dimensional and a three-dimensional discrete element software using the platform LMGC90 to study the ballast behaviour (Saussine et al, 2004). The first step is to provide validations through comparisons with some existing experiments or developed for this purpose. It is a prerequisite before an application to the industrial problem. Next, we present some research orientations: on the one hand the behaviour of a granular media subjected to a great number of loading cycles, which is strongly connected to the problems of the ballast settlement and its densification and on the other hand a three-dimensional application of the software to the study of the lateral resistance of the track (Saussine et al, 2004).

Keywords: ballasted track, discrete element approach, cyclic loading, lateral resistance

1. Les enjeux des recherches sur l'infrastructure

Depuis les débuts du ferroviaire, les voies ballastées sont très largement employées en raison de leur flexibilité à la fois pour la construction et la maintenance mais également pour leur capacité à répondre aux exigences mécaniques fixées par le transport de lourdes charges sur une grande distance. Le ballast est un des éléments support de la voie. Il est capital pour la SNCF et RFF de conserver une bonne qualité de voie pour assurer le confort et la sécurité des circulations.

En dépit de l'expérience acquise sur les infrastructures ferroviaires, la durée de vie d'une voie ferrée reste difficile à estimer et les mécanismes à l'origine de sa dégradation sont mal connus. Les réponses aux questions posées proviennent essentiellement de l'expérience terrain ou d'une approche empirique. Une des activités de recherche de la SNCF est d'étudier le comportement du ballast de manière à intégrer ces connaissances dans une stratégie de maîtrise des coûts pour la conception et la maintenance préventive, en particulier concernant le renouvellement de la voie ferrée : les renouvellements sur ligne nouvelle (au bout de 15 ans sur la ligne Paris Sud Est) ont posé de réelles difficultés d'exploitation et par voie de conséquence des coûts très importants.

Dans ce contexte, il nous apparaît nécessaire de mieux comprendre le fonctionnement interne de la voie et des mécanismes qui conduisent à plus ou moins long terme à l'apparition des défauts de voie. Les perspectives offertes par les approches granulaires permettent d'étudier les phénomènes liés à la microstructure du ballast. L'intérêt de ces méthodes a été largement démontré par les travaux du Professeur J.J. Moreau, notamment sur les avalanches et les monuments anciens (Moreau, 2000).

Dans cet article, nous nous intéresserons spécifiquement au comportement du ballast mais des recherches prenant en compte une approche plus globale de la voie sont également en cours (Quétin et al, 2003).

2. Construction et entretien d'une voie ferrée ballastée

L'objectif de cette partie n'est pas de donner une description exhaustive de la voie mais de présenter les principes généraux qui gouvernent la construction, le fonctionnement et la maintenance de la voie ferrée ballastée (Alias, 1984).

2.1 Structure de la voie

Les principaux éléments constitutifs de la voie ferrée ballastée sont présentés sur la figure 19.1. L'armement constitue la partie supérieure de la voie : il est formé par le rail tenu par l'intermédiaire d'attaches de semelles caoutchouc cannelées de 9 mm et enfin de traverses (en bois, en béton, monobloc ou bibloc). La structure d'assise est constituée par la couche de ballast, la sous-couche et repose sur la plate-forme de terrassement dont la partie supérieure est compactée en couche de forme. Les

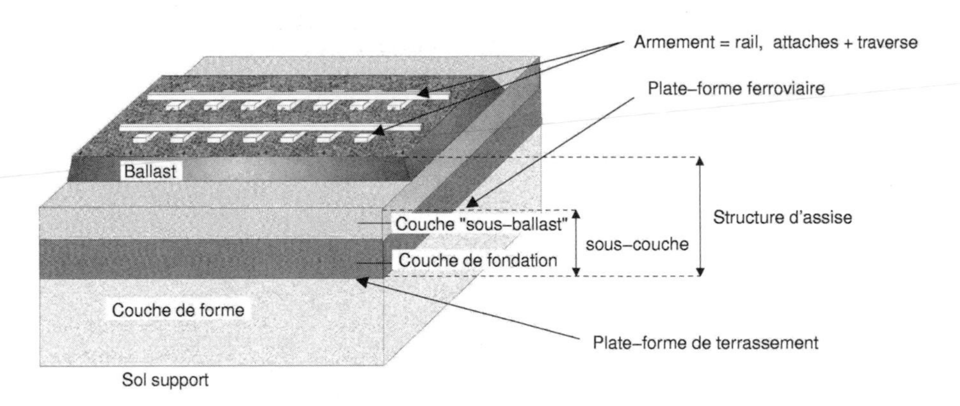

Figure 19.1. Constitution d'une voie ferrée ballastée — Constitution of the ballasted track.

règles de dimensionnement des structures d'assises ont été réexaminées à l'occasion des projets de lignes nouvelles et peuvent être appliquées pour la réfection des lignes classiques. Elles consistent à choisir l'épaisseur des couches d'assise de façon à ce que les contraintes induites sur la plate-forme ne dépassent pas une valeur limite. L'épaisseur varie en fonction de la qualité des sols support en place.

L'épaisseur de la couche de ballast est d'au moins 30 cm sur Ligne à Grande Vitesse (LGV). Sur voie classique, l'épaisseur sous traverse béton varie entre 15 et 25 cm selon le classement des lignes.

2.2 Le Ballast

2.2.1 Origine. Le lit de ballast est composé d'une couche d'agrégats minéraux grossiers concassés répondant à des critères de qualité portant essentiellement sur la pétrographie, la morphologie des grains ainsi que la dureté et les propriétés d'altérabilité. Par le passé, on a pu utiliser d'autres matériaux beaucoup plus friables (comme le calcaire ou des matériaux roulés...) mais ces matériaux ont été exclus de toute utilisation ferroviaire en raison de leurs mauvaises performances du point de vue mécanique et durabilité. Aux débuts de l'exploitation des premières lignes à grande vitesse, les normes de dureté et de résistance à la fragmentation ont ainsi été 'renforcées' pour limiter la détérioration du ballast.

La granulométrie du ballast utilisé en voie est fixée à 25/50 mm. Ce fuseau est amené à évoluer (31.5/63 mm) depuis la mise en application d'une nouvelle norme européenne sur les granulats.

2.2.2 Les fonctions du ballast. Le ballast assure un certain nombre de fonctions essentielles dans la voie :

- l'ancrage latéral et longitudinal de la voie,
- la transmission et la répartition des charges statiques et dynamiques du rail vers la plate-forme,
- le drainage des eaux pluviales,
- et une fonction environnementale, de plus en plus reconnue, qui est de moindres émissions acoustiques qu'une voie sur dalle et l'amortissement des vibrations engendrées par les circulations.

Le ballast présente un avantage supplémentaire en raison de sa flexibilité du point de vue de sa mise en oeuvre et de la maintenance ainsi que son faible coût de construction par rapport à une voie sur dalle classique.

2.3 Géométrie d'une voie ballastée

Il existe entre la voie et le véhicule des interactions très étroites en raison des efforts exercés par le véhicule sur la voie. Les imperfections de cette dernière réagissent sur la stabilité du véhicule avec des conséquences importantes sur le confort des passagers voire sur la sécurité des circulations, d'où la nécessité de conserver une bonne qualité de voie.

Le suivi de la géométrie est un des critères essentiel de qualité de la voie : au cours du temps des défauts de nivellement ou "défauts de voie" apparaissent dus à un tassement différentiel du ballast (voir partie suivante). On distingue plusieurs types de défauts :

- Les défauts de nivellement longitudinal qui mesurent (par file de rail), la différence de hauteur entre un point et la moyenne d'altitude du rail calculée sur une base de 15 m,

- les défauts de dressage qui mesurent (par file de rail) la variation de la position transversale par rapport à la moyenne calculée sur une base de 10 m,

- Les défauts de nivellement transversal. Le dévers mesure la différence d'altitude entre les deux files de rail. On calcule les écarts de dévers en chaque point du rail en référence à une moyenne calculée sur 10 m. Le gauche mesure la torsion de la voie (écart de dévers ramené sur une base de 3 ou 9 m).

Les défauts de voie sont relevés périodiquement grâce à des voitures d'auscultation (de type voiture Mauzin). A titre indicatif, sur LGV, pour un défaut de nivellement (défaut isolé) au delà de 15 mm, une restriction des vitesses de circulation est imposée.

2.4 Maintenance d'une voie ferrée ballastée

La maintenance consiste essentiellement à agir sur le ballast ou sur l'armement de la voie. Ces opérations sont plus ou moins complexes selon la nature des défauts et peuvent aller jusqu'à un renouvellement partiel ou complet d'une partie ou de la totalité de la voie.

Le bourrage-dressage mécanisé est une des opérations de nivellement les plus couramment pratiquées. Elle permet par un apport de matériau neuf et par une action de serrage et de vibration des grains situés sous la traverse (à une fréquence d'environ 35 Hz) de redonner un niveau géométrique satisfaisant à la voie et de reconstituer les "moules" c'est à dire les zones unies et compactes situées sous les traverses (Alias, 1984).

Le renouvellement complet de la voie (ballast et armement) intervient lorsque les opérations de maintenance deviennent inefficaces pour assurer une bonne qualité de voie et surtout garantir la durabilité de cette géométrie dans le temps. C'est cette analyse qui a conduit au renouvellement de tronçons de la LGV Paris Sud Est après 15 années d'exploitation.

Il est en revanche beaucoup plus délicat d'effectuer des opérations sur la sous-couche ou la plate-forme ferroviaire. Sur les LGV, il n'est en effet pas envisageable de déposer la voie en raison des très fortes contraintes d'exploitation, d'où l'importance du dimensionnement et de la réception de la plate-forme ferroviaire lors de la construction d'une ligne nouvelle pour limiter sa dégradation.

3. Les problématiques rencontrées

Le ballast est largement tributaire du trafic écoulé depuis sa mise en place. La vie de la voie ferrée est ponctuée par différentes opérations de maintenance, de renouvellement partiel ou total de ses composants. Tout remaniement du ballast est le point de départ d'un nouveau cycle. La dégradation de la qualité du ballast et/ou de ses interfaces peut cependant conduire à accélérer considérablement le rythme des opérations de maintenance.

Tassement et mise en place du ballast : le maintien de la géométrie de la voie est lié pour une grande part au comportement du ballast. Lors de la construction ou après une opération de relevage, on constate un tassement très rapide de la voie caractéristique de sa mise en place. Lorsque la voie est dite "stabilisée", le ballat continue de se tasser mais de manière beaucoup plus lente. C'est un processus naturel pour tout granulat mais la difficulté est que ce tassement n'est pas homogène d'un point à l'autre de la voie ce qui peut être à l'origine de l'apparition des défauts de voie. L'enjeu est donc de mieux comprendre les mécanismes à l'origine de l'apparition des tassements, d'étudier le rôle des interfaces (descentes de charge vers la sous-couche, traverse...) et d'identifier les facteurs aggravants.

L'augmentation des **vitesses de circulation** est un point important en raison des enjeux économiques pour l'entreprise. En effet, depuis les débuts de l'exploitation des lignes TGV, la vitesse commerciale des trains est passée de 270 km/h à 300 km/h et bientôt 320 km/h (sur la ligne TGV Est en construction). Des expériences de laboratoire (Bodin, 2001) ont mis en évidence que le tassement et la réponse élastique de la voie augmentent avec la vitesse de circulation. Au delà d'une certaine "fréquence" critique, le phénomène s'accélère. D'autres facteurs ont également une incidence, comme la **charge à l'essieu** des matériels ferroviaires en perspective d'une augmentation du tonnage à l'essieu des trains de fret sur les lignes classiques.

Résistance latérale : une des fonctions principales du ballast est d'assurer l'ancrage latéral et longitudinal de la voie. Cette propriété est particulièrement importante pour assurer la stabilité de la voie en latéral et ainsi contrôler les effets dus au flambement thermique du rail en période estivale. Il est donc souhaitable de bien comprendre quels sont les facteurs qui contribuent à la résistance latérale de la voie.

Bourrage : le bourrage est une opération complexe. Les paramètres du bourrage sont fixés par les constructeurs de machine (fréquence de sollicitation, serrage, descente des bourroirs dans la voie...). La qualité du bourrage est également très dépendante du savoir faire technique des opérateurs de terrain.
Les performances du bourrage sont estimées à partir de la qualité géométrique de la voie immédiatement après l'opération mais n'est pas évaluée en fonction de la tenue de la géométrie dans le temps. Il est donc très important de bien définir les spécifications fonctionnelles du bourrage pour mieux maîtriser les contraintes opérationnelles de la maintenance.

Stabilisation : la stabilisation de la voie est étroitement liée à sa mise en place. La résistance du ballast est largement tributaire de l'état de stabilisation de la voie : tout remaniement de la voie entraîne une déconsolidation du milieu et particulièrement une diminution de sa résistance latérale. Une voie récemment posée ou remaniée n'acquiert que progressivement sa stabilité définitive : un écoulement de trafic de 100000 tonnes est considéré comme nécessaire. La stabilisation peut être effectuée au moyen d'un stabilisateur dynamique. La stabilisation dynamique n'est cependant pas considérée comme suffisante pour restituer complètement le niveau de consolidation souhaité. Le couplage entre la stabilisation et le bourrage et sa répercussion sur la durabilité de la tenue de la géométrie dans le temps est donc un sujet particulièrement intéressant.

Usure du ballast : pour comprendre le comportement de la voie (tassement, mise en place, résistance latérale...) à ses différents stades, il est également nécessaire de connaître les modes de dégradation des grains de ballast. La vie du ballast dans la voie entraîne des phénomènes plus ou moins lents d'attrition ou de fragmentation qui conduisent à l'apparition d'éléments fins. Ces éléments fins vont peu à peu venir colmater les interstices entre les grains. Le ballast perd alors sa fonction drainante. Des facteurs extérieurs peuvent être également à l'origine du colmatage de la voie comme des remontées de boue en provenance de la plate-forme, la détérioration des traverses... Ces effets accélèrent encore les processus de dégradation il est donc nécessaire de les prendre en compte dans toute analyse du comportement de la voie.

4. Modélisation du ballast par l'approche Contact Dynamics

4.1 Principe

Le travail réalisé à la SNCF en collaboration avec le LMGC porte sur le développement d'un code par éléments discrets 2D et 3D dans la plate-forme logicielle LMGC90 en vue d'une application au ballast (Saussine et al, 2004). L'ensemble des simulations numériques présentées ont été réalisées avec LMGC90.

Une première étude, appliquée au bourrage et réalisée à la SNCF (Oviedo, 2001), a montré la pertinence d'une telle approche pour étudier le comportement du ballast. La méthode Contact Dynamics a été retenue en raison de ses performances numériques : le principe de résolution implicite de la méthode rend la résolution sur un pas de temps plus coûteuse mais ce même pas de temps est beaucoup plus grand que dans les autres approches, d'où un gain de temps significatif constaté dans l'ensemble de nos applications. C'est un avantage d'un point de vue numérique pour traiter un grand nombre de grains comme c'est le cas pour des géométrie 3D ou pour simuler un grand nombre de cycles de chargement. Une description détaillée de l'approche Contact Dynamics développée par Moreau et Jean pourra être trouvée dans (Moreau et Jean, 1992).

La démarche adoptée consiste à valider les simulations numériques par confrontation avec des résultats expérimentaux, expériences existantes (tambours tournants...) ou développées à cet effet. C'est un préalable avant de passer aux applications industrielles de la méthode.

Nous présentons quelques axes de recherche : d'une part le comportement d'un assemblage granulaire soumis à un grand nombre de cycles de chargement, ce qui rejoint la problématique du tassement du ballast et de sa mise en place et d'autre part une application du code 3D à l'étude de la résistance latérale du ballast (Saussine et al, 2004).

4.2 Comportement de la voie sous un grand nombre de cycles de chargement

Un dispositif expérimental bidimensionnel, inspiré du principe des matériaux de Schneebeli, a été développé avec pour objectif de valider le code de calcul bidimensionnel sur un exemple simplifié mais représentatif du ferroviaire.

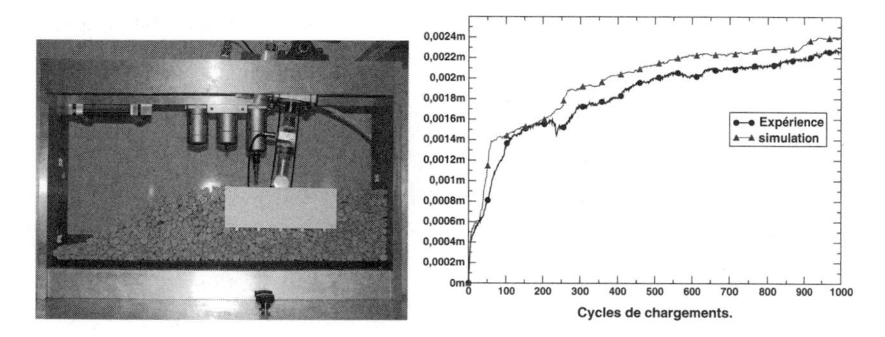

Figure 19.2. Dispositif expérimental bidimensionnel : Comparaison des tassements expérimentaux et numériques pour un échantillon — Two-dimensional experimental set-up: Comparison of experimental and numerical settlement for one sample.

Le dispositif expérimental 19.2, développé au LCPC (Laboratoire Central des Ponts et Chaussées) représente une portion de voie ferroviaire (Cholet et al, 2003). Il est constitué d'un bloc en aluminium représentant une portion de traverse, de prismes à sections polygonales modélisant les grains de ballast. Les prismes, en ciment haute performance et de trois diamètres différents (1 cm, 1.5 cm et 2 cm) reposent sur une couche d'élastomère modélisant la couche sous-ballast. L'ensemble est disposé dans un bâti rectangulaire sur lequel est fixé un vérin asservi en force. Celui-ci permet d'appliquer des chargements cycliques sur le blochet avec différentes inclinaisons. Des photos sont prises au cours du chargement pour suivre le déplacement de quelques grains. La position du blochet est enregistrée.

Sur une expérience, nous avons confronté les tassements mesurés pour un échantillon numérisé et les simulations numériques. Une des difficultés de cette opération a porté sur la modélisation de la sous-couche déformable et le recalage dans l'échantillon numérique (la sous-couche n'est pas numérisée). On constate que le tassement évolue de manière similaire dans la simulation et l'expérience et les mêmes phénomènes de recirculation des grains autour du blochet ont pu être observés. Nous avons pu ainsi démontrer la faisabilité d'une telle étude. Les comparaisons réalisées entre les simulations et les résultats expérimentaux sont encourageantes et donnent un bon indice de confiance quant à l'aptitude de la méthode Contact Dynamics à reproduire les phénomènes observés.

La compréhension des phénomènes de tassement et de mise en place de la voie nécessite de simuler un grand nombre de cycles de chargement : sur LGV, un mois d'exploitation correspond à environ 1 million de passages d'essieu. Les simulations ont pu être 'poussées' sur plusieurs milliers de cycles de chargement (35000 cycles) tout en concervant une bonne qualité de calcul : le cumul des erreurs d'interpénétration entre les grains reste contrôlé. C'est peu en comparaison des situations réelles mais cette étude nous a permis de réaliser des avancées significative dans ce domaine qui reste largement ouvert.

4.3 Résistance latérale de la voie

Comme nous l'avons vu à la partie précédente, la résistance latérale de la voie est un facteur important pour la sécurité des circulations.

L'étude de la résistance latérale à l'aide de LMGC90 implique l'élaboration d'échantillons et d'un protocole d'expérimentation : nous nous sommes inspirés d'essais existants réalisés à la SNCF.

Figure 19.3. Essai de résistance latérale sur un massif de ballast avec un blochet (à gauche) - Profil de vitesse : seuls les grains de ballast ayant une vitesse entre 10% et 50% de la vitesse maximale sont représentés (à droite) — Lateral resistance test on a ballast bed with a half sleeper (on the left) - Profile speed: only the ballast grains having a speed between 10% and 50% maximum speed are represented (on the right)

Une petite portion de voie comportant un seul blochet (demi-traverse) a été modélisée et en respectant la granulométrie réelle du ballast. Les grains utilisés pour les simulations sont issus de la numérisation de 1000 grains de ballast réel. Une fois l'échantillon déposé (figure 1.3) il est possible de réaliser plusieurs types d'essais sur le même échantillon, par exemple des essais de traction sur le blochet pour différentes charges verticales, des cycles de chargement en nombres limités. Ces essais de traction peuvent être exploités pour homologuer les traverses ou pour établir les référentiels sur la résistance latérale de la voie.

Nous avons commencé par étudier l'influence du coefficient de frottement entre le blochet et les grains de ballast sur le déripage du blochet soumis à une force incrémentale transversale (figure 1.4) mais les résultats obtenus demandent à être calibrés (Saussine et al, 2004).

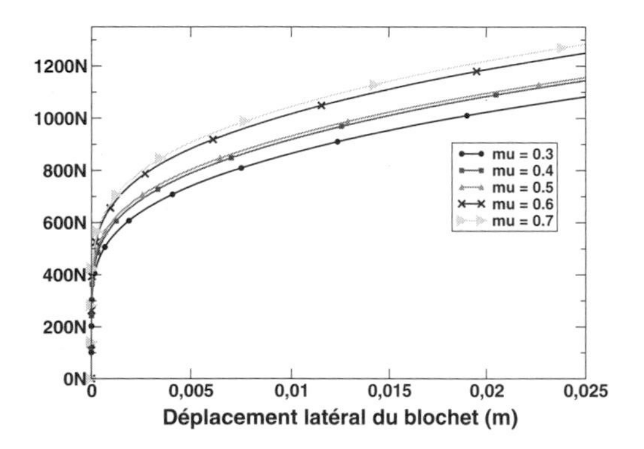

Figure 19.4. Champs de vitesse dans l'échantillon sous l'action de la force - Influence des différents coefficients de frottement. — Fields speed in the loaded sample - Influence of the coefficient of friction.

D'autres applications sont envisagées comme l'étude de l'influence du poids, de la forme de la traverse...

5. Conclusion

Les problèmes de conception ou d'exploitation de la voie ferrée ballastée sont nombreux et variés. Il est nécessaire d'étudier le fonctionnement du ballast afin d'optimiser la conception de la voie mais également de comprendre les mécanismes de dégradation pour mieux appréhender la maintenance.

Les méthodes par éléments discrets nous permettent d'appréhender la complexité des phénomèmes rencontrés. La démarche adoptée consiste à valider les simulations numériques par confrontation avec des résultats expérimentaux existants. C'est un préalable avant toute application industrielle de ce type d'approche. Nous présentons les premiers résultats issus des simulations numériques : simulation d'un grand nombre de cycles de chargement et résistance latérale de la voie.

References

J. Alias, *La voie ballastée - Techniques de construction et d'entretien*, 1984 (2ème édition).

V. Bodin, *Comportement du ballast des voies ferrées soumises à un chargement vertical et latéral*, Thèse de doctorat de l'Ecole Nationale des Ponts et Chaussées. 2001.

C. Cholet, G. Saussine, G. Combe, C. Bohatier, F. Dubois and K. Sab, Mechanical behaviour of the ballast using discrete element methods dans *Proceedings of The World Congress on the Railway Research.* 2003

M. Jean and J. J. Moreau, Unilaterality and dry friction in the dynamics of rigid body collections dans *Contact Mechanics.* Presses Polytechniques et Universitaires Romandes. 1992

J. J. Moreau, Contact et frottement en dynamique des systèmes de corps rigides dans *Revue Européenne des Eléments Finis.* Vol 9. 2000.

X. Oviedo. *Etude du comportement du ballast par un modèle micro-mécanique - Application aux opérations de maintenance de la voie ferrée ballastée.* Thèse de doctorat de l'Ecole Nationale des Ponts et Chaussées. 2001.

F. Quetin, L. Schmitt, K. Sab and D. Duhamel. Modelling of the dynamical behaviour of railway ballasted track including its platform dans*Proceedings of the world congress on railway research.* 2003.

G. Saussine, C. Cholet and P.E. Gautier. Modelling ballast under cyclic loading using discrete element method dans *Proceedings of International conference on cyclic behaviour of soils and liquefaction phenomena.* 2004 (à paraitre).

Chapter 20

SCALING BEHAVIOUR OF VELOCITY FLUCTUATIONS IN SLOW GRANULAR FLOWS

Farhang Radjaï
LMGC, CNRS-Université Montpellier II, Place Eugène Bataillon, 34095 Montpellier Cedex, France
radjai@lmgc.univ-montp2.fr

Stéphane Roux
Laboratoire Surface du Verre et Interfaces, CNRS-Saint Gobain, 39 Quai Lucien Lefranc, 93303 Aubervilliers Cedex, France.
stephane.roux@saint-gobain.com

Abstract Steric exclusions among particles lead to strong velocity fluctuations in a granular flow. Modelling the effective behaviour of granular materials depends on the extent and scaling properties of these fluctuations. We consider here slow granular flows of rigid particles simulated by a discrete element method. Bi-periodic boundary conditions allow for macroscopically homogeneous shearing up to large strains. We obtain thus reliable statistics for an accurate analysis of particle velocity fluctuations. We find that the probability distribution function of velocity components, evaluated from particle displacements, crucially depends on time resolution. It varies from stretched exponential to gaussian as the integration time is increased. On the other hand, the spatial power spectrum of the fluctuating velocity field is a power law, reflecting long range correlations and the self-affine nature of the fluctuations. Finally, by considering individual particle displacements, we show that the particles have a superdiffusive motion with respect to the mean background flow. These scaling behaviours bear a close analogy with the known scaling properties of turbulent fluid flows although the underlying physics is drastically different.

Keywords: granular media, discrete element method, plastic deformation, steric exclusions, probability density function, space correlations, superdiffusion, turbulence.

Introduction

Granular media have proved to be a rich source of intriguing phenomena at all scales: from the anisotropic dilatant stress-strain behaviour (Hicher, 2000) to particle-scale phenomena such as inhomogeneous force transmission (Moreau, 1997, Radjai et al., 1996) or size segregation (Moreau, 1993, Moreau, 1994a). Generally, one has little intuition about what happens, and quite often theoretical understanding remains a challenge (de Gennes, 1999, Jaeger et al., 1996, Roux and Radjai, 2001).

In analogy with molecular fluids, when a granular material flows, a large number of degrees of freedom interact in a disordered way. The complexity of granular flows arises from the dissipative nature of these interactions (friction, inelastic collisions) and geometrical frustrations induced by steric exclusions of particles (Moreau, 1988). These excluded-volume effects result in large force and velocity fluctuations in time. The quasistatic behaviour is of elasto-plastic type with hardening variables which are not yet well appreciated from particle-scale arguments. In the conventional approach, it is assumed that the size of the representative elementary volume is sufficiently small that the fluctuations in local variables fully average out at scales of engineering interest.

Recently, several authors have reported on strong velocity fluctuations in typical shear tests where it is often assumed that the velocity field is homogeneous (Misra and Jiang, 1997, Kuhn, 1999, Miller et al., 1996). They also remarked that the fluctuations occur in a correlated fashion, but they did not analyze the length and time scales involved in these fluctuations. An accurate evaluation of the statistics of fluctuations requires, indeed, long-time homogeneous and steady shearing, a condition that seems hardly accessible from experiments.

In this paper, we present a numerical investigation of velocity fields in a slow two-dimensional granular flow where the homogeneity of shearing is ensured by means of bi-periodic boundary conditions. We will briefly introduce the numerical procedures used for these simulations. Then, we study the distribution functions of the velocities, calculated from particle displacements, as a function of the integration time. In order to evaluate the space correlations, we consider also the spatial power spectra of the velocities. The temporal behaviour will be analyzed by considering the fluctuating trajectories of the particles in the bulk. Finally, we discuss the relevance of our findings to the macroscopic modelling of granular media and we briefly depict the striking analogy suggested by these findings with the scaling features of fluid turbulence.

1. Numerical procedures

The investigated granular system is a two-dimensional assembly of 4000 discs with diameters uniformly distributed between D_{min} and D_{max} with $D_{max} = 3D_{min}$. The particles interact through a stiff linear repulsive force as a function of mutual overlaps and the Coulomb friction law. The coefficient of friction is 0.5. The equations of motion for particle displacements and rotations are integrated by means of a predictor-corrector scheme (Allen and Tildesley, 1987).

In numerical simulations performed up to large cumulative deformation, the boundary conditions tend to introduce spurious effects in the vicinity of the walls, and may even lead to a strongly inhomogeneous strain field. The latter should, however, be as homogeneous as possible to extract a meaningful statistics for intrinsic fluctuations. In order to reduce these edge effects, we have introduced bi-periodic boundary conditions: it consists in imposing such a periodicity along both directions of the simulation cell for the stress and the strain fields. The displacement field, however, is not periodic if the mean strain is non-zero. It contains an affine part $\delta \boldsymbol{r}^i \equiv (\delta r^i_x, \delta r^i_y)$ in addition to a bi-periodic (fluctuating) field $\delta \boldsymbol{s}^i \equiv (\delta s^i_x, \delta s^i_y)$ of zero mean ($\langle \delta \boldsymbol{s} \rangle = 0$). This requires a special treatment of these two components with additional characteristics associated with the non-periodic part of the displacement, such as a fictive inertia (Parrinello and Rahman, 1980). However, the latter can be shown to play a negligible role in the results. The origin of space being now an arbitrary parameter, the stress-strain and more generally any geometrical observable has to be translationally invariant, and hence the strain remains spatially homogeneous on a large scale.

In our simulations, the gravity was set to zero and a confining pressure was applied along the y direction. The width L of the simulation cell was kept constant. The system is driven by imposing the affine component ${\delta r_x}^i = \delta t \gamma r^i_y$, where γ is a constant shear rate and δt is the time step. In other words, the Fourier mode $k = 0$ of the total strain is imposed, corresponding to a large scale forcing. This driving mode was applied on a dense packing prepared by isotropic compaction. The height of the packing increases (dilation) in the initial stages of shearing before a homogeneous steady state is reached where volume changes fluctuate around zero, so that $\langle {\delta r_y}^i \rangle \simeq 0$. The solid fraction in the steady state fluctuates in the range $[0.79, 0.81]$ and the average coordination number is 3.8. Our focus here is on the fluctuating components $(\delta s^i_x, \delta s^i_y)$ in the steady state.

2. Time scales

Although our dynamic simulations involve the physical time, the inertial effects are negligibly small and the contact network evolves quasistatically at time scales well below γ^{-1}. We normalize all times by γ^{-1} so that the dimensionless time t in what follows will actually represent the cumulative shear strain. We will also use the mean particle diameter D to scale displacements. As a result, the velocities will be scaled by γD and the power spectra in space by $(D^2\gamma)^2$. In our simulations the time step is $\delta t \simeq 10^{-7}$, and more than 2.10^7 steps are simulated, corresponding to a total strain larger than 2.

We may distinguish two separate time scales involved in a quasistatic granular flow: 1) contact time, corresponding to the elastic response time; and 2) shear time that corresponds to the time imposed by shearing. The latter is of the order of γ^{-1}. The contact time for a multicontact assembly is of the same order of magnitude as the duration of a binary collision. Up to a numerical time discretization, the particle motions below the contact time scale are smooth. We are interested here in times longer than the contact time scale. In other words, we consider the rigid relative displacements of the particles that underly the plastic deformations of the material. At such scales, the collisions can be considered as instantaneous so that the particle velocities evolve discontinuously in time (Moreau, 1994b). In our simulations, the elastic response time is around 10^{-5} in dimensionless units.

At contact time scales, it is physically plausible to rely on particle velocities calculated for one time step since at those times the motion is smooth. But beyond the elastic time, the physically relevant particle velocities should be evaluated from particle displacements. We consider here the periodic part of the velocity field and we define the fluctuating velocities $\boldsymbol{v}^i$ as a function of the integration time τ by

$$\boldsymbol{v}^i(t, t+\tau) = \frac{1}{\tau} \int_t^{t+\tau} \delta \boldsymbol{s}^i(t')\, dt' \qquad (20.1)$$

Since we are concerned with steady flow, the statistical properties of $\boldsymbol{v}$ are independent of t although, as shown below, they crucially depend on τ. Hence, accurate statistics can be obtained by cumulating the data from different time slices of a single simulation running for a long time. As we shall see below, the scaling of the fluctuations with τ reveals a

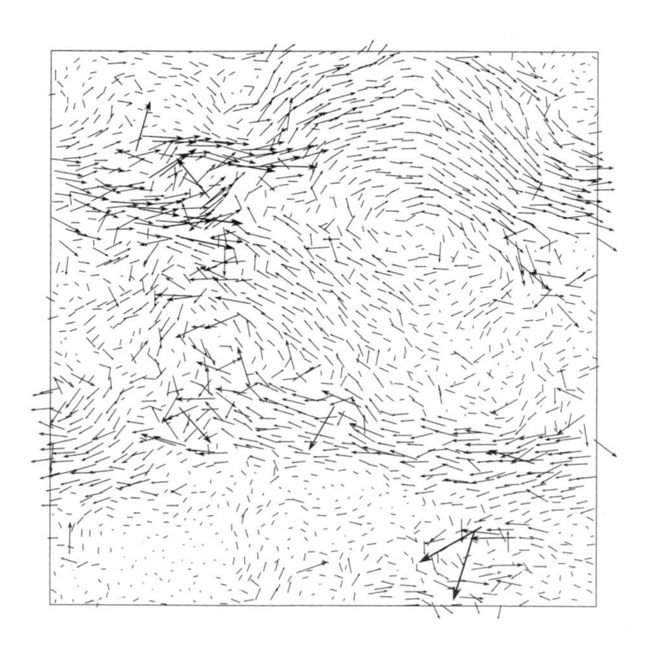

Figure 20.1. A snapshot of particle displacements $\delta \boldsymbol{s}^i$ with respect to the mean background flow.

new characteristic time corresponding to a transition in the shape of the probability density function of fluctuating velocities.

Fig. 20.1 shows a snapshot of fluctuating velocities $\boldsymbol{v}^i$ for a short time lag $\tau = 10^{-5}$. We see that large-scale well-organized displacements coexist with a strongly inhomogeneous distribution of amplitudes and directions on different scales. Convection rolls appear quite frequently, but they survive typically for strains τ less than 10^{-3}. After such short times, large-scale rolls break down and new statistically uncorrelated structures appear. This behaviour is radically different from turbulence eddies which survive long enough to undergo a significant distortion due to fluid motion (Frisch, 1995).

3. Probability density functions

We consider here the probability density functions (pdf's) of the components $v_x^i(t, t+\tau)$ and $v_y^i(t, t+\tau)$ as a function of time resolution τ. The pdf's of v_y^i are shown in Fig.20.2 for a short integration time $\tau = 10^{-3}$, and for a long integration time $\tau = 10^{-1}$. The pdf has changed from a nearly Gaussian shape at large τ to a non-Gaussian shape with broad

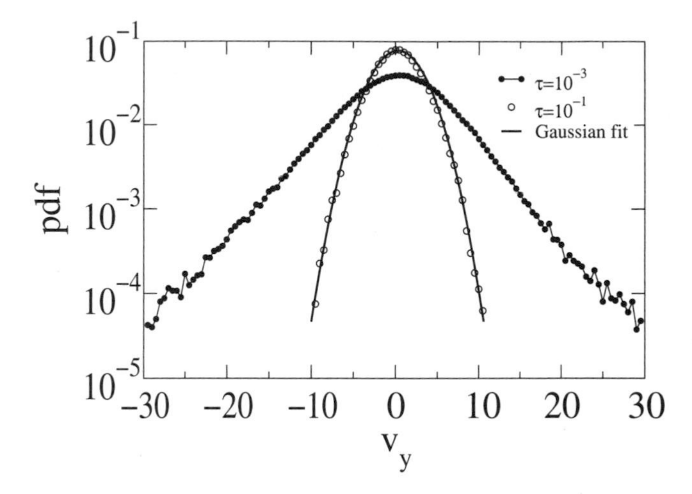

Figure 20.2. The pdf's of the y-components of fluctuating velocities for two different integration times: 10^{-3} (broad curve) and 10^{-1} (narrow curve). The latter is fitted by a Gaussian. The error bars are too small to be shown.

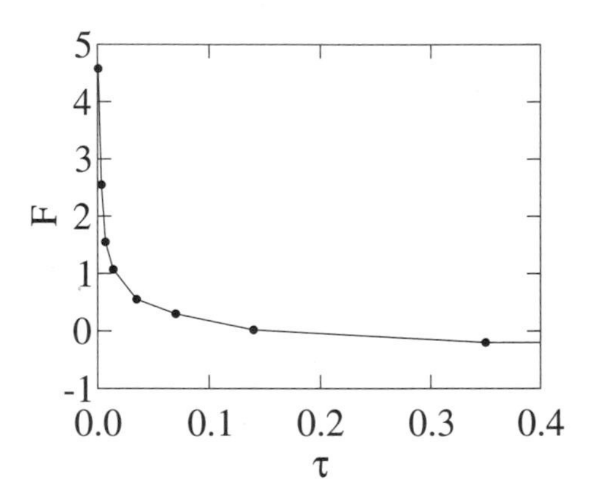

Figure 20.3. Flatness F of the distribution of velocity fluctuations as a function of the integration time τ.

stretched exponential tails extending nearly to the center of the distribution at small τ.

In order to characterize this non-Gaussian broadening of the pdf's as a function of τ, we calculated the flatness $F = \langle v_y^4 \rangle / \langle v_y^2 \rangle - 3$, which is zero for a Gaussian distribution and 3 for a purely exponential distribution. The values of F as a function of τ, shown in Fig.20.3, are consistent with zero at large τ ($\tau > 10^{-1}$) and rises to 5 for our finest time resolution ($\tau = 10^{-7}$). A strictly similar behaviour was observed for the component v_x.

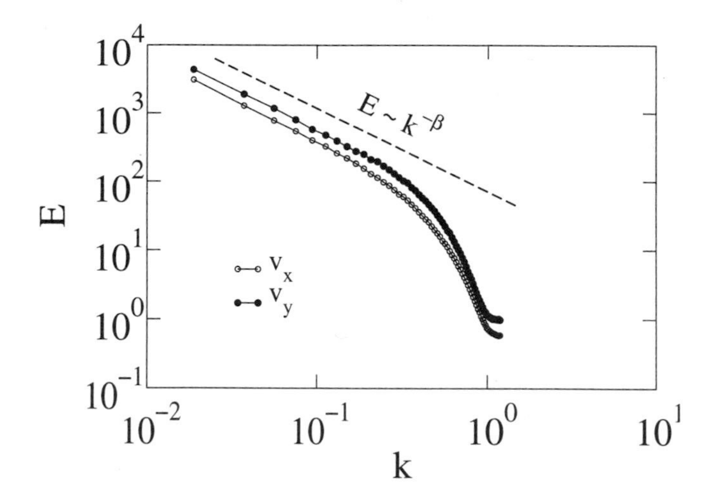

Figure 20.4. Averaged power spectrum of the x and y components of the fluctuating velocity field with $\tau = 10^{-7}$ for one-dimensional cross sections along the mean flow. For the units, see text.

The basic physical mechanism underlying the fluctuations is the mismatch of the uniform strain field, applied here in the bulk, with mutual exclusions of the particles. As a result, the local strains deviate from the mean (global) strain. The observation of a transition toward a Gaussian distribution for large time lags is a sign of partial loss of correlation and/or exhaustion of large fluctuations in the increment of displacement which occur at different times. Fig. 20.3 allows us to define a new time scale corresponding to the value of τ at transition to gaussian distribution for which $F = 0$. This occurs for $\tau \simeq 10^{-1}$. Beyond this point, as we shall see below, the velocity fluctuations are basically reflected in the diffusive motions of individual particles.

4. Space correlations

In order to quantify the extent of correlations in space, we estimated the power spectrum E of velocity fluctuations both along and perpendicular to the flow and at different times. The Fourier transform was performed over the fluctuating velocity field defined on a fine grid by interpolating the velocities from particle centers. The power spectra were quite similar along and perpendicular to the flow, and for different snapshots of the flow.

The averaged spectrum on one-dimensional cross sections is shown in Fig.20.4. It has a clear power-law shape $k^{-\beta}$ ranging from the smallest wavenumber $k = D/L$, corresponding to the system size L, up to a cut-off around $k = 0.5$, corresponding to nearly two particle diameters.

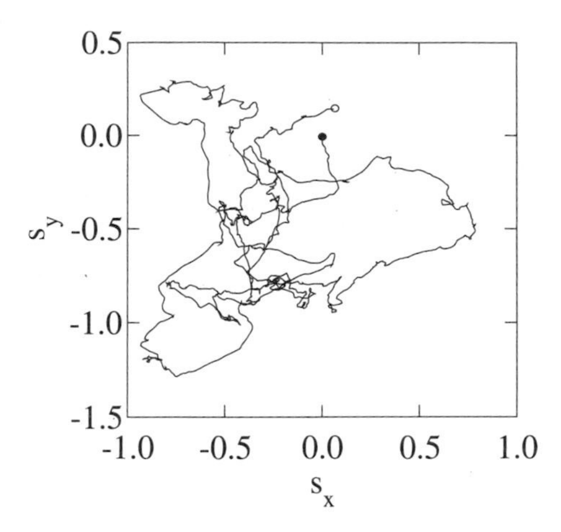

Figure 20.5. Diffusive trajectory of one particle with respect to the mean background flow expressed in the unit of mean particle diameter D for a cumulative shear strain of 2. The displacements may be compared with the approximative cell dimensions $60DX70D$.

The exponent is $\beta \simeq 1.24 \simeq 5/4$ over nearly one decade. This means in practice that the fluctuating velocity field is self-affine with a Hurst exponent $H = (\beta - 1)/2 = 0.12$ (Feder, 1988).

Due to the peculiar behavior of the velocity field (discontinuous in time), one might expect that the power spectrum is sensitive to the time resolution τ. However, we checked that the value of β is independent of τ. It is also noteworthy that, the presence of long range correlations in displacements, reflected in the value of β, is in strong contrast with the observed correlation lengths of nearly $10D$ for contact forces (Radjai et al., 1996, Miller et al., 1996).

5. Kinematic diffusion

The long-time behaviour may be analyzed by considering the effective diffusive motion of the particles. Normal diffusion implies that the root-mean-square (rms) relative displacements λ in a given direction varies in proportion to the square root of time. Let us remark that the random motions of the particles in a quasistatic granular flow are of *kinematic* origin. In this sense, they differ from dynamic diffusion in a fluid where the diffusivity of the particles is induced by incoherent velocity fluctuations and molecular interactions.

We analyzed the diffusive motions of single particles in our quasistatic granular flow. One example of a single particle trajectory with respect to the background strain is shown in Fig.20.5. We see that the fluctuating displacement is of the order of the mean particle diameter for a strain

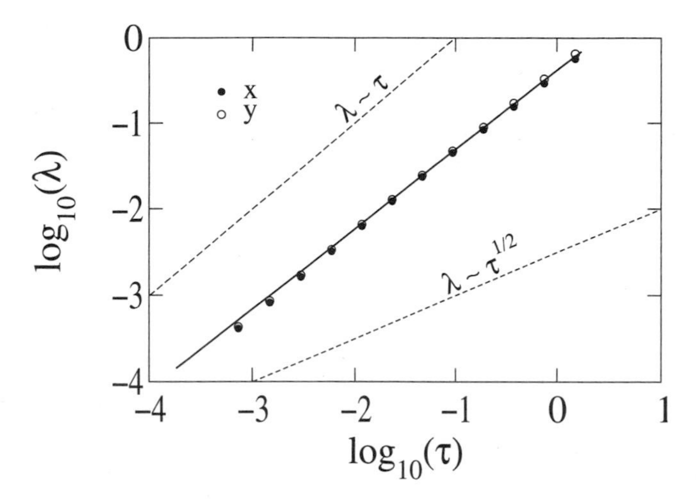

Figure 20.6. Root-mean-square relative displacements λ along x and y directions as a function of time τ fitted by a power law of exponent 0.9 (straight solid line). As a guide to the eyes, a power law of exponent 1/2, corresponding to normal diffusion, and the line $\lambda \propto \tau$ are shown as well (dashed lines).

of the order of unity. Fig.20.6 shows the rms relative displacement $\lambda(\tau)$ of all particle pairs initially in contact, as a function of time along x and y directions. This clearly corresponds to a superdiffusion behaviour $\lambda \propto \tau^\alpha$ with $\alpha \simeq 0.9$ for both components over nearly 3 decades of strain. Particle self-diffusivities exhibit a similar law. Since large-scale structures are short lived, this anomalous diffusion scaling can not be solely attributed to velocity correlations. It reveals, above all, the long-time configurational memory of a granular medium in quasistatic flow.

We note that Fig.20.6 shows no anisotropy for the diffusion. The lengths being in units of mean particle diameter, the relative diffusive displacements of the particles are quite weak (below one particle diameter). At such scales, the steric exclusion effects dominate over the large-scale strain field, and hence the anisotropy may be weak. Moreover, we observe no crossover to normal diffusion within the investigated strain range. We cannot exclude that for larger strains, when two particles are more widely separated, a normal diffusion law is recovered.

6. Analogy with turbulence

The fluctuating particle displacements with respect to the background shear flow were shown to have the following scaling characteristics:

1 The probability distribution functions undergo a transition from stretched exponential to gaussian as the time lag is increased.

2 The spatial power spectrum of the velocity field obeys a power law, reflecting long range correlations and the self-affine nature of the fluctuations.

3 The fluctuating displacements have a superdiffusive character.

There appears an evident analogy with the scaling features of fluid turbulence (Radjai and Roux, 2002). Turbulence studies focus mainly on velocity differences δv measured at a fixed point of a fluid over a time interval τ or between two points separated by a distance r (Frisch, 1995). This is in contrast with granular flow which involves a discrete displacement field that is carried by individual particles. Up to this difference in framework, the scaling properties discussed above are shared by turbulent fluid flows. The broadening of the exponential tails of the pdf's at increasingly smaller scales is a well-known property of fully developed turbulence (for velocity differences). It is attributed to the phenomenon of intermittency, i.e. strong localized energy transfers at small scales (Frisch, 1995). On the other hand, the power-law scaling $k^{-\beta}$ for the spectrum of velocity differences is a hallmark of 3D turbulence with $\beta \simeq 5/3$ (to be compared with $\beta \simeq 5/4$ in our granular flow). Finally, in 3D fluid turbulence at high Reynolds numbers, the long-time pair diffusivity of suspended particles is anomalous, following the Richardson law $\lambda \propto \tau^{3/2}$ which reflects the Kolmogorov-Obukhov velocity spectrum (Frisch, 1995).

A strict analogy with fluid turbulence makes certainly not much sense because of a drastically different physics that underlies granular fluctuations (hard-core interactions) as compared to turbulent fluctuations (Navier-Stokes equations, inertia regime). But, the observed analogy in terms of scaling characteristics is consistent enough to upgrade kinematic fluctuations in quasistatic granular flows to the rank of a systematic phenomenology which could be coined by the term "granulence" as compared to "turbulence" in fluid dynamics!

Interestingly, this analogy works with three-dimensional turbulence although our data concern a two-dimensional granular flow. The energy cascade in turbulence is governed by inertia, and two- and three-dimensional systems do differ significantly in this respect. In quasistatic granular flow, the fluctuating velocity field is a consequence of the geometrical compatibility of the strain with particle arrangements, and dissipation is mainly governed by friction at the particle scale. The difference between two- and three-dimensional systems may thus be less crucial, but it was not investigated in the present work. Quite independently of its physical origins, this analogy is suggestive enough to be used as a promising strategy towards a refined probabilistic description of granular flow in the plastic regime.

The conventional approach disregards kinematic fluctuations in macroscopic modelling of plastic flow in granular media. If the effective plastic properties (yield surface, flow rule and hardening variables) are dependent only on the the mean strain, this approach is not in contradiction with the above nontrivial scaling of fluctuating velocities. But, if the effective properties involve higher-order moments of the velocities, the scaling should be taken into account. For example, the hardening behaviour might involve higher order moments since the evolution of the microstructure (typically, the contact network) is locally controlled by steric exclusions (Troadec et al., 2002). Schematically, in larger systems the long-range correlations in particle motions induce stronger randomness in local motions. As a result, we expect that longer shearing is required to reach the steady (critical) state which is characterized by a well-defined organization of the microstructure in terms of density and anisotropy (Wood, 1990, Rothenburg and Bathurst, 1989). On the other hand, if some effective properties are indeed controlled by velocity fluctuations, and if a representative elementary volume exists (which is possible if the power-law scaling in space breaks down for some reason for larger systems than ours), its dimension should be much larger than 60 particle diameters (linear size of our system). In all cases, the velocity fluctuations play a key role in modelling the effective bahaviour from the particle scale.

Acknowledgments

Personal note from F.R.: I met Jean Jacques Moreau for the first time at the "Powders and Grains" meeting held in Birmingham in 1993. Using his laptop computer during a video presentation, Jean Jacques showed us how his "contact dynamics method", new then but widely used today, was able to simulate efficiently granular flows and reproduce puzzling phenomena such as size segregation. He then explained, by means of a rigourous analysis of the data, a physical mechanism at the origin of size segregation in a vibrated granular material. After such a delightful theoretical demonstration, I was really impressed when he declared that in his view the numerical results should be used "in interaction with experiments". Only rarely a mathematician talks of experiments! I had the priviledge to meet Jean Jacques quite often later in France and I enjoy discussing frequently with him since I joined the LMGC in 1997. His limitless enthusiasm for science, his ability to break through "cultural" walls between research fields and to associate rigourous reasoning with experimental evidence have always been a source of lively research on granular materials in the LMGC. I am greatly indebted to Jean Jacques for his encouraging attitude which has allowed me to initiate my research work on granular materials on solid grounds and to benefit in many ways from his vast competence during the last ten years.

References

Allen, M. P. and Tildesley, D. J. (1987). *Computer Simulation of Liquids*. Oxford University Press, Oxford.

de Gennes, P.-G. (1999). Granular matter: a tentative review. *Reviews of Modern Physics*, 71:S374–S382.

Feder, J. (1988). *Fractals*. Plenum press, NewYork.

Frisch, U. (1995). *Turbulence, the Legacy of A. N. Kolmogorov*. Cambridge University Press, Cambridge, England.

Hicher, P.-Y. (2000). Experimental behaviour of granular materials. In Cambou, B., editor, *Behaviour of Granular Materials*, pages 1–98, Wien. Springer.

Jaeger, H. M., Nagel, S. R., and Behringer, R. P. (1996). Granular solids, liquids, and gases. *Reviews of Modern Physics*, 68(4):1259–1273.

Kuhn, M. (1999). *Mechanics of Materials*, 31:407.

Miller, B., O'Hern, C., and Behringer, R. P. (1996). Stress fluctuations for continously sheared granular materials. *Phys. Rev. Lett.*, 77:3110–3113.

Misra, A. and Jiang, H. (1997). Measured kinematic fields in the biaxial shear of granular materials. *Computers and Geotechnics*, 20(3/4):267–285.

Moreau, J. J. (1988). *Unilateral contact and dry friction in finite freedom dynamics*, pages 1–82. Number 302.

Moreau, J. J. (1993). New computation methods in granular dynamics. In *Powders & Grains 93*, page 227, Rotterdam. A. A. Balkema.

Moreau, J. J. (1994a). Some numerical methods in multibody dynamics: Application to granular materials. *European Journal of Mechanics A/Solids*, supp.(4):93–114.

Moreau, J. J. (1994b). Some numerical methods in multibody dynamics: application to granular materials. *Eur. J. Mech. A*, 13:93.

Moreau, J. J. (1997). Numerical investigation of shear zones in granular materials. In Wolf, D. E. and Grassberger, P., editors, *Friction, Arching and Contact Dynamics*, Singapore. World Scientific.

Parrinello, M. and Rahman, A. (1980). *Phys. Rev. Lett.*, 45:1196.

Radjai, F., Jean, M., Moreau, J. J., and Roux, S. (1996). Force distribution in dense two-dimensional granular systems. *Phys. Rev. Lett.*, 77(2):274.

Radjai, F. and Roux, S. (2002). Turbulentlike fluctuations in quasistatic flow of granular media. *Phys. Rev. Lett.*, 89:064302.

Rothenburg, L. and Bathurst, R. J. (1989). Analytical study of induced anisotropy in idealized granular materials. *Geotechnique*, 39:601–614.

Roux, S. and Radjai, F. (2001). Statistical approach to the mechanical behavior of granular media. In Aref, H. and Philips, J., editors, *Mechanics for a New Millennium*, pages 181–196, Netherlands. Kluwer Acad. Pub.

Troadec, H., Radjai, F., Roux, S., and Charmet, J.-C. (2002). Model for granular texture with steric exclusions. *Phys. Rev. E*, 66:041305.

Wood, D. (1990). *Soil behaviour and critical state soil mechanics.* Cambridge University Press, Cambridge, England.

V

TOPICS IN NONSMOOTH SCIENCE

Chapter 21

MORPHOLOGICAL EQUATIONS AND SWEEPING PROCESSES*

Jean-Pierre Aubin

Centre de Recherche Viabilité, Jeux et Contrôle, Université Paris-Dauphine

J.P.Aubin@wanadoo.fr

José Alberto Murillo Hernández*

Departamento de Matemática Aplicada y Estadística, Universidad Politécnica de Cartagena

alberto.murillo@upct.es

This paper is dedicated to Jean Jacques Moreau, with respectful friendship.

Abstract In this paper the existence of viable evolutions governed by joint evolutionary-morphological systems is analyzed. For "nonviably" consistent evolutionary-morphological systems, it is shown that viable solutions can be obtained correcting the dynamics by means of viability multipliers. Moreau's sweeping processes arise in a natural way in the framework of correction of dynamical systems by viability multipliers for adapting the state to evolving constraints governed by morphological equations.

Keywords: Morphological equations, viability multipliers, sweeping process.

*Work supported in part by the European Community's Human Potential Programme under contract HPRN-CT-2002-00281, Evolution Equations.

*Alberto Murillo Hernández acknowledges the financial support provided through the European Community's Human Potential Programme under contract HPRNCT-2002-00281 (Evolution Equations for Deterministic and Stochastic Systems). This paper is also supported by grant PI-53/00809/FS/01 (Nonlinear Differential Equations: Analysis, Simulation and Control) from the Fundación Séneca (Gobierno Regional de Murcia, Spain).

1. Introduction

Viability theory deals with evolutions $t \mapsto x(t)$ which are viable in "tubes" (time-dependent subsets of constraints) $t \rightsquigarrow K(t)$, that is, which satisfy

$$\forall\, t \geq 0, \;\; x(t) \in K(t)$$

This problem was first studied by Jean Jacques Moreau in some seminal papers (Moreau, 1971, Moreau, 1972, Moreau, 1973) and (Moreau, 1977) where the *sweeping process*

$$x'(t) \in -N_{K(t)}(x(t)) \tag{21.1}$$

was introduced for governing such viable evolutions. This process and some related evolution problems are still today the object of active mathematical researches (see for instance (Castaing et al., 1993, Thibault, 2003) or (Moreau, 2003) for recent references).

These tubes were first assumed to be given, and the characterization of this viability property led to the introduction of a class of derivatives of set-valued maps, called *graphical derivatives.* Yet, biological and economical considerations led us not only to characterize and regulate viable evolutions of contingent (i.e., multivalued) dynamical systems to given tubes $K(\cdot)$, but also to "endogenize" the evolution of tubes. The issue quickly arose of having the evolution of tubes governed itself by a kind of differential equation, called *morphological equation.* It enabled us to study the necessary and sufficient conditions linking the dynamics governing the evolution of the state $x(\cdot)$ and the dynamics ruling the evolution of the closed subset $K(\cdot)$ in such a way that the above viability condition is satisfied. Unfortunately, the concepts of graphical derivatives of tubes are no longer adequate for defining the velocities of a tube, needed to design morphological equations governing the evolution of subsets. This required the construction of another "differential calculus" in the metric space of nonempty compact sets in order to study these morphological equations. The concept of *mutation of a tube* introduced in (Aubin, 1999) leads us to define tube velocities in an adequate manner.

We shall revisit the viability problem when the evolution of **both the state and the tube** is governed by a joint evolutionary-morphological system of the form

$$\begin{cases} (i) & x'(t) \;\in\; F(t, x(t), K(t)) \\ (ii) & \overset{\circ}{K}(t) \ni \Phi(t, x(t), K(t))(\cdot) \end{cases}$$

required to be "viably consistent" in the sense that for all $t \geq 0$, $x(t) \in K(t)$. The notations are explained later of the form (21.9)-(21.10) required to be "viably consistent" in the sense that we shall define below.

2. Characterization of Viable Tubes

Let us denote by $\mathcal{S}_F(T,x) \subset \mathcal{C}(0,\infty;X)$ the set of evolutions governed by differential inclusion $x'(t) \in F(t,x(t))$ starting at state x at initial time T (the set-valued map $\mathcal{S}_F$ is called the evolutionary system associated with F).

We first consider the case when the tube $K(\cdot)$ is given a priori.

DEFINITION 21.1 *Let us consider two tubes $t \rightsquigarrow K(t)$ and $t \rightsquigarrow C(t)$, taking values in $X := \mathbb{R}^d$. We shall say that the tube $K(\cdot)$ is* viable outside the tube $C(\cdot)$ *under a time-dependent evolutionary system $\mathcal{S}_F$ if for any $(T,x) \in$ Graph(K), there exist an evolution $x(\cdot) \in \mathcal{S}(T,x)$ and a time $t^\star := t^\star_{x(\cdot)} \geq T$ such that*

$$\begin{cases} i) & \text{if } t^\star < +\infty, \quad x(t^\star) \in K(t^\star) \cap C(t^\star) \\ ii) & \forall\, t \in [T, t^\star[, \quad x(t) \in K(t) \end{cases} \tag{21.2}$$

In order to characterize viable tubes under time-dependent differential inclusions, we introduce the (contingent) derivative at $(t,x) \in \text{Graph}(K)$ of $K(\cdot)$ regarded as a set-valued map from $\mathbb{R}_+$ to X: Given $\tau \in \mathbb{R}$, then $v \in DK(t,x)(\tau)$ if and only if (τ, v) belongs to the Bouligand-Severi tangent cone to the graph of $K(\cdot)$ at (t,x), $T_{\text{Graph}(K)}(t,x)$. We observe that it is enough to know this contingent derivative only in the directions $1, 0$ and -1 because $\tau \rightsquigarrow DK(t,x)(\tau)$ is positively homogeneous. Moreover $T_{K(t)}(x) \subset DK(t,x)(0)$. We recall the observation:

LEMMA 21.2 *The* forward derivative *of $K(\cdot)$ at $(t,x) \in$ Graph(K) is defined by*

$$DK(t,x)(1) = \left\{ v \in X \mid \liminf_{h \to 0+} d\left(v, \frac{K(t+h) - x}{h}\right) = 0 \right\} \tag{21.3}$$

We deduce from Viability Theorem 11.1.3 in (Aubin, 1991) the following characterization of viable tubes $K(\cdot)$ outside a target tube $C(\cdot)$:

THEOREM 21.3 *Assume that the set-valued map $F : \mathbb{R}_+ \times X \rightsquigarrow X$ is Marchaud (i.e., upper semicontinuous with compact convex values and linear growth) and the tubes $K(\cdot)$, $C(\cdot)$ are closed (i.e., with closed graph). Then $K(\cdot)$ is viable outside $C(\cdot)$ under the evolutionary system $\mathcal{S}_F$ if and only if*

$$\forall\, t \geq 0,\ \forall\, x \in K(t) \setminus C(t),\ \ 0 \in F(t,x) - DK(t,x)(1) \tag{21.4}$$

where $\mathcal{S}_F(t,x) \subset \mathcal{C}(0,\infty;X)$ *refers to the set of solutions of the Cauchy problem* $x'(\tau) \in F(\tau, x(\tau))$, $x(t) = x$.

3. Mutations of Tubes

The underlying idea behind the concept of mutation of a tube is to regard the limit v of the usual differential quotients $\dfrac{\mathbf{u}(t+h)-\mathbf{u}(t)}{h}$ measuring the variation of a differentiable map $\mathbf{u} : \mathbb{R} \to X$ on half-lines $t+h$ as elements v satisfying

$$\lim_{h \to 0+} \frac{d(\mathbf{u}(t) + hv, \mathbf{u}(t+h))}{h} = 0. \tag{21.5}$$

The half lines $h \mapsto t+h$ and $h \mapsto x+hv$ in vector spaces $\mathbb{R}$ and X are regarded as *transitions* mapping each element to a neighboring one in a given "direction", depicted by vectors 1 and v in the case of vector spaces.

Since only the distance is involved in this reformulation of the directional derivative, we can extend this concept to the case of tubes $t \mapsto K(t) \in \mathcal{K}(X)$ regarded as maps from $\mathbb{R}$ to the metric space $\mathcal{K}(X)$ of nonempty compact subsets of X supplied with the Pompeiu-Hausdorff distance defined by

$$\mathbf{dl}(L, M) := \max\left(\sup_{y\in L} d_M(y),\ \sup_{z\in M} d_L(z)\right)$$

where $d_M(y) := \inf_{z\in M}\|y - z\|$ is the distance to M. We recall the classical result:

THEOREM 21.4 *Let* $\mathcal{X}$ *be a complete metric space. Supplied with this Pompeiu-Hausdorff distance, the space* $\mathcal{K}(\mathcal{X})$ *of nonempty compact subsets of* $\mathcal{X}$ *is a complete metric space.*

We replace in (21.5) the half line $h \mapsto \mathbf{u}(t) + hv$ in the vector space X by the "half curve" $h \mapsto \vartheta_\varphi(h, K(t))$ where $\varphi : X \to X$ is a Lipschitz map, $\vartheta_\varphi(h,x)$ denotes the value at time h of the solution to the differential equation $z' = \varphi(z)$ starting at $z(0) = x$ at time h and $\vartheta_\varphi(h, M) := \{\vartheta_\varphi(h,x)\}_{x\in M}$ is the *reachable set* from $M \subset X$ at time h of φ.

It is then tempting to say that either $\varphi(\cdot) \in \mathrm{Lip}(X,X)$ or its associated transition ϑ_φ belongs to the *mutation* of $K(\cdot)$ at t in the forward direction 1 if

$$\lim_{h\to 0+} \frac{\mathbf{dl}(\vartheta_\varphi(h, K(t)), K(t+h))}{h} = 0 \tag{21.6}$$

We can thus interpret definition (21.6) of the mutation $\overset{\circ}{K}(t)(1)$ by saying that *the transition* $\vartheta_{\varphi}(h, K(t))$ *of the image* $K(t)$ *and the image* $K(t+h)$ *of the transition* $t + h1$ *are "equivalent"*.

Consequently, in the case of tubes $t \rightsquigarrow K(t)$ with nonempty compact values, we define the *velocity of the tube* $K(\cdot)$ at time t as the mutation $\overset{\circ}{K}(t) := \overset{\circ}{K}(t)(1)$ of $K(\cdot)$ at t in the forward direction 1.

In other words, the "half curve" $h \mapsto \vartheta_{\varphi}(h, K(t))$ plays the role of the half line $h \mapsto x + hv$ in vector spaces. Both $h \mapsto t + h$ and $h \mapsto \vartheta_{\varphi}(h, K(t))$ can be regarded as *transitions* mapping each element to a neighboring one in a given "direction", depicted by a vector in the case of vector spaces and by a Lipschitz map in the case of $\mathcal{K}(X)$. They share enough common properties for allowing us to extend to metric spaces the classical differential calculus on vector spaces, called *mutational calculus.* This is the idea used by Jean Céa and Jean-Paul Zolésio in 1976 for defining shape derivatives and gradients of shape maps (mapping subsets of a vector space to another vector space) in the framework of shape optimization or optimal design.

Here, this idea allows us to find the velocities of tubes among Lipschitz maps $\varphi \in \text{Lip}(X, X)$ (associated with the "transitions" $h \mapsto \vartheta_{\varphi}(h, K)$), which contain the usual vectors regarded as constant maps. They play the role of directions when one defines directional derivatives of usual maps.

4. Morphological Transitions on Compact Sets

By the way, we shall enrich this structure by associating transitions not only with Lipschitz single-valued maps, but also with bounded Lipschitz set-valued maps with closed convex values, that we call *morphological maps*, as candidates to be mutations of a tube. Enlarging the class of transitions increases the chances of a tube to have mutations.

We thus associate with any morphological map $\Phi \in \text{LIP}(X, X)$ the differential inclusion

$$x'(t) \in \Phi(x(t)) \tag{21.7}$$

and its *reachable map* defined by $\vartheta_{\Phi}(h, x) := \{x(h)\}_{x(\cdot) \in \mathcal{S}_{\Phi}(x)}$ and

$$\vartheta_{\Phi}(h, B) := \bigcup_{x \in B} \vartheta_{\Phi}(h, x)$$

$\mathcal{S}_{\Phi}(x)$ being the set of solutions to (21.7) starting at x, with $B \subset X$ a nonempty set. The map ϑ_{Φ} defined by $(h, B) \mapsto \vartheta_{\Phi}(h, B)$ is called the *morphological transition* (on the space $\mathcal{K}(X)$ of all the compact subsets of X) associated with Φ.

DEFINITION 21.5 *Let $K(\cdot)$ be a tube. We shall say that the set-valued map $\Phi \in$ LIP(X,X) (or its associated transition ϑ_Φ) belongs to the* mutation of $K(\cdot)$ at t in the forward direction 1 *if*

$$\lim_{h \to 0^+} \frac{\mathbf{dl}(\vartheta_\Phi(h, K(t)), K(t+h))}{h} = 0$$

Then we shall write $\Phi \in \overset{\circ}{K}(t)(1)$.

We shall regard this forward mutation $\overset{\circ}{K}(t) := \overset{\circ}{K}(t)(1)$ as the velocity of the tube $K(\cdot)$ at t for defining mutational equations governing the evolution of the tube.

Example 1. Dilations by Structuring Elements

For instance, when we consider a constant map $\Phi \in$ LIP(X,X), equal to a convex compact B, i.e., $\Phi(x) \equiv B$, regarded as a *structuring element*, the associated *structuring transition* is equal to what is called in mathematical morphology the *dilation* $\vartheta_\Phi(h,C) \rightsquigarrow C + hB$ of C by the convex compact subset B.

Example 2. Mutations of Level Sets of Smooth Functions

When the tube $t \rightsquigarrow K(t) := \{x \in X \mid u(t,x) = \lambda\}$ is given by the level sets of a smooth nondegenerate function u, we provide an analytical formula of the mutation of the tube providing a link between mutational equations and the "level set approach" describing the evolution of level sets through the evolution of the functions governed by partial differential equations:

THEOREM 21.6 (TH. 3.6.1 IN (AUBIN, 1999)) *Let us consider a differentiable function $u : \mathbb{R}_+ \times X \to \mathbb{R}$ satisfying*

$$\begin{cases} i) & \forall\, t \geq 0,\ u'_x(t,\cdot) \text{ and } u'_t(t,\cdot) \text{ are Lipschitz on a neighborhood} \\ & \mathcal{N}(t) \text{ of } K(t) \\ ii) & \exists\, \alpha > 0,\ M \geq \alpha \text{ such that } \forall\ (s,x) \in [t,t+1] \times \mathcal{N}(t), \\ & \alpha \leq \|u'_x(s,x)\| \leq M \text{ and } \|u'_t(t,x)\| \leq M \\ iii) & \forall\, t \geq 0,\ K(t) \text{ is compact} \end{cases}$$

Then, if $K(t) := \{x \in X \mid u(t,x) = \lambda\}$ is not empty, the Lipschitz map $\varphi(t)$ defined by

$$\varphi(t)(x) := -\frac{u'_t(t,x)}{\|u'_x(t,x)\|} \frac{u'_x(t,x)}{\|u'_x(t,x)\|}$$

belongs to the mutation $\overset{\circ}{K}(t)$ *of* $K(\cdot)$ *at* t *in the forward direction.*

The map

$$x \mapsto a(t,x) := -\frac{u_t'(t,x)}{\|u_x'(t,x)\|}$$

is called the *normal velocity* of the level set of the function $u(\cdot)$. Since $\boldsymbol{\eta}(t,x) := u_x'(t,x)/\|u_x'(t,x)\|$ is the unit normal to $K(t)$ at $x \in K(t)$, the restriction of the mutation $\varphi(t)$ to $K(t)$ can thus be written as

$$\forall\, x \in K(t),\ \varphi(t)(x) \ = -\frac{u_t'(t,x)}{\|u_x'(t,x)\|}\,\boldsymbol{\eta}(t,x)$$

So that the level set approach, when valid, relates naturally to the mutational approach.

5. Mutations and Derivatives of Tubes

Therefore, the question of comparing contingent derivatives and mutations of tubes arises. "Graphical derivatives" are defined in a local way at each point of the graph of a set-valued map. However "mutations" are *global*, in the sense that they are defined on the domain of the set-valued map. Furthermore, graphical derivatives are set-valued maps while mutations are bounded Lipschitz set-valued maps with compact convex images (morphological maps). One can deduce from Frankowska's formulas of infinitesimal generators of set-valued semigroups a formula linking these two radically different concepts.

THEOREM 21.7 (COROLLARY 5.4.3 IN (AUBIN, 1999)) *Let* $K(\cdot)$ *be a tube with closed values and let* $\Phi \in \overset{\circ}{K}(t)$ *a forward mutation of* $K(\cdot)$ *at* t. *Then:*

$$DK(t,x)(1) = \Phi(x) + T_{K(t)}(x) \tag{21.8}$$

for any $x \in K(t)$.

6. Viability of Tubes Governed by Morphological Equations

Let X be a finite dimensional vector space and $\mathcal{K}(X)$ be the metric space of nonempty compact subsets of X. We assume that the evolutions $x(\cdot) \in \mathcal{S}(\tau,x)$ are given by all the solutions to the differential inclusion

$$x'(t) \in F(t,x(t),K(t)) \tag{21.9}$$

starting at time τ at point x, where $F : \mathbb{R}_+ \times X \times \mathcal{K}(X) \rightsquigarrow X$. Notice that differential inclusion (21.9) includes control systems

$$\begin{cases} i) & x'(t) = f(t, x(t), u(t), K(t)) \\ ii) & u(t) \in U(t, x(t), K(t)) \end{cases}$$

(by setting $F(t, x, K) := f(t, x, U(t, x, K), K))$ and also Moreau sweeping processes (21.1).

Given a map $\Phi : \mathbb{R}_+ \times X \times \mathcal{K}(X) \to \mathrm{LIP}(X, X)$ and the associated morphological equation

$$\overset{\circ}{K}(t) \ni \Phi(t, x(t), K(t))(\cdot) \tag{21.10}$$

we shall consider the evolutions governed by the joint evolutionary-morphological system (21.9)-(21.10), that is, $(x(\cdot), K(\cdot)) \in \mathcal{S}(T, x, K)$ will be a solution of (21.9)-(21.10) satisfying initial conditions $x(T) = x$ and $K(T) = K$.

DEFINITION 21.8 *We set*

$$\mathcal{V}(X) := \{(x, P) \in X \times \mathcal{K}(X) \mid x \in P\}$$

and consider a tube $C(\cdot)$. The joint evolutionary-morphological system (21.9)-(21.10) is said to be viably consistent with respect to $C(\cdot)$ *if for any $(T, x, K) \in \mathbb{R}_+ \times \mathcal{V}(X)$, there exist a joint evolution $(x(\cdot), K(\cdot))$ in $\mathcal{S}(T, x, K)$ and a time $t^\star := t^\star_{x(\cdot)} \geq T$ such that (21.2) holds.*

We deduce from Theorems 21.3 and 21.7 the following characterization of viably consistent joint evolutionary-morphological systems:

THEOREM 21.9 *Let $C(\cdot)$ be a closed tube. Assume that the set-valued map F is Marchaud and that Φ is continuous and satisfies*

$$\begin{cases} i) & M := \sup_{t \in \mathbb{R}_+,\, (x,K) \in \mathcal{V}(X)} \|\Phi(t, x, K)\|_\Lambda < +\infty, \\ & \text{where } \|\Phi(t, x, K)\|_\Lambda \text{ denotes the Lipschitz constant associated} \\ & \text{with } \Phi(t, x, K) \in \mathrm{LIP}(X, X) \\ ii) & c := \sup_{t \in \mathbb{R}_+,\, (x,K) \in \mathcal{V}(X)} \|\Phi(t, x, K)\|_\infty < +\infty, \\ & \|\Phi(t, x, K)\|_\infty := \sup_{z \in X} \sup_{y \in \Phi(t,x,K)(z)} \|y\|, \end{cases}$$

Then the joint evolutionary-morphological system (21.9)-(21.10) is viably consistent if and only if for every $t \geq 0$, every compact $K \in \mathcal{K}(X)$, for any $t \geq 0$ and $x \in K \setminus C(t)$

$$0 \in F(t, x, K) - \Phi(t, x, K)(x) - \overline{\mathrm{co}}(T_K(x)) \tag{21.11}$$

7. Viability Multipliers for Joint Evolutionary-Morphological Systems

There is no reason why an arbitrary joint evolutionary-morphological system should be viably consistent. We can, therefore, introduce the *viability kernel*, as the largest closed subset of initial states from which starts at least one viable evolution, in the sense that (21.2) is satisfied, see (Aubin, 2001, Saint-Pierre, 2002). This does not, however, exhaust the problem of restoring viability. One can imagine several methods for this purpose:

1. Correct initial dynamics by introducing "viability multipliers" as control variables.
2. Change the initial conditions by introducing a *reset map* $\mathcal{R}$ taking any pair $(x, K) \in \mathcal{V}(X)$ into a (possibly empty) subset $\mathcal{R}(x, K) \subset \mathcal{V}(X)$ of new "initialized states" (for more details about *impulsive evolution* we refer to (Aubin et al., 2002)).

In this section we briefly sketch the first approach, that is, the use of viability multipliers to correct the dynamics governing the evolution of the state in order to obtain a new viably consistent joint evolutionary-morphological system .

Definition 21.10 *We call the function $\varphi^\circ : \mathbb{R}_+ \times \mathcal{V}(X) \to \mathbb{R}_+$ defined by*

$$\varphi^\circ(t, x, K) := \min_{y \in F(t,x,K)} d_{\Phi(t,x,K)(x)+\overline{co}(T_K(x))}(y) \tag{21.12}$$

viability discrepancy. *Given a joint evolution $t \mapsto (x(t), K(t)) \in \mathcal{V}(X)$, every map $p(\cdot)$ satisfying*

$$p(t) \in F(t, x(t), K(t)) - \Phi(t, x(t), K(t))(x(t)) - \overline{co}(T_{K(t)}(x(t))) \tag{21.13}$$

is said to be a viability multiplier *for the system (21.9)-(21.10).*

Under the assumptions made on F and Φ in Theorem 21.9, we deduce that $\varphi^0(\cdot)$ has linear growth.

Theorem 21.11 *Let $\varphi : \mathbb{R}_+ \times \mathcal{V}(X) \to \mathbb{R}_+$ be an upper semicontinuous function with linear growth such that $\varphi^0(\cdot) \le \varphi(\cdot)$. Also assume that F, Φ and $C(\cdot)$ satisfy the same assumptions as in Theorem 21.9. Then the corrected joint evolutionary-morphological system*

$$\begin{cases} i) & x'(t) \in F(t, x(t), K(t)) - p(t) \\ ii) & \overset{\circ}{K}(t) \ni \Phi(t, x(t), K(t))(\cdot) \\ iii) & \|p(t)\| \le \varphi(t, x(t), K(t)) \end{cases} \tag{21.14}$$

is viably consistent with respect to $C(\cdot)$.

Sketch of proof. Set $G(t,x,K) := F(t,x,K) - c(t,x,K)B_X$, where B_X is the unit closed ball in X. Since $\varphi(\cdot)$ is an upper semicontinuous. function with linear growth, it is clear that G is Marchaud. Thus, by Theorem 21.9, it suffices to show that $0 \in G(t,x,K) - \Phi(t,x,K)(x) - \overline{\text{co}}\,(T_K(x))$, for every $t \geq 0$, $K \in \mathcal{K}(X)$ and $x \in K \setminus C(t)$, but this claim follows from the very definition of $\varphi^0(\cdot)$. □

REMARK 21.12 *Given a solution* $(x(\cdot), K(\cdot))$ *to the corrected system (21.14), if* $t \mapsto p(t) \in \varphi(t, x(t), K(t))B_X$ *is the associated control, from Theorem 21.7 we obtain that* $p(\cdot)$ *is a viability multiplier.*

REMARK 21.13 *When the viability multiplier* $p^0(\cdot)$ *is the projection of zero onto* $F(\cdot, x(\cdot), K(\cdot)) - \Phi(\cdot, x(\cdot), K(\cdot))(x(\cdot)) - \overline{\text{co}}\left(T_{K(\cdot)}(x(\cdot))\right)$ *or, equivalently,* $\varphi^0(t, x(t), K(t)) = \|p^0(t)\|$, *then the Moreau Projection Theorem onto closed convex cones applied to the set* $\overline{\text{co}}\left(T_{K(t)}(x(t))\right)$ *implies that* $p^0(t) \in N_{K(t)}(x(t))$. *Therefore, in such a case, the evolution* $x(\cdot)$ *governed by the the perturbed sweeping process*

$$x'(t) \in F(t, x(t), K(t)) - N_{K(t)}(x(t)) \tag{21.15}$$

is viable. This happens, for instance, when $\varphi^0(\cdot)$ *is upper semicontinuous. In this case, we conclude that sweeping process (21.15) provides a way to restore viability in the joint system (21.9)-(21.10).*

References

J.-P. Aubin. *Viability Theory*, Birkhäuser, 1991.

J.-P. Aubin. *Mutational and Morphological Analysis. Tools for Shape Evolution and Morphogenesis*, Birkhäuser, 1999.

J.-P. Aubin. Viability Kernels and Capture Basins of Sets under Differential Inclusions, *SIAM J. Control*, **40** (2001), 853–881.

J.-P. Aubin, A. Bayen, N. Bonneuil and P. Saint-Pierre. *Viability, Control and Games: Regulation of Complex Evolutionary Systems Under Uncertainty and Viability Constraints*, Springer-Verlag, to appear.

J.-P. Aubin and H. Frankowska. *Set-Valued Analysis*, Birkhäuser, 1990.

J.-P. Aubin, J. Lygeros, M. Quincampoix, S. Sastry and N. Seube. Impulse Differential Inclusions: A Viability Approach to Hybrid Systems, *IEEE Trans. Automatic Control*, **47** (2002), 2–20.

C. Castaing, T.X. Dúc Ha, and M. Valadier. Evolution Equations Governed by the Sweeping Process, *Set-Valued Anal.*, **1** (1993), 109–139.

J.J. Moreau. Rafle par un convexe variable (Première partie), *Sém. Anal. Convexe*, (exposé 15, 43 pages), (1971).

J.J. Moreau. Rafle par un convexe variable (Deuxième partie), *Sém. Anal. Convexe*, (exposé 3, 36 pages), (1972).

J.J. Moreau. Problème d'évolution associé à un convexe mobile d'un espace hilbertien, *C.R. Acad. Sci. Paris*, **276** (1973), 791–794.

J.J. Moreau. Evolution problem associated with a moving set in a Hilbert space, *J. Differential Equations*, **26** (1977), 347–374.

J.J. Moreau. An introduction to Unilateral Dynamics, in *Novel approaches in Civil Engineering*, (M. Brémond and F. Maceri eds.), Sprinver-Verlag, 2003.

P. Saint-Pierre. Hybrid Kernels ans Capture Basins for Impulse Constrained Systems, in *Proceedings of Hybrid Systems: Computation and Control, HSCC 2002*, (C. Tomlin and M. Greenstreet eds.), 2002.

L. Thibault. Sweeping process with regular and nonregular sets, *J. Differential Equations*, **193** (2003), 1–26.

Chapter 22

HIGHER ORDER MOREAU'S SWEEPING PROCESS

Formulation and numerical simulation

Vincent Acary
and Bernard Brogliato
INRIA-BipOp project, ZIRST, 655 avenue de l'Europe, 38334 Saint-Ismier, France
Vincent.Acary@inrialpes.fr
Bernard.Brogliato@inrialpes.fr

Abstract In this chapter we present a mathematical formulation of complementarity dynamical systems with arbitrary dimension and arbitrary relative degree between the complementary slackness variables. The proposed model incorporates the state jumps via high-order distributions through the extension of Moreau's sweeping process, which is a special type of differential inclusion. The time-discretization of these nonsmooth systems, which is non-trivial, is also presented. Applications of such high-order sweeping processes can be found in dynamic optimization under state constraints and electrical circuits with ideal diodes, where it may be helpful for a better understanding of the closed-loop dynamics induced by some feedback laws.

Keywords: Complementarity systems, Hybrid systems, Convex analysis, Differential inclusions, Variational inequalities, Numerical simulation, Zero dynamics, Relative degree.

Introduction

The general objective of this chapter is the study of complementarity dynamical systems with arbitrary relative degree. As we shall briefly see below, such systems possess complex dynamics and their well-posedness, numerical time integration, and analysis for control, have not yet been understood except in some particular cases. It is proposed here to settle a general dynamical framework for such higher relative degree complementarity systems, using the concept of differential inclusions and Moreau's sweeping process. Besides showing the coherence of the presented dynamics and its usefulness in designing a numerical time-stepping scheme (which in particular paves the way for well-posedness studies), numerous problems like optimal control with state inequality constraints and feedback control of circuits can benefit from the approach.

1. The ZD canonical representation

In this section several state space representations are derived, which will prove to be useful to formalize the extended sweeping process.

1.1 Canonical state space representations

Let us consider the following linear complementarity system:

$$\begin{cases} \dot{x}(t) = Ax(t) + B\lambda(t), \quad x(0^-) = x_0 \\ 0 \leq \lambda(t) \perp w(t) = Cx(t) \geq 0 \end{cases} \tag{22.1}$$

where $x \in I\!R^n$, $\lambda \in I\!R$. If the pair (A, B) is controllable, then the system in (22.1) has a relative degree $r^{w\lambda} \leq n$ (Sontag, 1998). Let us note that it is implicitly assumed in (22.1) that the relative degree is strictly larger than zero. Actually, the framework that is presented next is essentially linked to systems with $r^{w\lambda} \geq 1$. As a consequence it is of little interest for so-called relay systems, whose relative degree $r^{w\lambda}$ is always zero (Camlibel, 2001, Chapter 7). This allows one to perform a state space transformation, with new state vector $z = Wx$, W square full-rank, and $z^T = (w, \dot{w}, \ddot{w}, ..., w^{(r^{w\lambda}-1)}, \xi^T) = (\bar{z}^T, \xi^T)$, $\xi \in I\!R^{n-r^{w\lambda}}$ such that the new state space representation is (Sannuti, 1983)

$$
\begin{cases}
\dot{z}_1(t) = z_2(t) \\
\dot{z}_2(t) = z_3(t) \\
\dot{z}_3(t) = z_4(t) \\
\\
\vdots \\
\\
\dot{z}_{r^{w\lambda}-1}(t) = z_{r^{w\lambda}}(t) \\
\dot{z}_{r^{w\lambda}}(t) = CA^{r^{w\lambda}}W^{-1}z(t) + CA^{r^{w\lambda}-1}B\lambda(t) \\
\\
\dot{\xi}(t) = A_\xi \xi(t) + B_\xi z_1(t) \\
\\
0 \leq z_1(t) \perp \lambda(t) \geq 0, \quad z(0^-) = z_0
\end{cases}
\tag{22.2}
$$

In Systems and Control theory, the dynamics $\dot{\xi} = A_\xi \xi$ is called the *zero dynamics*, so we shall denote the state space form in (22.2) the ZD representation. The notion of relative degree is quite similar to that of index in DAE theory (Campbell and Gear, 1995), or to what is called the state constraint order in optimal control (Hartl et al., 1995). We note that the formalism in (22.2) continues to hold if $\lambda, w \in I\!R^m$. The system has a uniform relative degree if all the Markov parameter $C_i A^{r^{w\lambda}-1} B_j$ are non zero for some $r^{w\lambda}$ and all integers $i, j \in \{1, m\}$, where C_i is the ith row of C while B_j is the jth column of B. Then $CA^{r^{w\lambda}-1}B$ is an $m \times m$ matrix. One has in the multivariable case $z_1^i(\cdot) = w_i(\cdot)$, $1 \leq i \leq m$ and

$$
\begin{cases}
\dot{z}_1^i(t) = z_2^i(t) \\
\dot{z}_2^i(t) = z_3^i(t) \\
\vdots \\
\dot{z}_{r^{w\lambda}}^i(t) = C_i A^{r^{w\lambda}} W^{-1} z(t) + C_i A^{r^{w\lambda}-1} B\lambda \\
\\
\dot{\xi}_i(t) = A_{\xi_i} \xi_i(t) + B_{\xi_i} z_1^i(t), \quad 1 \leq i \leq m \\
\\
0 \leq z_1(t) \perp \lambda(t) \geq 0
\end{cases}
\tag{22.3}
$$

Grouping terms together $z_1 = (z_1^1, \ldots, z_1^m)^T$, and so on, one gets the same expression as in (22.2) but all z_i, $1 \leq i \leq r^{w\lambda}$, are $m-$dimensional. One has also $mr^{w\lambda} \leq n$.

1.2 Distributional dynamics

Until now possible state $x(\cdot)$ jumps have not been introduced. It is of utmost importance to notice that in general, the solutions of (22.2) (equivalently of (22.1)) will not be differentiable. Consider for instance the initial data $z_i(0^-) \leq -\delta$ for some $\delta > 0$ and all $1 \leq i \leq r^{w\lambda}$. Then obviously all the z_i, $1 \leq i \leq r^{w\lambda}$, need to jump to some non-negative value so that the unilateral constraint $z_1(t) \geq 0$ is satisfied on $(0, \epsilon)$ for some $\epsilon > 0$. At this stage we can just say that a jump mapping is needed. Its form will depend on the type of system one handles (in Mechanics, this is the realm of impact mechanics (Brogliato, 1999)). If one considers (22.2) as an equality of distributions, then we can rewrite it as

$$\left\{\begin{array}{l} Dz_1 = z_2 \\ Dz_2 = z_3 \\ Dz_3 = z_4 \\ \\ \vdots \\ \\ Dz_{r^{w\lambda}-1} = z_{r^{w\lambda}} \\ Dz_{r^{w\lambda}} = CA^{r^{w\lambda}} W^{-1} z + CA^{r^{w\lambda}-1} B\lambda \\ \\ D\xi = A_\xi \xi + B_\xi z_1 \end{array}\right. \tag{22.4}$$

where D denotes the distributional derivative (Ferreira (1997)). At a reinitialization time one has $z(t_k^+) = \mathcal{F}[z(t_k^-)]$, where $\mathcal{F}(\cdot)$ is an operator that will be defined later. Let us denote the jump of a function $f(\cdot)$, with right and left limits at time t, as $\sigma_f(t) = f(t^+) - f(t^-)$. Consider the above initial conditions on $z(\cdot)$. Then Dz_1 is a distribution of degree 2 and we get $Dz_1 = \{\dot{z}_1\} + \sigma_{z_1}(0)\delta_0 = z_2$. Consequently Dz_2 is a distribution of degree 3 (Ferreira (1997), Theorem 1.1), and $Dz_2 = D^2 z_1 = D\{\dot{z}_1\} + \sigma_{z_1}(0)\dot{\delta}_0 = \{\dot{z}_2\} + \sigma_{\{\dot{z}_1\}}(0)\delta_0 + \sigma_{z_1}(0)\dot{\delta}_0 = z_3$, and $\{\dot{z}_1\} = \{z_2\}$. Then Dz_3 is a distribution of degree 4, and we get $Dz_3 = D\{\dot{z}_2\} + \sigma_{\{\dot{z}_1\}}(0)\dot{\delta}_0 + \sigma_{z_1}(0)\ddot{\delta}_0 = \{\dot{z}_3\} + \sigma_{\{\dot{z}_2\}}(0)\delta_0 + \sigma_{\{\dot{z}_1\}}(0)\dot{\delta}_0 + \sigma_{z_1}(0)\ddot{\delta}_0 = z_4$, and $\{\dot{z}_2\} = \{z_3\}$. Thus $\sigma_{\{\dot{z}_1\}}(0) = z_2(0^+) - z_2(0^-)$, and $\sigma_{\{\dot{z}_2\}}(0) = z_3(0^+) - z_3(0^-)$. And so on. Until now we have decomposed only the left hand side of the dynamics as distributions of some degrees. Now let us get back to the distributional dynamics in (22.4). Starting from $Dz_1 = z_2$, one deduces that the right hand side has to be of the same degree than the left hand side. This means that the right hand side is equal to $\{z_2\} + \mu_1$, where μ_1 is a distribution of degree 2, i.e. a measure. Similarly from $Dz_2 = z_3$ one deduces that $z_3 = \{z_3\} + \mu_2'$, where μ_2' has degree 3 and can therefore further be decomposed as $\mu_2 + \mu_1'$, with $\deg(\mu_2) = 2$ and $\deg(\mu_1') = 3$. It is not difficult to see that $\mu_1' = \dot{\mu}_1$, using similar arguments as in (Brogliato,

1999, §1.1). Therefore $Dz_2 = \{z_3\} + \mu_2 + \dot{\mu}_1$. The variables μ_1 and μ_2 are slack variables (or Lagrange multipliers), and are measures of the form $\mu_i = g_i(t)dt + d\mu_i$, $g_i(\cdot)$ being Lebesgue integrable function, and $d\mu_i$ an atomic measure. We will see later that one cannot merge $g_i(t)$ with $\{z_{i+1}\}$, because $\mathrm{supp}(g_i(t)dt) \subset \{t|\ z_1(t) = 0\}$, and the measure μ_i will obey specific sign conditions. It happens that when no external inputs (functions of time) act on the system, the non-atomic part of μ_i, $1 \le i \le r^{w\lambda} - 1$, will always be zero, see lemma 22.10. Continuing the reasoning until $Dz_{r^{w\lambda}}$, we obtain $Dz_{r^{w\lambda}} = CA^{r^{w\lambda}}W^{-1}\{z\} + CA^{r^{w\lambda}-1}B\lambda$ where $\deg(\lambda) = \deg(Dz_{r^{w\lambda}}) = r^{w\lambda}+1$. Consequently from (22.4) one gets

$$\left\{\begin{array}{l} Dz_1 = \{z_2\} + \mu_1 \\ Dz_2 = \{z_3\} + \dot{\mu}_1 + \mu_2 \\ Dz_3 = \{z_4\} + \ddot{\mu}_1 + \dot{\mu}_2 + \mu_3 \\ \\ \vdots \\ Dz_i = \{z_{i+1}\} + \mu_1^{(i-1)} + \mu_2^{(i-2)} + \ldots + \dot{\mu}_{i-1} + \mu_i \\ \vdots \\ Dz_{r^{w\lambda}-1} = \{z_{r^{w\lambda}}\} + \mu_1^{(r^{w\lambda}-1)} + \ldots + \mu_{r^{w\lambda}-1} \\ Dz_{r^{w\lambda}} = CA^{r^{w\lambda}}W^{-1}\{z\} + CA^{r^{w\lambda}-1}B\lambda \end{array}\right. \tag{22.5}$$

We keep the notation λ for the multiplier which appears in the last line, and whose expression will be given in the next section. It is important at this stage to realize that λ is the unique source of higher degree distributions in the system, which will allow the state to jump. Therefore the measures μ_i have themselves to be considered as "sub-multipliers". In (22.5) we have separated the regular (functions) parts denoted as $\{\cdot\}$ and the atomic distributional parts. Notice that $\{z_{i+1}\} = \{\dot{z}_i\}$. Also $D\{z_i\} = \{z_{i+1}\} + \mu_i$. From this last fact it is convenient to extract the "measure" part of (22.5) as

$$\left\{\begin{array}{l} Dz_1 = \{z_2\} + \mu_1 \\ D\{z_2\} = \{z_3\} + \mu_2 \\ D\{z_3\} = \{z_4\} + \mu_3 \\ \\ \vdots \\ D\{z_i\} = \{z_{i+1}\} + \mu_i \\ \vdots \\ D\{z_{r^{w\lambda}-1}\} = \{z_{r^{w\lambda}}\} + \mu_{r^{w\lambda}-1} \\ D\{z_{r^{w\lambda}}\} = CA^{r^{w\lambda}}W^{-1}\{z\} + CA^{r^{w\lambda}-1}B\mu_{r^{w\lambda}} \end{array}\right. \tag{22.6}$$

where the terms $D\{z_i\}$ can now be interpreted as the differential measures of $\{z_i\}$ (Marques, 1993). It is noteworthy that (22.6) is not at all equivalent to (22.4). It will be quite useful, however, for the characterization of the extended sweeping process and some of its properties, as well as for time-discretization. Roughly speaking, (22.6) represents the system before and after a state reinitialization, whereas (22.4) intends to also represent the dynamics at jump times.

The measures μ_i and the distribution λ in (22.5) play a similar role to the Lagrange multiplier in Mechanics with unilateral contact. Viewing the dynamics as an equality of distributions as in (22.5) paves the way for time-discretization with time-stepping algorithms, i.e. numerical schemes working without event detection procedures and constant time-step.

Positivity of λ: Only the Dirac measures μ_i and time functions are signed. Consequently writing $\lambda \geq 0$ is meaningless. The correct writing of the complementarity $0 \leq z_1 \perp \lambda \geq 0$ (see corollary 22.9 below) requires to reformulate the dynamics under a suitable representation and will be done through several steps. Another point of view is to assert that $\lambda \geq 0$ implies that λ **is** a measure. However as we shall see this is not sufficient to assure $z_1(t) \geq 0$. Consequently one has to resort to higher degree distributions to give a reasonably general meaning to the dynamics in (22.2).

2. The extended Moreau's sweeping process

In order to simplify the presentation we shall assume in many places that $m = 1$. When the statements or results also obviously hold for $m \geq 2$ and uniform relative degree (see (22.3)) this will be pointed out. Starting from (22.4) (22.5) the extended sweeping process is written as follows (in order to lighten the writing we will denote the non-singular part of a distribution z as $z(t)$)

$$\mu_i \in -\partial\psi_{T_\Phi^{i-1}(z_1(t^-),\ldots,z_{i-1}(t^-))}(z_i(t^+)) \text{ for all } 1 \leq i \leq r^{w\lambda} \tag{22.7}$$

where $T_\Phi^i(z_1, \ldots, z_i) = T_{T_\Phi^{i-1}(z_1,\ldots,z_{i-1})}(z_i)$, $T_\Phi^1(z_1) = T_\Phi(z_1)$, $T_\Phi^0(z_1) = \Phi$ and $T_\Phi(x) = \{v \mid v \geq 0 \text{ if } x \leq 0, v \in I\!R \text{ if } x > 0\}$ is the tangent cone to $\Phi = I\!R^+$ at x (extended outside Φ), defined as in (Moreau, 1988) to take into account constraint violations. We shall keep the notation Φ noting that in general when one starts from the ZD dynamics, one gets $\Phi = (I\!R^+)^m$. Moreover we also keep in mind the extension of the material that follows towards formalisms involving convex sets $\Phi(t)$ and not necessarily being the ZD dynamics of a given system. The sets

$\partial\psi_{T_\Phi^{i-1}(z_1(t^-),\ldots,z_{i-1}(t^-))}(z_i(t^+))$ are cones and therefore the inclusions in (22.7) make sense: since $\mu_i = g_i(t)dt + d\mu_i$, (22.7) means that either $g_i(t) \in -\partial\psi_{T_\Phi^{i-1}(z_1(t^-),\ldots,z_{i-1}(t^-))}(z_i(t^+))$ for all $1 \le i \le r^{w\lambda}$, or that $\sigma_{\mu_i}(t) \in -\partial\psi_{T_\Phi^{i-1}(z_1(t^-),\ldots,z_{i-1}(t^-))}(z_i(t^+))$ for all $1 \le i \le r^{w\lambda}$, where $\sigma_{\mu_i}(t)$ is the density of μ_i with respect to δ_t. Thus one sees that λ in (22.5) can be given a meaning as

$$\lambda = (CA^{r^{w\lambda}-1}B)^{-1}[\mu_1^{(r^{w\lambda}-1)} + \ldots + \dot{\mu}_{r^{w\lambda}-1}] + \mu_{r^{w\lambda}} \tag{22.8}$$

provided $CA^{r^{w\lambda}-1}B$ is invertible. Then λ is uniquely defined as in (22.8). The positivity of λ is now understood as the positivity of $\mu_{r^{w\lambda}}$. It is then important to see that the distributional dynamics

$$D^{r^{w\lambda}}z_1 = Dz_{r^{w\lambda}} = CA^{r^{w\lambda}}W^{-1}\{z\} + CA^{r^{w\lambda}-1}B\lambda \tag{22.9}$$

with λ in (22.7) (22.8), is equivalent to (22.5) (22.7). In (22.9) we used the standard notation for distributional derivatives (Ferreira (1997)). Notice that (22.9) (22.8) (22.7) is the same as (22.4) (22.8) (22.7).

DEFINITION 22.1 *The higher order (or extended) sweeping process is the dynamical system represented in (22.4)-(22.5)-(22.7). This is a particular measure differential inclusion.*

ASSUMPTION 22.2 *Let the test functions $\phi(\cdot)$ be with compact support and n times differentiable with continuous $(n+1)$th derivative. The solutions of the higher order sweeping process in (22.4)-(22.5)-(22.7) are distributions T such that $\langle T, \phi\rangle = \sum_{i=0}^{n} \int_{supp(\phi)} \phi^{(i)}(t)dh_i$, where h_i, $1 \le i \le n$ are RCLSBV (Right Continuous Locally Special Bounded Variation), and $n \le r^{w\lambda} + 1$.*

Thus $dh_i = \dot{h}_i(t)dt + \mu_{h_ia}$, where $\dot{h}_i(\cdot)$ is Lebesgue integrable and μ_{h_ia} is an atomic measure with countable set of atoms. Distributions of the form $f(t) + \sum_{i=0}^{l} a_i\delta_{t_k}^{(i)}$, $l < +\infty$, belong to the above set, with $g_i(t) = a_iH_i(t)$, $H_i(\cdot)$ is the Heaviside function with jump at $t = t_k$. The proposed framework permits atomic distributions with support a set of times $\{t_k\}_{k\ge 0}$, $0 \le t_k < +\infty$, with possible accumulations, and that may even not be orderable. *We also note that assumption 22.2 implies that outside the state jump times, right and left limits exist so that $z(t^+) = \lim_{s\to t, s>t} z(s)$ and $z(t^-) = \lim_{s\to t, s<t} z(s)$ have a meaning for all t in the interval of existence of solutions, which is a crucial property in the developments which follow.* This framework also allows us to recover the case of Mechanics which involves signed distributions of degree ≤ 2, i.e. measures (Ballard, 2000). The framework proposed in assumption

22.2 is thought to be large enough to encompass the non-autonomous and nonlinear cases as well. One sees that from assumption 22.2, the measures μ_i are the derivatives of some functions $\nu_i(\cdot) \in RCLSBV$. We recall that the continuous part of the derivative of special functions of bounded variation is absolutely continuous.

From assumption 22.2, (22.6) (22.7) and (22.8) the extended sweeping process can be formulated as the evolution Variational Inequality

$$\begin{cases} \langle D\{z_i\} - \{z_{i+1}\}, v - z_i(t^+)\rangle \geq 0, \\ \forall\ v \in T_\Phi^{i-1}(z_1(t^-), ..., z_{i-1}(t^-)), 1 \leq i \leq r^{w\lambda} - 1 \\ \\ \langle (CA^{r^{w\lambda}-1}B)^{-1}[D\{z_{r^{w\lambda}}\} - CA^{r^{w\lambda}}W^{-1}\{z\}], v - z_{r^{w\lambda}}(t^+)\rangle \geq 0, \\ \forall\ v \in T_\Phi^{r^{w\lambda}-1}(z_1(t^-), ..., z_{r^{w\lambda}-1}(t^-)) \\ \\ \langle \dot{\xi}(t) - A_\xi \xi(t) - B_\xi z_1(t), v - \xi(t)\rangle \geq 0, \ \ \forall\, v \in \mathbb{R} \\ \\ z(0^-) = z_0 \in \mathbb{R}^n \end{cases} \tag{22.10}$$

REMARK 22.3 *Assumption 22.2 is not too stringent. Especially the fact that solutions admit right and left limits everywhere seems to be a basic requirement. Many of the technical results that follow use it.*

Starting from (22.2) one is tempted to write the inclusion

$$Dz_{r^{w\lambda}} - CA^{r^{w\lambda}}W^{-1}z(t) \in -CA^{r^{w\lambda}-1}B\ \partial\psi_\Phi(z_1(t))$$

which makes sense only if λ is a measure since $\partial\psi_\Phi(z_1(\cdot))$ is a cone. This inclusion is replaced by

$$\mu_{r^{w\lambda}} \in -\ \partial\psi_{T_\Phi^{r^{w\lambda}-1}(z_1(t^-),...,z_{r^{w\lambda}-1}(t^-))}(z_{r^{w\lambda}}(t^+))$$

in (22.7). The following is true

LEMMA 22.4 *The inclusion* $\partial\psi_{T_\Phi^{r^{w\lambda}-1}(z_1,...,z_{r^{w\lambda}-1})}(z_{r^{w\lambda}}) \subseteq \partial\psi_\Phi(z_1)$ *holds for all* $z_1, ..., z_{r^{w\lambda}}$. *Also from (22.7) one has* $z_1 > 0 \Longrightarrow \mu_{r^{w\lambda}} = 0$, *and if* $z_1 = 0$ *then* $\mu_{r^{w\lambda}} \geq 0$.

Proof: If $z_1 > 0$ then $T_\Phi(z_1) = \mathbb{R}$ and $\partial\psi_\Phi(z_1) = \{0\}$, and since $T_\Phi^{i-1}(z_1, ..., z_{i-1}) = \mathbb{R}$ then $\partial\psi_{T_\Phi^{i-1}(z_1,...,z_{i-1})}(z_i) = \{0\}$ for all $1 \leq i \leq r^{w\lambda}$. In particular from (22.7) $\mu_{r^{w\lambda}} = 0$. If $z_1 = 0$ then $\partial\psi_{\mathbb{R}^+}(z_1) = \mathbb{R}^-$. Depending on the values of $z_2, ..., z_i$ being positive or non-positive, one may have $\partial\psi_{T^i_\Phi(z_1,\dots,z_i)}(z_{i+1}) = I\!R^-$ or $\partial\psi_{T^i_\Phi(z_1,\dots,z_i)}(z_{i+1}) = \{0\}$ for all $1 \le i \le r^{w\lambda}-1$. Indeed assume that $z_1 = z_2 = \dots = z_j = 0$ and $z_{j+1} > 0$ (this implies that $z \succeq 0$). Then $(\Phi) = T^0_\Phi(z_1) = T_\Phi(z_1) = T^2_\Phi(z_1, z_2) = T^3_\Phi(z_1, z_2, z_3) = \dots = T^j_\Phi(z_1, \dots, z_j) = I\!R^+$. And $T^{j+1}_\Phi(z_1, \dots, z_{j+1}) = T^{j+2}_\Phi(z_1, \dots, z_{j+2}) = \dots = T^{r^{r\lambda}-1}_\Phi(z_1, \dots, z_{r^{w\lambda}-1}) = I\!R$. This can be seen since for instance $T^{j+2}_\Phi(z_1, \dots, z_{j+2}) = T_{T^{j+1}_\Phi(z_1,\dots,z_{j+1})}(z_{j+2}) = T_{I\!R}(z_{j+2}) = I\!R$ because also $T^{j+1}_\Phi(z_1, \dots, z_{j+1}) = T_{I\!R^+}(z_{j+1}) = I\!R$. Consequently $\partial\psi_{T^i_\Phi(z_1,\dots,z_i)}(z_{i+1}) = I\!R^-$ for all $0 \le i \le j$, whereas $\partial\psi_{T^i_\Phi(z_1,\dots,z_i)}(z_{i+1}) = \{0\}$ for $j+1 \le i \le r^{w\lambda}-1$. We conclude that under such conditions $\mu_i \ge 0$ for all $1 \le i \le j+1$, and $\mu_i = 0$ for all $j+2 \le i \le r^{w\lambda}$. Consequently $\mu_{r^{w\lambda}} \ge 0$ when $z_1 = 0$. The inclusion is also proved. ∎

LEMMA 22.5 *The distribution Dz_1 in (22.5) (22.7) is of degree ≤ 2, so that $z_1(\cdot)$ is a function of time and the zero-dynamics is an ODE (i.e. $D\xi = \frac{d\xi}{dt}(t)$). Also λ has degree $\le r^{w\lambda}+1$.*

Proof: From (22.7) μ_1 has degree ≤ 2 so from (22.5) and (22.8) the result follows. ∎

The following lemmas prove that the extended sweeping process inclusion defines a well-posed state jump mapping.

LEMMA 22.6 *From (22.4) (22.5) (22.7) we get for all $1 \le i \le r^{w\lambda}-1$ and all $t \ge 0$,*

$$
\begin{array}{c}
z_{i+1}(t^+) - z_{i+1}(t^-) \in -\partial\psi_{T^i_\Phi(z_1(t^-),\dots,z_i(t^-))}(z_{i+1}(t^+))\,(\mathbf{a}) \\
\Updownarrow \\
z_{i+1}(t^+) = prox\left[T^i_\Phi(z_1(t^-),\dots,z_i(t^-)); z_{i+1}(t^-)\right] \qquad (\mathbf{b})
\end{array}
\tag{22.11}
$$

Proof: The proof starts by noting that (22.5) is an equality of distributions, and that at state jump times, Dz_i is a distribution of degree strictly larger than $z_{i+1}(\cdot)$ which is a function (Ferreira (1997), Theorem 1.1). Equaling distributions of same degree results in (22.11) (a) (see (Brogliato, 1999, §1.1) for such a reasoning in the case of distributions of degree ≤ 2). The rest of the proof is a direct consequence of the equivalence $-x + y \in \partial\psi_K(x) \Longleftrightarrow x = \text{prox}[K; y]$ (Rockafellar and Wets, 1998, Example 10.2) (Brezis, 1973, Example 2.8.2), where K is a nonempty convex set, and $\text{prox}[K; y]$ denotes the closest vector to y in K (i.e. the projection of y on K). ∎

REMARK 22.7 *The proximation operation in (22.11) (b) is used here in the sense of (Moreau, 1963), i.e. $prox_f z$ is the point where the function $u \mapsto \frac{1}{2}||z-u||^2 + f(u)$ attains its minimum value. When $f(\cdot) = \psi_K(\cdot)$, then $prox_f z = prox[K; z]$.*

This shows that jumps are automatically taken into account by the dynamics as it is written in (22.4)-(22.5)-(22.7). We notice also that the lower triangular structure of the tangent cones which appear in (22.7) merely reflects the way the measures μ_i appear in (22.5). We also have

LEMMA 22.8 *Let $m \geq 1$.*
Let us assume that $CA^{r^{w\lambda}-1}B = (CA^{r^{w\lambda}-1}B)^T > 0$. Then for all $t \geq 0$,

$$-z_{r^{w\lambda}}(t^+) + z_{r^{w\lambda}}(t^-) \in CA^{r^{w\lambda}-1}B\;\partial\psi_{T_\Phi^{r^{w\lambda}-1}(z_1(t^-),\ldots,z_{r^{w\lambda}-1}(t^-))}(z_{r^{w\lambda}}(t^+))$$

$$\Updownarrow$$

$$z_{r^{w\lambda}}(t^+) = prox_{(CA^{r^{w\lambda}-1}B)^{-1}}\left[T_\Phi^{r^{w\lambda}-1}(z_1(t^-),\ldots,z_{r^{w\lambda}-1}(t^-)); z_{r^{w\lambda}}(t^-)\right] \tag{22.12}$$

and these generalized equations possess the same unique solution $z_{r^{w\lambda}}(t^+)$ for any $z_{r^{w\lambda}}(t^-)$.

COROLLARY 22.9 *Let the solution of (22.4) (22.5) (22.7) exist on a time interval $[\tau, \tau+\epsilon]$ and assumption 22.2 holds. Then $z_i(t^+) \in T_\Phi^{i-1}(z_1(t^-), \ldots, z_i$ for all $2 \leq i \leq r_{w\lambda}$, and $0 \leq z_1(t^+) \perp \mu_{r^{w\lambda}}(t) \geq 0$ for all $t \in [\tau, \tau+\epsilon]$.*

Proof: From lemmas 22.4, 22.6 and 22.8, and also from assumption 22.2 which implies that solutions have right and left limits everywhere in $[\tau, \tau+\epsilon]$. Indeed from lemma 22.4 we have that $z_1 > 0 \Longrightarrow \mu_{r^{w\lambda}} = 0$, and if $z_1 = 0$ then $\mu_{r^{w\lambda}} \geq 0$. Let us assume now that $\mu_{r^{w\lambda}}(t) > 0$. This implies that $\partial\psi_{T_\Phi^{r^{w\lambda}-1}(z_1(t^-),\ldots,z_{r^{w\lambda}-1}(t^-))}(z_{r^{w\lambda}}(t^+)) = \mathbb{R}^+$ which in turn implies that $z_1(t^-) \leq 0$ from the definition of the tangent cones (see also the developments in the proof of lemma 22.4). Now from lemma 22.6 it follows that $z_1(t^+) \geq 0$ for all $t \in [\tau, \tau+\epsilon]$ since $z_1(t^+) = prox[\Phi; z_1(t^-)]$. We deduce that in fact if $\mu_{r^{w\lambda}} > 0$ then $z_1(t^+) = 0$. Thus $0 \leq z_1(t^+) \perp \mu_{r^{w\lambda}}(t) \geq 0$. The assertion $z_i(t^+) \in T_\Phi^{i-1}(z_1(t^-, \ldots, z_{i-1}(t^-))$ is a consequence of (22.11) and (22.12): right limits, which by assumption exist, belong to the tangent cones. ■

This result (which is also a consequence of the inclusion in lemma 22.4 and of (22.7) with $i = r^{w\lambda}$) implies that on intervals $[\tau, \tau+\epsilon)$, $\epsilon > 0$, on which $z_1(t) = 0$, the inclusion (22.4)-(22.5)-(22.7) implies the existence of a multiplier $\lambda(t) = \mu_{r^{w\lambda}}(t)$ that belongs to $-\partial\psi_{\mathbb{R}^+}(0) = \mathbb{R}^+$, equivalently which satisfies $0 \le z_1(t^+) \perp \lambda(t) \ge 0$ and is the solution of the linear complementarity problem (LCP)

$$0 \le \lambda(t) \perp CA^{r^{w\lambda}}W^{-1}z(t) + CA^{r^{w\lambda}-1}B\lambda(t) \ge 0 \qquad (22.13)$$

Equivalently, $\mu_{r^{w\lambda}}$ is the sum of an atomic measure $d\mu_{r^{w\lambda}}$, $\text{supp}(d\mu_{r^{w\lambda}}) \subset \{t|\ z_1(t) = 0\}$, that corresponds to jumps in $z_{r^{w\lambda}}(\cdot)$ and whose magnitude is the solution of the LCP in (22.12), and of a Lebesgue measure $g_{r^{w\lambda}}(t)dt$ where $g_{r^{w\lambda}}(t)$ is the solution of the LCP in (22.13). In corollary 22.9, the complementarity condition $0 \le z_1(t^+) \perp \mu_{r^{w\lambda}}(t) \ge 0$ can equivalently be written as $0 \le z_1(t^+) \perp g_{r^{w\lambda}}(t) \ge 0$, since the complementarity holds at the right limit $z_1(^+)$. Similarly $\mu_i = g_i(t)dt + d\mu_i$ for all $1 \le i \le r^{w\lambda}$, with $\text{supp}(d\mu_i) \subset \{t|\ z_1(t) = 0\}$. On $[\tau, \tau + \epsilon)$ one has $g_i(t) = 0$ for all $1 \le i \le r^{w\lambda} - 1$, as can easily be deduced from $z_1(t) \equiv 0$. We thus have proved the following

LEMMA 22.10 *Consider the extended sweeping process dynamics (22.4) (22.5) (22.7) and let assumption 22.2 hold. Then $g_i(t) = 0$ for all $1 \le i \le r^{w\lambda} - 1$ and almost all $t \ge 0$, whereas $g_{r^{w\lambda}}(t)$ is the solution of the LCP in (22.13).*

LEMMA 22.11 *The following inclusion holds*

$$\partial\psi_{T_\Phi^{i-1}(z_1,\ldots,z_{i-1})}(z_i) \subseteq \partial\psi_{T_\Phi^{i-2}(z_1,\ldots,z_{i-2})}(z_{i-1}) = N_{T_\Phi^{i-2}(z_1,\ldots,z_{i-2})}(z_{i-1}) \qquad (22.14)$$

for all $1 \le i \le r^{w\lambda}$.

Proof: Let $z_1 = z_2 = \ldots = z_j = 0$ and $z_{j+1} > 0$. Then as already shown in the proof of lemma 22.4, one has $\partial\psi_{T_\Phi^i(z_1,\ldots,z_i)}(z_{i+1}) = I\!R^-$ for all $0 \le i \le j$, and $\partial\psi_{T_\Phi^i(z_1,\ldots,z_i)}(z_{i+1}) = \{0\}$ for all $j+1 \le i \le r^{w\lambda} - 1$. So one sees that in particular it always holds that $\partial\psi_{T_\Phi^k(z_1,\ldots,z_k)}(z_{k+1}) \subseteq \partial\psi_{T_\Phi^{k-1}(z_1,\ldots,z_{k-1})}(z_k)$ for any $1 \le k \le r^{w\lambda} - 1$. ■

COROLLARY 22.12 *The operators $z_i(t^+) \mapsto -\mu_i$, $1 \le i \le r^{w\lambda}$, in (22.7) are maximal monotone.*

Proof: These operators can be rewritten as the following cone CP

$$T_\Phi^{i-1}(z_1(t^-), \ldots, z_{i-1}(t^-)) \ni z_i(t^+) \perp$$
$$- \mu_i \in \partial\psi_{T_\Phi^{i-1}(z_1(t^-),\ldots,z_{i-1}(t^-))}(z_i(t^+)) \qquad (22.15)$$

Since from lemma 22.11 we have

$$\partial\psi_{T_\Phi^{i-1}(z_1(t^-),\ldots,z_{i-1}(t^-))}(z_i(t^+)) \subseteq N_{T_\Phi^{i-2}(z_1(t^-),\ldots,z_{i-2}(t^-))}(z_{i-1}(t^+)$$

the cones in both sides of (22.15) are closed polar convex cones. Consequently the operators that correspond to these cone CPs are maximal monotone. ■

We notice that from (22.15) we get

$$T_\Phi(z_1(t^-)) \ni z_2(t^+) \perp -\mu_2 \in \partial\psi_{T_\Phi(z_1(t^-))}(z_2(t^+)) \subseteq N_\Phi(z_1(t^+))$$

It is also noteworthy that from corollary 22.9 one gets that the operator $-\mu_{r_w\lambda} \mapsto z_1(t^+)$ is also maximal monotone. Hence our framework contains that of the sweeping process for Lagrangian systems (Moreau, 1988).

3. Numerical time integration scheme

This section addresses the problem of the numerical time integration of complementarity dynamical systems with arbitrary relative degree. Particularly, it is shown how one can take advantage of the formalism in (22.4)-(22.7) to design a time-stepping scheme, i.e, a time integration scheme without explicit event handling procedure.

A naive way to design a time-stepping scheme for non smooth systems is to apply a backward Euler method to the dynamical equation and to discretize the complementarity condition in a fully implicit way. One obtains for the system (22.1) the following discretized system:

$$\begin{cases} \dfrac{x_{k+1} - x_k}{h} = Ax_{k+1} + B\lambda_{k+1} \\ w_{k+1} = Cx_{k+1} + D\lambda_{k+1} \\ 0 \leq \lambda_{k+1} \perp w_{k+1} \geq 0 \end{cases} \tag{22.16}$$

where h is the constant time step of a subdivision $\{t_k\}$ of the time interval $[0, T]$ and the subscript k denotes the approximation of a value at time t_k.

A straightforward substitution of w_{k+1} in the complementarity condition leads to solve the following complementarity problem at each step:

$$0 \leq \lambda_{k+1} \perp C(I - hA)^{-1}x_k + hC(I - hA)^{-1}B\lambda_{k+1} \geq 0 \tag{22.17}$$

For the linear complementarity systems, some sufficient conditions for consistency and convergence of this backward Euler scheme have been given in (Camlibel, 2002). They also exhibit several examples for which the scheme does not work at all. Let us consider one of these examples, introduced in (Camlibel, 2001, Example 6.3.3), where the system (22.1) is defined by

$$A = \begin{bmatrix} 0 & 1 & 0 \\ 0 & 0 & 1 \\ 0 & 0 & 0 \end{bmatrix}; B = \begin{bmatrix} 0 \\ 0 \\ 1 \end{bmatrix}; C = \begin{bmatrix} 1 & 0 & 0 \end{bmatrix}; x_0 = \begin{bmatrix} 0 \\ -1 \\ 0 \end{bmatrix} \tag{22.18}$$

In this example, the relative degree $r^{w\lambda}$ is equal to 3 ($D = 0, CB = 0, CAB = 0, CA^2B \neq 0$). Using the time discretization defined above, we can remark that:

$$\lim_{h \longrightarrow 0} hC(I - hA)^{-1}B = 0 \tag{22.19}$$

It is clear that if the time step h vanishes, which may be needed in many practical cases or for the convergence analysis purpose, then the LCP matrix in (22.17) has little chance to be well conditioned due to the fact that $CB = 0$. Furthermore, the numerical solution is given by

$$x_k = \begin{bmatrix} \frac{k(k+1)}{h} \\ k \\ \frac{1}{h} \end{bmatrix}, \forall k \geq 1; \quad \lambda_1 = \frac{1}{h^2}; \quad \lambda_k = 0, \forall k \geq 2 \tag{22.20}$$

which cannot converge towards a solution when h vanishes.

In fact, the backward Euler scheme is only consistent for the systems of relative degree, $r_{w\lambda} \leq 1$. Indeed, we can construct easily many examples of inconsistency with systems of relative degree equal to two.

A similar time-stepping method has been introduced in (Moreau, 1977) for a mathematical analysis purpose of the existence of solution for the first order sweeping process. For systems of relative degree equal to two, such as Lagrangian mechanical systems with unilateral constraints, Moreau (Moreau, 1988) introduces the Contact Dynamics method which extends the simple Euler method when some discontinuities may be encountered in the first time derivative of the state. Although the resulting scheme seems to be very close to a standard Euler scheme, there are several slight, but fundamental differences based on a sound mathematical analysis of Moreau about the nature of the solution. For the Lagrangian systems, majors lessons of this seminal work may be stated as follows:

- the use of a differential measure associated with a function of bounded variation leads to a first order approximation given directly by the integration of the differential measure,
- the use of finite values, as primary unknowns, such as velocity and impulse. This feature allows one to capture the discontinuities when the time step vanishes,

- the reformulation of constraints in terms of velocity associated with a viability lemma to ensure the satisfaction of the position constraint.

In the next section, we propose a time-stepping scheme based on these remarks, which is able to integrate in time linear complementarity problems of any relative degree.

3.1 The proposed numerical scheme

The proposed numerical scheme is based on the ZD dynamical form written in terms of measures (22.6). Let us consider a subdivision $\{t_k\}$ of the interval $[0, T]$. The time integration of the differential measure on $]t_k, t_{k+1}]$ is given by

$$\int_{]t_k,t_{k+1}]} D\{z_i\} = z_i(t_{k+1}^+) - z_i(t_k^+) \tag{22.21}$$

and we pose as a primary unknown the right value of the non singular part of z_i such that $z_{i,k} = z_i(t_k^+)$ (recall that $z(t)$ or $\{z\}$ denote the non singular part of the distribution z). These remarks leads to the following numerical integration rule for a generic line of the system (22.6):

$$\int_{]t_k,t_{k+1}]} D\{z_i\} = z_{i,k+1} - z_{i,k} = \int_{]t_k,t_{k+1}]} \{z_{i+1}\}\, dt + \int_{]t_k,t_{k+1}]} \mu_i \tag{22.22}$$

$$\approx h z_{i+1,k+1} + \int_{]t_k,t_{k+1}]} \mu_i \tag{22.23}$$

In order to manipulate only finite values, this second unknown, which is the multiplier corresponding to μ_i, is defined as:

$$r_{i,k+1} = \int_{]t_k,t_{k+1}]} \mu_i \tag{22.24}$$

and we assume that:

$$r_{i,k+1} \in -\partial\psi_{T_\Phi^{i-1}(z_{1,k},\ldots,z_{i-1,k})}(z_{i,k+1}) \text{ for all } 1 \leq i \leq r^{w\lambda} \tag{22.25}$$

The approximation of $z_i(t^-)$ in the inclusion (22.7) by $z_{i,k}$ is a basic choice. The operation can be viewed as a prediction of the state before a discontinuity. More accurate prediction may be performed using higher order derivatives, if exist, of $z_i(t^-)$. Finally, the proposed discretization scheme may be summarized as follows:

$$
\left\{
\begin{array}{l}
z_{1,k+1} - z_{1,k} = hz_{2,k+1} + r_{1,k+1} \\
z_{2,k+1} - z_{2,k} = hz_{3,k+1} + r_{2,k+1} \\
\vdots \\
z_{i,k+1} - z_{i,k} = hz_{i+1,k+1} + r_{i,k+1} \\
\vdots \\
z_{r^{w\lambda},k+1} - z_{r^{w\lambda},k} = hCA^{r^{w\lambda}} W^{-1} z_{k+1} + CA^{r^{w\lambda}-1} B r_{r^{w\lambda},k+1} \\
\xi_{k+1} - \xi_k = hA_\xi \xi_{k+1} + hB_\xi z_{1,k+1} \\
r_{i,k+1} \in -\partial\psi_{T_\Phi^{i-1}(z_{1,k},\ldots,z_{i-1,k})}(z_{i,k+1}) \text{ for all } 1 \le i \le r^{w\lambda}
\end{array}
\right.
\tag{22.26}
$$

If $\Phi = I\!R^+$, the inclusion (22.25) implies a sequence of unilateral constraints on $z_{i,k}$ to be satisfied. If the integer j is the first for which $z_{j,k}$ is positive, then the system (22.26) is reduced to a linear complementarity problem involving $z_{i,k+1}, r_{i,k+1}$ for all $i, 1 \le i \le j$.

3.2 Numerical examples

We illustrate in this part the ability of the scheme to solve the preliminary example. The numerical solution given by the scheme (22.26) for the initial condition $z_0 = [0; -1; 0]^T$ is given by $z_k = [0; 0; 0]^T, \forall k \ge 1$ and $r_2, 1 = 1, r_2, k = 0, \forall k \ge 2$. The Figure 22.1 depicts a similar result with the initial condition $z_0^T = (1, -1, 0)$.

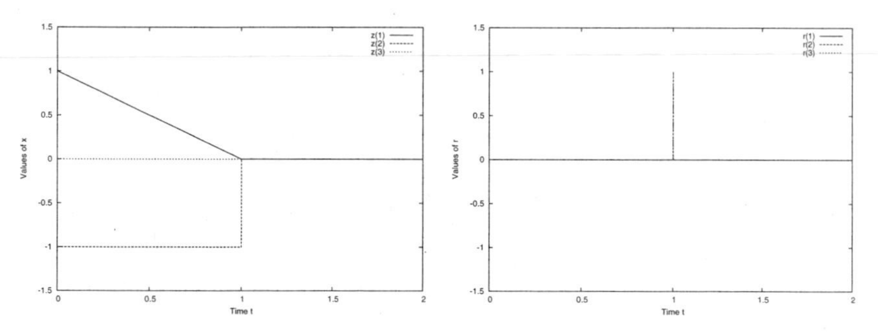

Figure 22.1. EMSP scheme – Initial data $z(0^-)^T = (1, -1, 0)$

Acknowledgments

The authors wish to thank Daniel Goeleven (IREMIA, university of La Réunion, France) for useful comments on their work. This work was partially supported by the European project SICONOS IST-2001-37172.

References

J. Campos Ferreira, *Introduction to the Theory of Distributions*, Pitman Monographs and Surveys in Pure and Applied Mathematics, 87, Addison Wesley Longman Limited, 1997.

H. Brezis, 1973 *Opérateurs Maximaux Monotones et Semi-groupes de Contraction dans les Espaces de Hilbert*, North-Holland Publishing, Amsterdam.

J.J. Moreau, "Les liaisons unilatérales et le principe de Gauss", CRAS Paris, t.256 (1963), pp.871-874.

J.J. Moreau, *Evolution problem associated with a moving convex set in a Hilbert space*, J. Differential Equations, 26 (1977), pp.347-374.

J.J. Moreau, "Unilateral contact and dry friction in finite freedom dynamic", CISM Courses and Lectures no 302 (1988), International Center for Mechanical Sciences, J.J. Moreau and P.D. Panagiotopoulos (Eds.), Springer-Verlag, pp.1-82.

M.D.P. Monteiro Marques, *Differential Inclusions in Nonsmooth Mechanical Problems, Shocks and Dry Friction*, Birkhauser, Progress in Nonlinear Differential Equations and Their Applications, 1993.

P. Sannuti, "Direct singular perturbation analysis of high-gain and cheap control problems", Automatica, vol.19 (1983), no 1, pp.41-51.

K. Camlibel, *Complementarity Methods in the Analysis of Piecewise Linear Dynamical Systems*, PhD thesis, Katholieke Universiteit Brabant, ISBN: 90 5668 073X (2001).

K. Camlibel, W.P.M.H. Heemels, J.M. Schumacher, *Consistency of a time-stepping method for a class of piecewise-linear networks*, IEEE Transactions on Circuits and Systems I, vol.49 (2002), pp.349-357.

S.L. Campbell and W. Gear. *The index of general nonlinear DAEs*. Numerische Mathematik, 72:173–196, 1995.

R.F. Hartl, S.P. Sethi, R.G. Vickson, *A survey of the maximum principles for optimal control problems with state constraints*, SIAM Review, vol.37 (1995), no 2, pp.181-218.

E.D. Sontag, *Mathematical Control Theory; Deterministic Finite Dimensional Systems*, Springer Texts in Applied Mathematics 6, second edition, 1998.

R.T. Rockafellar, R.J.B. Wets, 1998 *Variational Analysis*, Springer, Grundlehren der mathematischen Wissenschaften, vol.317.

P. Ballard, "The dynamics of discrete mechanical systems with perfect unilateral constraints", Arch. Rational Mech. Anal., vol.154 (2000), no 3, pp.199-274.

B. Brogliato, *Nonsmooth Mechanics*, Second edition, Springer Verlag London, 1999.

Chapter 23

AN EXISTENCE RESULT IN NON-SMOOTH DYNAMICS

Manuel D.P. Monteiro Marques

CMAF and Faculdade de Ciências, Universidade de Lisboa, 2, Av Prof. Gama Pinto, 1649-003 Lisboa, Portugal

mmarques@ptmat.fc.ul.pt

Laetitia Paoli

Université de Saint-Etienne, 23 rue Michelon, 42023 Saint Etienne Cedex 2, France

laetitia.paoli@univ-st-etienne.fr

Abstract Starting in 1983, Jean Jacques Moreau gave remarkable new formulations of the dynamics of mechanical systems submitted to inelastic or frictional contact as measure-differential inclusions. This paved the way to proofs of existence of solutions, by the first author, in 1985 for the case of standard inelastic shocks and in 1988 for frictional dynamics, both for one contact problems. These mathematical theoretical results left out many other situations, that could be handled numerically by J.J. Moreau and collaborators in a very effective way, as well as the Painlevé example. Here we present a generalization of such existence results to variable mass matrices and non isotropic friction.

Keywords: Unilateral constraint, friction, non trivial mass matrix, measure differential inclusion, velocities with bounded variation, time-discretization scheme.

Introduction

We consider a mechanical system with a finite number of degrees of freedom subjected to a single unilateral constraint. Using generalized coordinates, the system is represented by a point $q \in \mathbb{R}^d$ and the motion is described by a function $q : I \to \mathbb{R}^d$ with $I = [0, T]$, $T > 0$. The unilateral constraint is expressed by an inequality $f(q) \leq 0$ and thus

$$q(t) \in L = \{q \in \mathbb{R}^d; f(q) \leq 0\} \quad \forall t \in I \tag{23.1}$$

where $f \in C^1(\mathbb{R}^d; \mathbb{R})$ such that ∇f does not vanish in a neighbourhood of the hypersurface $S = \{q \in \mathbb{R}^d; f(q) = 0\}$.

Since we expect discontinuous velocities when $f\big(q(t)\big) = 0$, an appropriate functional framework is to look for a motion q such that $\dot{q}$ is a function of bounded variation, see (Schatzman, 1978, Moreau, 1983, Moreau, 1985, Moreau, 1986, Moreau, 1988). It follows that $\dot{q}$ has right and left limits, $\dot{q}^+$ and $\dot{q}^-$, for all $t \in (0, T)$. Moreover, denoting $u = \dot{q}^+$, we can introduce the Stieltjes measure $du = d\dot{q}$ and the dynamics is described by

$$q(t) = q_0 + \int_0^t u(s)\, ds \quad \forall t \in I \tag{23.2}$$

and

$$M(q)du = g\big(t, q, M(q)u\big)dt + dr \tag{23.3}$$

where M is the mass matrix of the system and dr is a measure describing the reaction when the constraint is active. Using Radon-Nikodym's theorem we can rewrite (23.3) as follows

$$M(q)u'_\mu = g\big(t, q, M(q)u\big)t'_\mu + r'_\mu \quad d\mu \text{ a.e.}$$

where μ is a positive measure such that du, dt and dr are absolutely continuous with respect to $d\mu$ (we can choose $d\mu = |du| + dt$ for instance) and u'_μ, t'_μ and r'_μ are the densities of du, dt and dr with respect to $d\mu$, see (Moreau, 1985) and (Moreau, 1988) for instance.

The reaction dr and the transmission of the velocities at impacts (i.e when $f\big(q(t)\big) = 0$) can be described more precisely if we know whether the constraint is perfect or not. In both cases, there exists a subset $\mathcal{R}(q)$ of $\mathbb{R}^d$ such that

$$r'_\mu \in \mathcal{R}\big(q(t)\big) \quad d\mu \text{ a.e.} \tag{23.4}$$

and an impact law relating $\dot{q}^+$, $\dot{q}^-$ and dr.

In order to obtain existence results for this kind of problem, we study the convergence of a sequence of approximants $(q_n, u_n)_{n \geq 0}$ of the form

$$\begin{cases} q_n(t) = q_0 + \displaystyle\int_0^t u_n(s)\, ds \quad \forall t \in I \\ u_n(t) = u_n^i \quad \forall t \in [t_n^i, t_n^{i+1}[, \quad 0 \leq i \leq 2^n - 1 \end{cases} \tag{23.5}$$

with

$$t_n^i = i\frac{T}{2^n} \quad 0 \leq i \leq 2^n, \quad \forall n \geq 0$$

where the u_n^i's are defined by a time-discretization scheme derived from the description of the dynamics.

In the next two sections we will focus on the case of inelastic shocks without friction (section 1) and with friction (section 2).

1. Inelastic shocks without friction

When the unilateral constraint is frictionless the reaction belongs to the opposite of the normal cone to L at q, see (Moreau, 1983, Moreau, 1985, Moreau, 1986), i.e

$$\mathcal{R}(q) = -N_L(q) = \begin{cases} \{\lambda \nabla f(q); \lambda \leq 0\} & \text{if } f(q) \geq 0, \\ \{0\} & \text{if } f(q) < 0. \end{cases}$$

Equations (23.3) and (23.4) can be rewritten as a measure differential inclusion

$$g\big(t, q, M(q)u\big)t'_\mu - M(q)u'_\mu \in N_L\big(q(t)\big) \quad d\mu \text{ a.e.} \tag{23.6}$$

Let us describe now the transmission of velocities at impacts. The condition (23.1) implies that

$$\dot{q}^+(t) \in V\big(q(t)\big), \quad \dot{q}^-(t) \in -V\big(q(t)\big) \quad \forall t \in (0, T) \tag{23.7}$$

with $V(q) = \{v \in I\!R^d; \nabla f(q) \cdot v \leq 0\}$ if $f(q) \geq 0$ and $V(q) = I\!R^d$ otherwise. Here, as above, $\nabla f(q)$ denotes the gradient with respect to the canonical scalar product in the reference coordinates space $I\!R^d$. Integrating (23.6) on the set $\{t\}$ we obtain

$$M\big(q(t)\big)\big(\dot{q}^+(t) - \dot{q}^-(t)\big) \in I\!R^- \nabla f\big(q(t)\big) \tag{23.8}$$

if $f\big(q(t)\big) = 0$.

We can decompose the impulsion $p = M(q)\dot{q}$ into normal and tangential components with respect to the scalar product defined by the inverse of the mass matrix. To be precise,

$$p = p_N + p_T; \quad p_N \in I\!R \nabla f(q), \quad \nabla f(q)^T M^{-1}(q) p_T = 0.$$

Relations (23.8) and (23.7) imply that (with obvious notation for right and left limits):

$$p_T^+ = p_T^-, \quad p_N^- \in I\!R^+ \nabla f(q), \quad p_N^+ \in I\!R^- \nabla f(q)$$

if $f\big(q(t)\big) = 0$. If we assume moreover that the kinetic energy does not increase during impacts (energetic consistency of the model), we get

$$\begin{aligned} 2E_c^- &= p_T^{-T} M^{-1}(q) p_T^- + p_N^{-T} M^{-1}(q) p_N^- \\ \geq 2E_c^+ &= p_T^{+T} M^{-1}(q) p_T^+ + p_N^{+T} M^{-1}(q) p_N^+ \end{aligned}$$

and thus the impact law is given by

$$p_N^+ = -ep_N^-, \quad p_T^+ = p_T^-$$

with a restitution coefficient $e \in [0, 1]$, which is equivalent to

$$\dot{q}^+(t) = \dot{q}^-(t) - (1+e) \frac{\nabla f\big(q(t)\big) \cdot \dot{q}^-(t)}{\nabla f\big(q(t)\big)^T M^{-1}\big(q(t)\big) \nabla f\big(q(t)\big)} M^{-1}\big(q(t)\big) \nabla f\big(q(t)\big) \tag{23.9}$$

if $f\big(q(t)\big) = 0$.

Following J.J.Moreau (Moreau, 1983, Moreau, 1985), we consider now the case of inelastic shocks i.e $e = 0$. The impact law (23.9) can then be rewritten as

$$\dot{q}^+(t) = \text{prox}_{q(t)}\big(\dot{q}^-(t), V\big(q(t)\big)\big) \quad \text{if } f\big(q(t)\big) = 0 \tag{23.10}$$

where $\text{prox}_q\big(v, V(q)\big)$ denotes the projection (or "proximal point" of the vector v on the convex set V, relatively to the kinetic metric at q defined by the mass matrix $M(q)$. Observing that $V\big(q(t)\big)$ is the set of admissible right velocities for the system at time t, equation (23.10) can be interpreted as a minimization principle.

Moreover, taking into account the impact law, we can rewrite equation (23.6) as follows

$$-r'_\mu = g\big(t, q, M(q)u\big)t'_\mu - M(q)u'_\mu \in \partial\psi_{V(q(t))}\big(u(t)\big) \quad d\mu \text{ a.e.} \tag{23.11}$$

where $\psi_{V(q)}$ denotes the indicator function of $V(q)$ and $\partial\psi_{V(q)}$ is its subdifferential.

Indeed, given the definition of $V(q)$, we have $\partial\psi_{V(q(t))}\big(u(t)\big) \subset N_L\big(q(t)\big)$ for all t and, if the velocity is discontinuous at time t, we have $q(t) \in \partial L$ and

$$-dr\big(\{t\}\big) = M\big(q(t)\big)\big(\dot{q}^+(t) - \dot{q}^-(t)\big) \in \partial\psi_{V(q(t))}\big(\dot{q}^+(t)\big),$$

hence

$$\dot{q}^+(t) = \text{prox}_{q(t)}\big(\dot{q}^-(t), V\big(q(t)\big)\big),$$

which gives again equations (23.6) and (23.9). For a more rigourous study of the equivalence of the two formulations see (Paoli, 2001).

Starting from (23.11) we propose the following time-discretization scheme:

$$q_n^{i+1} = q_n^i + hu_n^i, \tag{23.12}$$

$$u_n^{i+1} = \text{prox}_{q_n^{i+1}}\Big(u_n^i + hM^{-1}(q_n^{i+1})g\big(t_n^{i+1}, q_n^{i+1}, M(q_n^{i+1})u_n^i\big), V(q_n^{i+1})\Big) \tag{23.13}$$

for all $i \in \{0, \ldots, 2^n - 1\}$, $n \geq 0$, with an appropriate initialization for q_n^0 and u_n^0, see (Monteiro Marques, 1985, Monteiro Marques, 1993).

The construction of this scheme is based on the following idea. Without constraint, the dynamics is described by the second order ODE

$$M(q)\ddot{q} = g\big(t, q, M(q)\dot{q}\big)$$

and the velocity at t_n^{i+1} can be approximated by

$$u_n'^{i+1} = u_n^i + hM^{-1}(q_n^{i+1})g\big(t_n^{i+1}, q_n^{i+1}, M(q_n^{i+1})u_n^i\big).$$

This velocity can be considered as a left velocity at t_n^{i+1} for the constrained system and (23.13) is obtained by applying a discretized formulation of the impact law (23.10).

The convergence of the approximate motions $(q_n, u_n)_{n \geq 0}$ given by (23.5) and (23.12)-(23.13) has been proved in (Monteiro Marques, 1985), see also (Monteiro Marques, 1993), when the mass matrix is trivial (i.e $M(q) = \text{Id}_{\mathbb{R}^d}$ for all q), ∇f is Lipschitz continuous and g is a continuous function of t and q, uniformly bounded on $I \times \mathbb{R}^d$. A generalization of this result to the case of partially or totally elastic shocks (i.e $e \neq 0$) has been obtained by M.Mabrouk (Mabrouk, 1998). In this paper the function g may depend also on $\dot{q}$ but the assumption of a trivial mass matrix still holds.

2. Inelastic shocks with friction

2.1 Isotropic Coulomb friction

In this subsection we assume for simplicity that the mass matrix of the system is equal to $\text{Id}_{\mathbb{R}^d}$ and that the classical Coulomb's law of friction holds. When the constraint is active, the set of reactions $\mathcal{R}(q)$ is then a revolution cone $C(q)$ revolving about the inward normal to L at q, denoted $n(q)$, with an angle $\alpha(q) \in (0, \pi/2)$ i.e

$$\mathcal{R}(q) = \begin{cases} C(q) = \{v \in \mathbb{R}^d; v \cdot n(q) \geq \|v\| \cos\alpha(q)\} & \text{if } f(q) \geq 0, \\ \{0\} \quad \text{if } f(q) < 0. \end{cases}$$

Moreover, if $f\big(q(t)\big) = 0$, we have the following alternative, see (Moreau, 1986, Moreau, 1988)

- if $u(t) \cdot \nabla f\big(q(t)\big) < 0$ there is no reaction,
- if $u(t) \cdot \nabla f\big(q(t)\big) = 0$ then, if the reaction belongs to $\text{Int}\Big(C\big(q(t)\big)\Big)$ we have $u(t) = 0$, otherwise $u(t)$ is in the opposite direction of the tangential part of the reaction.

Using the density r'_μ, we obtain $d\mu$-a.e.

$$\bullet \text{ if } f\big(q(t)\big) < 0 \text{ then } r'_\mu = 0, \tag{23.14}$$

$$\bullet \text{ if } f\big(q(t)\big) = 0 \text{ and } u(t) \cdot \nabla f\big(q(t)\big) < 0, \text{ then } r'_\mu = 0, \tag{23.15}$$

$$\bullet \text{ if } f\big(q(t)\big) = 0 \text{ and } u(t) \cdot \nabla f\big(q(t)\big) = 0, \text{ then}$$

$$-u(t) \in \mathrm{proj}_{T\left(q(t)\right)} N_{C\left(q(t)\right)} \big(r'_\mu(t)\big) \tag{23.16}$$

where $T\big(q(t)\big) = V\big(q(t)\big) \cap \big(-V\big(q(t)\big)\big)$. When the velocity is discontinuous at time t, the measure dr has a Dirac equal to $\big(\dot{q}^+(t) - \dot{q}^-(t)\big)\delta_t$ and (23.16) implies that

$$-u(t) \in \mathrm{proj}_{T\left(q(t)\right)} N_{C\left(q(t)\right)} \big(u(t) - \dot{q}^-(t)\big)\big)$$

which is equivalent to

$$u(t) = \mathrm{prox}\big(0, \big(\dot{q}^-(t) + C\big(q(t)\big)\big) \cap T\big(q(t)\big)\big). \tag{23.17}$$

Using the same ideas as in the frictionless case, we propose the following discretization:

$$\begin{aligned} q_n^{i+1} &= q_n^i + h u_n^i \\ u_n'^{i+1} &= u_n^i + h g\big(t_n^{i+1}, q_n^{i+1}, u_n^i\big) \\ u_n^{i+1} &= \mathrm{prox}\big(0, \big(u_n'^{i+1} + C(q_n^{i+1})\big) \cap T(q_n^{i+1})\big), \end{aligned}$$

for all $i \in \{0, \ldots, 2^n - 1\}$, $n \geq 0$, with an appropriate initialization for q_n^0 and u_n^0, see (Monteiro Marques, 1993).

The convergence of this discretization has been proved in (Monteiro Marques, 1993) under the assumption that ∇f is Lipschitz continuous, g is a continuous function of t and q, uniformly bounded on $I \times I\!R^d$ and α is a continuous function with values in $(0, \pi/2)$.

2.2 Generalization

We consider now the case of a non trivial mass matrix and non isotropic friction. The set of reactions $\mathcal{R}(q)$ is then a cone given by

$$\mathcal{R}(q) = \begin{cases} C(q) = I\!R^+\big(n(q) + D_1(q)\big) & \text{if } f(q) \geq 0, \\ \{0\} \quad \text{if } f(q) < 0, \end{cases}$$

where $n(q)$ denotes once again the inward normal to L at q and $D_1(q)$ is a closed, bounded, convex subset of $I\!R^d$ such that $n(q) \notin \mathrm{span}\big(D_1(q)\big)$. We assume moreover that

• $q \mapsto M(q)$ is locally Lispschitz continuous with values in the set of symmetric, definite positive $d \times d$ matrices,

• $q \mapsto D_1(q)$ is continuous with respect to Hausdorff distance,

• $r \cdot \nabla f(q) < 0$ for all $r \in \mathcal{R}(q) \setminus \{0\}$, for all $q \in L$.

In order to generalize relations (23.14)-(23.17), we observe that (23.14)-(23.15)-(23.16) imply that, for all $t > 0$ such that $f\big(q(t)\big) = 0$, we have $u(t) \in T\big(q(t)\big)$ (otherwise $u(t) \cdot \nabla f\big(q(t)\big) < 0$ and $u(t) = \dot{q}^-(t)$ which is impossible) and (23.16) is equivalent to

$$u(t) \cdot r'_\mu(t) = \min\big\{u(t) \cdot r;\ r \in C\big(q(t)\big) \text{ s.t. } n\big(q(t)\big) \cdot \big(r - r'_\mu(t)\big) = 0\big\}. \tag{23.18}$$

Moreover, when the velocity is discontinuous at time t, dr has a Dirac at t and the right velocities v which are kinematically admissible satisfy

$$v = \dot{q}^-(t) + r, \quad r \in C\big(q(t)\big), \quad v \in T\big(q(t)\big).$$

Thus the set of admissible reactions at t is given by

$$C_{adm}\big(q(t)\big) = \big\{r \in C\big(q(t)\big);\ n\big(q(t)\big) \cdot \big(r - r'_\mu(t)\big) = 0\big\}.$$

Relation (23.18) can be interpreted as a maximal dissipation principle with respect to the set of admissible reactions. This interpretation leads to the following generalization:

$$\bullet \quad r'_\mu = M(q)u'_\mu - g\big(t, q, M(q)u\big)t'_\mu \in \mathcal{R}\big(q(t)\big) \quad d\mu \text{ a.e.}$$

and

$$\bullet \text{ if } f\big(q(t)\big) < 0 \text{ then } r'_\mu = 0, \tag{23.19}$$

$$\bullet \text{ if } f\big(q(t)\big) = 0 \text{ and } u(t) \cdot \nabla f\big(q(t)\big) < 0, \text{ then } r'_\mu = 0, \tag{23.20}$$

$$\bullet \text{ if } f\big(q(t)\big) = 0 \text{ and } u(t) \cdot \nabla f\big(q(t)\big) = 0, \text{ then}$$

$$u(t) \cdot r'_\mu(t) = \min\big\{u(t) \cdot r;\ r \in C_{adm}\big(q(t)\big)\big\}, \tag{23.21}$$

with

$$C_{adm}\big(q(t)\big) = \big\{r \in C\big(q(t)\big);\ n\big(q(t)\big)^T M^{-1}\big(q(t)\big)\big(r - r'_\mu(t)\big) = 0\big\}$$

if $r'_\mu(t) \neq 0$, and $C_{adm}\big(q(t)\big) = \{0\}$ if $r'_\mu(t) = 0$. We may observe that, if $t > 0$ and $\dot{q}^-(t) \cdot \nabla f\big(q(t)\big) > 0$, then the velocity is discontinuous at time t and (23.21) implies that

$$u(t) = \text{prox}_{q(t)}\Big(0, \big(\dot{q}^-(t) + M^{-1}\big(q(t)\big)C\big(q(t)\big)\big) \cap T\big(q(t)\big)\Big) \tag{23.22}$$

which is a generalization of (23.17).

For this problem we propose the following time-discretization:

$$q_n^{i+1} = q_n^i + hu_n^i, \tag{23.23}$$

$$u_n^{\prime i+1} = u_n^i + hM^{-1}(q_n^{i+1})g\big(t_n^{i+1}, q_n^{i+1}, M(q_n^{i+1})u_n^i\big), \tag{23.24}$$

$$u_n^{i+1} = \mathrm{prox}_{q_n^{i+1}}\big(0, \big(u_n^{\prime i+1} + M^{-1}(q_n^{i+1})C(q_n^{i+1})\big) \cap T(q_n^{i+1})\big), \tag{23.25}$$

for all $i \in \{0, \ldots, 2^n - 1\}$, $n \geq 0$, with an appropriate initialization for q_n^0 and u_n^0.

This scheme has already been implemented, see (Moreau, 1988, Jean and Moreau, 1992). It is an alternative to and it allows more general friction cones than the scheme proposed by D.Stewart (Stewart, 1998). The study of the convergence of the approximate motions $(q_n, u_n)_{n\geq 0}$ defined by (23.5) and (23.23)-(23.25) is not yet finished but we can describe here the main ideas. As in the previous cases, the proof is divided in three main steps. First we establish a priori estimates for the discrete velocities and accelerations and using Ascoli-Arzela and Helly theorems we pass to the limit. The ideas which yield these estimates are directly inspired by (Monteiro Marques, 1993) and (Stewart, 1998) but, since the kinetic metric depends on the position q, a quadratic term appears in the estimate of the discrete velocities and prevents us from directly obtaining global estimates. Hence, we just get the convergence of a subsequence of $(q_n, u_n)_{n\geq 0}$ on $[0, \tau]$, with $\tau \in (0, T]$, i.e.

$$\begin{array}{ll} q_n(t) \to q(t) & \text{uniformly on } [0, \tau], \\ u_n(t) \to v(t) & \text{for all } t \in [0, \tau], \end{array}$$

where $v \in BV(0, \tau; I\!R^d)$.

Then we define $u = v^+$ and, using the same techniques as in (Monteiro Marques, 1993), we prove in the next step that

$$q(t) = q_0 + \int_0^t u(s)\,ds \ \in L, \quad u(t) \in V\big(q(t)\big) \quad \forall t \in [0, \tau]$$

and the measure $dr = r'_\mu d\mu = u'_\mu d\mu - g\big(t, q, M(q)u\big)t'_\mu d\mu$ satisfies

$$r'_\mu(t) = 0 \quad \text{if } f\big(q(t)\big) < 0 \text{ or } f\big(q(t)\big) = 0 \text{ and } u(t) \cdot n\big(q(t)\big) > 0, \ \ d\mu \text{ a.e.}$$

Moreover the classical properties of convex analysis combined with the weak-* convergence of measures lead to

$$r'_\mu \in \mathcal{R}\big(q(t)\big) \quad d\mu \text{ a.e.}$$

and, by a global energy estimate, we can extend all these results to the whole interval $[0, T]$.

Finally, in the last step, we prove that the limit satisfies the friction law (23.21) when $f(q(t)) = 0$. We consider separately several cases.

Case 1: $\dot{q}^-(t) \notin V(q(t))$.

Then the right velocity should be determined uniquely by (23.22). The main argument relies on a continuity property for the projection operator involved in relation (23.25) in a neighbourhood of $\dot{q}^-(t)$.

Case 2: $\dot{q}^-(t) \in T(q(t))$.

Then we can distinguish two subcases according to whether the velocity is continuous at time t or not.

Case 2.1: $\dot{q}^-(t) = u(t)$.

The idea is to prove that if $r'_\mu(t) \neq 0$

$$-u(t) \in \mathrm{proj}_{T(q(t)),q(t)} N_{C(q(t))}\big(r'_\mu(t)\big),$$

(the orthogonal projection into $T(q(t))$ being performed with respect to the kinetic metric), which implies (23.21), by using the same "variational" techniques as in (Monteiro Marques, 1993).

Case 2.2: $\dot{q}^-(t) \neq u(t)$.

We observe first that this case can occur only if $M^{-1}(q(t))C(q(t)) \cap T(q(t)) \neq \{0\}$, which corresponds to Painlevé's type cases, see (Painlevé, 1905, Erdmann, 1994). Although the discretization is deterministic, it may produce different solutions for this case, see (Moreau, 1988).

When $\mathrm{span}(D_1(q(t)))$ is of dimension 1, which is the case studied in (Stewart, 1998), we can prove that the limit satisfies Coulomb's friction law. If the dimension of $\mathrm{span}(D_1(q(t)))$ is greater than or equal to 2, the question is still open.

Acknowledgments

This work has been partially supported by the European project SICONOS IST-2001 37172 and M. M. Marques was partially supported by FCT/POCTI/FEDER.

References

Erdmann M. *On a representation of friction in configuration space,* Int. J. Robotics Research, 13(3):240–271, 1994.

Mabrouk M. *A unified variational model for the dynamics of perfect unilateral constraints,* Eur. J. Mechanics A/Solids, 17:819–842, 1998.

Monteiro Marques M.D.P. *Chocs inélastiques standards : un résultat d'existence,* Séminaire d'Analyse Convexe, Univ. Sci. Tech. Languedoc 15, exp. 4, 1985.

Monteiro Marques M.D.P. *Differential inclusions in non-smooth mechanical problems: shocks and dry friction,* Birkhauser, Boston PNLDE 9, 1993.

Moreau J.J. *Liaisons unilatérales sans frottement et chocs inélastiques,* C.R. Acad. Sci. Paris, Série II, 296:1473–1476, 1983.

Moreau J.J. Standard inelastic shocks and the dynamics of unilateral constraints, In *Unilateral problems in structural analysis,* G.Del Piero & F.Maceri eds., pages 173–221, CISM courses and lectures 288, Springer-Verlag, New-York, 1985.

Moreau J.J. Dynamique de systèmes à liaisons unilatérales avec frottement sec éventuel, essais numériques, préprint 85-1, LMGC Montpellier, 1986.

Moreau J.J. Unilateral contact and dry friction in finite freedom dynamics, In *Nonsmooth Mechanics and Applications,* J.J.Moreau & P.D.Panagiotopoulos eds, pages 1–82, CISM courses and lectures 302, Springer-Verlag, New-York, 1988.

Jean M., and Moreau J.J. Unilaterality and dry friction in the dynamics of rigid body collections, In *Proceedings of Contact Mechanics International Symposium,* A.Curnier ed., pages 31–48, PPUR, Lausanne, 1992.

Painlevé P. *Sur les lois du frottement de glissement,* C.R. Acad. Sci. Paris, 121:112–115, 1895 and 141:401–405 and 141:546–552, 1905.

Paoli L. *Time-discretization of vibro-impact,* Phil. Trans. Roy. Soc. London A, 359:2405–2428, 2001.

Schatzman M. *A class of nonlinear differential equations of second order in time,* Nonlinear Anal.,Theory, Methods and Applications, 2:355–373, 1978.

Stewart D. *Convergence of a time-stepping scheme for rigid body dynamics and resolution of Painlevé's paradoxes,* Arch. Rational Mech. Anal., 145:215–260, 1998.

Chapter 24

FINITE TIME STABILIZATION OF NONLINEAR OSCILLATORS SUBJECT TO DRY FRICTION

Samir Adly
Laboratoire LACO, Université de Limoges, 123 avenue Albert Thomas, 87060 Limoges Cedex, France
samir.adly@unilim.fr

Hedy Attouch
Laboratoire ACSIOM, Université Montpellier II, Place Eugène Bataillon, 34095 Montpellier Cedex 05, France
attouch@math.univ-montp2.fr

Alexandre Cabot
Laboratoire LACO, Université de Limoges, 123 avenue Albert Thomas, 87060 Limoges Cedex, France
alexandre.cabot@unilim.fr

Abstract Given a smooth function $f : \mathbb{R}^n \to \mathbb{R}$ and a convex function $\Phi : \mathbb{R}^n \to \mathbb{R}$, we consider the following differential inclusion:
$(S) \qquad \ddot{x}(t) + \partial\Phi(\dot{x}(t)) + \nabla f(x(t)) \ni 0, \qquad t \geq 0,$
where $\partial\Phi$ denotes the subdifferential of Φ. The term $\partial\Phi(\dot{x})$ is strongly related with the notion of friction in unilateral mechanics. The trajectories of (S) are shown to converge toward a stationary solution of (S). Under the additional assumption that $0 \in \operatorname{int} \partial\Phi(0)$ (case of a dry friction), we prove that the limit is achieved in a finite time. This result may have interesting consequences in optimization.

Keywords: Dissipative dynamical system, differential inclusion, dry friction, finite time convergence.

1. Introduction and notation

Throughout the paper, $H = \mathbb{R}^n$ is equipped with the euclidean scalar product $\langle .,. \rangle$ and the corresponding norm $|\,.\,|$. Let $f : \mathbb{R}^n \to \mathbb{R}$ a function of class $\mathcal{C}^1$ that we wish to minimize over $\mathbb{R}^n$. A powerful method consists in following the trajectories of a gradient-based dynamical system. If the dynamics is dissipative, the trajectories will hopefully converge toward a critical point of the potential f. In the past ten years, much attention has been brought to the study of second-order in time dynamical systems. Because of the inertial aspects, the dynamics of such systems do not stop at each minimum point. These methods are endowed with better exploration properties than first-order ones. Among them, let us quote the "Heavy Ball with Friction" system:

$$(HBF) \qquad \ddot{x}(t) + \gamma\,\dot{x}(t) + \nabla f(x(t)) = 0, \qquad t \geq 0, \qquad \gamma > 0,$$

which has been initiated by Polyak (Polyack, 1964). Several authors have recently studied the properties of the (HBF) system (*cf.* Alvarez (Alvarez, 2000), Attouch-Goudou-Redont (Attouch et al., 2000)). The main numerical drawback of the (HBF) method comes from the oscillatory behaviour of the trajectory near the minimum. A first way to improve the (HBF) method consists in introducing a second-order information on f via the hessian matrix. In mechanical terms, this amounts to replacing the viscous friction $-\gamma\,\dot{x}$ by a hessian-driven one (and eventually combining both frictions). This approach has been studied by Attouch-Redont (Attouch and Redont, 2001) and Alvarez-Attouch-Bolte-Redont (Alvarez et al., 2002). Our point of view in this paper is quite different and consists in strengthening the friction when the velocity $\dot{x}$ vanishes. This approach is inspired by mechanical models involving dry friction. A very wide literature is devoted to the contact problems with Coulomb friction (see for example (Amassad and Fabre, 2003, Cadivel et al., 2000, Dumont et al., 2000, Eck and Jarusek, 2000, Han and Sofonea, 2002, Jean, 1988)). Let us recall that, in the one dimensional setting, the classical Coulomb friction is given by $-\alpha\,\mathrm{Sgn}\,(\dot{x})$, where $\alpha > 0$ and Sgn is the set-valued sign function: $\mathrm{Sgn}\,(x) = 1$ if $x > 0$, $\mathrm{Sgn}\,(x) = -1$ if $x < 0$ and $\mathrm{Sgn}\,(0) = [-1, 1]$. The multivalued function Sgn coincides with the subdifferential of the function $\mathbb{R} \ni x \to |x|$. These considerations lead us to study the following differential inclusion:

$$(S) \qquad \ddot{x}(t) + \partial\Phi(\dot{x}(t)) + \nabla f(x(t)) \ni 0, \qquad t \geq 0,$$

where $\partial\Phi$ denotes the subdifferential of the convex function $\Phi : \mathbb{R}^n \to \mathbb{R}$. Notice that in unilateral mechanics, the modelling by differential inclusions has been initiated by Moreau (Moreau, 1977, Moreau, 1985) and has been intensively studied by many authors (Mabrouk, Montei... Marques, 1993, Schatzman, 1978), to quote only some of them. The formulation (S) allows to recover a large variety of friction models. For example, the function $\Phi = \alpha\,|\,.\,|$ corresponds to the Coulomb friction and on the other hand, $\Phi = \gamma\,|\,.\,|^2/2$ gives the viscous friction and the associated (HBF) system. Notice also that the function $\Phi = \beta\,|\,.\,|^p/p$ ($p \in]1,2[$) generates an intermediate situation, which has been studied by Amann-Díaz (Amann and Díaz, 2003) and Díaz-Liñán (Díaz and Liñán, 2001) under the terminology of "strong friction".

The system (S) defined above may have interesting applications in cognitive sciences and decision sciences (mathematical economy, game theory,...). Indeed, the friction corresponds in this case to the cost of changing (cost of leaving a routine, cost of exploration by tests and errors, cost of dissimilarity,...). The viscous friction is associated to a quadratic cost with respect to the distance to the rest position. On the other hand, the dry friction corresponds to a linear cost so that the changes are relatively expensive. This last model may represent the inertial aspects of human behaviours. The introduction of costs of changing is an important characteristic of the theory of bounded rationality, *cf.* (Conlisk, 1996, Rubinstein, 1998, Sobel, 2000).

For any couple of initial conditions, the inclusion (S) is shown to admit a unique solution $x : [0, +\infty[\to \mathbb{R}^n$ of class $\mathcal{C}^1$ satisfying (S) almost everywhere. The behaviour at infinity of the (S) trajectory essentially depends on the nature of the friction term Φ near the origin. When $\partial\Phi(0) = \{0\}$, the gradient $\nabla f(x(t))$ tends to 0 when t tends to $+\infty$; however the convergence of the trajectories may not hold without further assumptions on f (like convexity or analyticity). On the other hand, when the friction is dry, *i.e.* $0 \in \operatorname{int}\partial\Phi(0)$ the convergence of the trajectories automatically holds. The main result of the paper consists in showing that the limit is achieved in a finite time (see Theorem 24.8). Such a limit x^* is a solution of the inclusion: $-\nabla f(x^*) \in \partial\Phi(0)$, *i.e.* x^* is a stationary solution of (S). Since our goal is to minimize f, we must choose a function Φ having a "small" subdifferential set $\partial\Phi(0)$, so as to force the trajectories to converge close to the "exact" critical points. In the various contexts recalled above (mechanics, cognitive sciences,...), the finite time stabilization of the system is meaningful. Such a result in finite dimension opens interesting perspectives concerning similar results for mechanical and physical systems with infinite degrees of freedom.

The paper is organized as follows. In section 2, we state the existence and uniqueness of the solution satisfying the Cauchy problem associated with (S). For the sake of readability, the proof of this result is postponed to section 4. The technique consists in using the Moreau-Yosida approxıates of Φ, then establishing uniform estimates and finally passing to the limit by means of compacity arguments. Section 3 is devoted to the asymptotic study of (S) in the case $0 \in \operatorname{int} \partial\Phi(0)$.

2. The second order differential inclusion (S)

2.1 Mechanical example

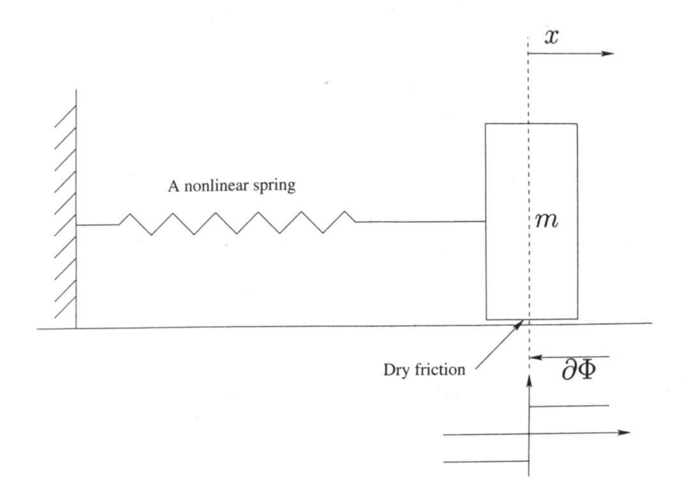

Figure 24.1. A nonlinear oscillator subject to dry friction.

Let us consider the nonlinear oscillator system whose dynamics equation is given by:

$$m\ddot{x} + g(x) \in -\partial\Phi(\dot{x}), \tag{24.1}$$

Equation (24.1) describes the motion of a mass m sliding on a surface and attached to a nonlinear spring. The term $-\partial\Phi(\dot{x})$ represents the dry frictional contact of the mass on the surface and $-g(x)$ corresponds to the force exerted by the spring on the mass. In large displacement operation springs are nonlinear: this is the case for example of the wool felt used for piano hammers which acts as a nonlinear hardening spring. Experimental measurements show that the nonlinear relationship between applied force and felt compression is of the form: $g(x) = k\,x^p \quad (x \geq 0)$, where k is a generalized stiffness coefficient and p is called the effective nonlinearity exponent. Static measurements typically produce values of p in the interval $[2.2, 3.5]$. For more details about the hardness of a piano hammer, we refer to (Russell, 1997). In general, for a nonlinear spring, the stiffness can be broken up into two parts: $g(x) = k_0 x + k_1 |x|^p \operatorname{sgn}(x)$, where k_0 and k_1 are the constants for the linear and nonlinear parts respectively. Here sgn denotes the classical sign function. The function g derives from the potential f given by:

$$f(x) = \frac{k_0}{2}|x|^2 + \frac{k_1}{p+1}|x|^{p+1}.$$

The dry friction is usually modelled by the Coulomb one: $\Phi_0(\dot{x}) = \gamma_0\,|\dot{x}|$, where γ_0 is a positive coefficient. The friction force is then given by $-\partial\Phi_0(\dot{x}) = -\gamma_0 \mathrm{Sgn}(\dot{x})$ where Sgn is the multivalued operator defined by: $\mathrm{Sgn}\,(\dot{x}) = 1$ if $\dot{x} > 0$, $\mathrm{Sgn}\,(\dot{x}) = -1$ if $\dot{x} < 0$ and $\mathrm{Sgn}\,(0) = [-1, 1]$. In many situations, the dry friction is combined with a viscous friction $-\gamma_1\,\dot{x}$, so that the total friction term equals:

$$\Phi(\dot{x}) = \gamma_0\,|\dot{x}| + \frac{\gamma_1}{2}\,|\dot{x}|^2.$$

For other examples of nonlinear oscillators involving dry friction, we refer the reader to (Adly and Goeleven, 2004, Cadivel et al., 2000, Dumont et al., 2000).

2.2 Global existence and uniqueness result

Consider two functions $\Phi : \mathbb{R}^n \to \mathbb{R}$ and $f : \mathbb{R}^n \to \mathbb{R}$ satisfying respectively the following assumptions:
$(\mathcal{H}_\Phi - i)$ Φ is convex;
$(\mathcal{H}_\Phi - ii)$ $\min_{x\in\mathbb{R}^n} \Phi(x) = \Phi(0) = 0$.
$(\mathcal{H}_f - i)$ f is of class $\mathcal{C}^1$ and ∇f is Lipschitz continuous on the bounded subsets of $\mathbb{R}^n$.
$(\mathcal{H}_f - ii)$ f is bounded from below.

Let us consider the problem of finding a continuous function x such that $\dot{x} \in L^\infty_{loc}([0, +\infty[, \mathbb{R}^n)$ and satisfying

$$(S) \qquad \ddot{x}(t) + \partial\Phi(\dot{x}(t)) + \nabla f(x(t)) \ni 0, \qquad t \geq 0.$$

We start with a general result of existence and uniqueness for the associated Cauchy problem.

THEOREM 24.1 (EXISTENCE AND UNIQUENESS) *Assume that hypotheses $(\mathcal{H}_f - i, ii)$ and $(\mathcal{H}_\Phi - i, ii)$ hold. Then, for every $(x_0, \dot{x}_0) \in \mathbb{R}^n \times \mathbb{R}^n$, there exists a unique solution $x : [0, +\infty[\to \mathbb{R}^n$ of (S) in the following sense:*

(a) $x \in \mathcal{C}^1([0, +\infty[, \mathbb{R}^n)$ and $x \in \mathcal{W}^{2,+\infty}([0, T], \mathbb{R}^n)$ for every $T > 0$.

(b) (S) is satisfied for almost every $t \in [0, +\infty[$.

(c) $x(0) = x_0$ and $\dot{x}(0) = \dot{x}_0$.

The proof of the uniqueness in Theorem 24.1 is classical and left to the reader. The problem of the existence is postponed to section 4.

REMARK 24.2 We note that the second-order system (S) can be written as a first-order one in $\mathbb{R}^n \times \mathbb{R}^n$:

$$(S^*) \qquad \dot{X} + \partial\varphi(X) + F(X) \ni 0,$$

with $X(t) = (x(t), \dot{x}(t)) = (x_1, x_2)$, $\varphi(x_1, x_2) = \Phi(x_2)$ and $F(x_1, x_2) = (-x_2, \nabla f(x_1))$. The system (S^*) can be seen as a first-order dynamical system governed by a perturbation of a maximal monotone operator $\partial\varphi$ with a locally Lipschitz operator F. We can not apply general existence and uniqueness results like Kato's Theorem (see Brezis (Brézis, 1972), Kato (Kato, 1970)) here since the perturbation function F is not Lipschitz continuous on the whole space $\mathbb{R}^n \times \mathbb{R}^n$.

3. Finite time convergence under dry friction

3.1 First asymptotical results

Once the global existence and uniqueness is acquired in the study of (S), we wish to investigate the asymptotical properties of (S). The key tool is the existence of a Lyapounov function E emanating from the mechanical interpretation of (S).

PROPOSITION 24.3 *Under the assumptions $(\mathcal{H}_f - i, ii)$ and $(\mathcal{H}_\Phi - i, ii)$, consider the unique solution of the (S) system and define the energy function by $E(t) := \frac{1}{2}|\dot{x}(t)|^2 + f(x(t))$. Then, the following holds:*
(i) For almost every $t \in \mathbb{R}_+$,

$$\dot{E}(t) \leq -\Phi(\dot{x}(t)) \leq 0, \tag{24.2}$$

and hence E is a Lyapounov function for the (S) system.
(ii) $\dot{x} \in L^\infty([0, +\infty[, \mathbb{R}^n)$ and $\Phi(\dot{x}) \in L^1([0, +\infty[, \mathbb{R})$.

Proof. (i) Let D be the subset of $\mathbb{R}_+$ on which the map $\dot{x}$ is derivable and the inclusion (S) is satisfied. Since the function $\dot{x}$ is absolutely continuous and taking into account Theorem 24.1 (b), it is clear that the set $\mathbb{R}_+ \setminus D$ is negligible. Let us prove that inequality (24.2) is satisfied for every $t \in D$. By differentiating the expression of E, we find:

$$\begin{aligned} \forall t \in D, \qquad \dot{E}(t) &= \langle \ddot{x}(t), \dot{x}(t)\rangle + \langle \nabla f(x(t)), \dot{x}(t)\rangle \\ &= -\langle -\ddot{x}(t) - \nabla f(x(t)), \dot{x}(t)\rangle. \end{aligned}$$

From the fact that $-\ddot{x}(t) - \nabla f(x(t)) \in \partial\Phi(\dot{x}(t))$, we deduce that

$$\dot{E}(t) \leq -(\Phi(\dot{x}(t)) - \Phi(0)) = -\Phi(\dot{x}(t))$$

and hence formula (24.2) holds for every $t \in D$.

(ii) In view of (i), the function E is non increasing and hence

$$E(t) = \frac{1}{2}|\dot{x}(t)|^2 + f(x(t)) \leq E(0).$$

We then deduce that $\frac{1}{2}|\dot{x}(t)|^2 \leq E(0) - \inf f$, *i.e.* $\dot{x} \in L^\infty([0, +\infty[, \mathbb{R}^n)$. Now integrate inequality (24.2) between 0 and t; we obtain

$$\int_0^t \Phi(\dot{x}(s))\, ds \leq E(0) - E(t) \leq E(0) - \inf f.$$

Taking the limit as $t \to +\infty$, we obtain $\Phi(\dot{x}) \in L^1([0, +\infty[, \mathbb{R})$. □

From now on, we will assume that the term Φ strictly dissipates the energy when the velocity $|\dot{x}|$ is positive, which amounts to suppose that: $(\mathcal{H}_\Phi - iii)$ $\operatorname{argmin} \Phi = \{0\}$.

An essential step in the asymptotical study of (S) consists in proving that the velocity $\dot{x}$ tends to 0 when $t \to +\infty$.

PROPOSITION 24.4 *Under hypotheses* $(\mathcal{H}_f - i, ii)$ *and* $(\mathcal{H}_\Phi - i, ii, iii)$, *let* x *be the unique solution of the* (S) *system. If* $x \in L^\infty([0, +\infty[, \mathbb{R}^n)$, *then we have* $\lim_{t\to+\infty} \dot{x}(t) = 0$.

Proof. Since ∇f is Lipschitz continuous on the bounded sets, it is clear that

$$t \mapsto \nabla f(x(t)) \quad \text{is bounded.} \tag{24.3}$$

From Proposition 24.3 (ii), the map $\dot{x}$ is bounded. In view of the boundedness of $\partial\Phi$ on bounded sets, this implies the existence of a bounded set $B \subset \mathbb{R}^n$ such that for every $t \in \mathbb{R}_+$,

$$\partial\Phi(\dot{x}(t)) \subset B. \tag{24.4}$$

As a consequence, we deduce from (24.3), (24.4) and (S) that $\ddot{x} \in L^\infty([0, +\infty[, \mathbb{R}^n)$, *i.e.* $\dot{x} \in \mathrm{Lip}([0, +\infty[, \mathbb{R}^n)$. Since the function Φ is Lipschitz continuous on the bounded sets, it is clear that $\Phi(\dot{x}) \in \mathrm{Lip}([0, +\infty[, \mathbb{R})$. This combined with the fact that $\Phi(\dot{x}) \in L^1([0, +\infty[, \mathbb{R})$ classically implies that $\lim_{t\to+\infty} \Phi(\dot{x})(t) = 0$.

Let $\bar{u} \in \mathbb{R}^n$ be a cluster point of the bounded set $\{\dot{x}(t),\ t \geq 0\}$. There exists a sequence (t_n) such that $\lim_{n\to+\infty} \dot{x}(t_n) = \bar{u}$. From the continuity of Φ, we infer that $\Phi(\bar{u}) = 0$, *i.e.* $\bar{u} \in \operatorname{argmin} \Phi = \{0\}$. Consequently, 0 is the unique limit point of $\{\dot{x}(t),\ t \geq 0\}$ and therefore $\lim_{t\to+\infty} \dot{x}(t) = 0$. □

To go further in the asymptotical results relative to (S), we have to specify our assumptions on the function Φ. Indeed, the behaviour of

x at infinity depends on the nature of the friction near the origin. A "strong" friction can force the trajectories of (S) to converge fastly and even in a finite time. These situations of "strong" friction are analyzed in the next section, where the adequate condition on Φ is shown to be $0 \in \operatorname{int} \partial\Phi(0)$.

3.2 Case of the dry friction: $0 \in \operatorname{int} \partial\Phi(0)$

Let us first remark that the condition $0 \in \operatorname{int} \partial\Phi(0)$ amounts to saying that the function Φ is minorized by $\alpha\,|\,.\,|$, for some $\alpha > 0$. Indeed, we have:

LEMMA 24.5 *Let $\Phi : \mathbb{R}^n \to \mathbb{R}$ a convex function satisfying $\Phi(0) = 0$. Then the following assertions are equivalent:*
(i) $\Phi \geq \alpha\,|\,.\,|$, for some $\alpha > 0$.
(ii) $0 \in \operatorname{int} \partial\Phi(0)$.

The proof of Lemma 24.5 is elementary and left to the reader. If $0 \in \operatorname{int} \partial\Phi(0)$, it is clear, in view of the previous lemma that assumption $(\mathcal{H}_\Phi - iii)$ is automatically satisfied. The following result shows the convergence of the (S) trajectories under the condition $0 \in \operatorname{int} \partial\Phi(0)$.

THEOREM 24.6 *Under the hypotheses $(\mathcal{H}_\Phi - i, ii)$ and $(\mathcal{H}_f - i, ii)$, assume moreover that $0 \in \operatorname{int} \partial\Phi(0)$. Then the unique solution x of (S) satisfies the following assertions:*
(i) $|\dot{x}| \in L^1([0, +\infty[, \mathbb{R})$ and therefore $x_\infty := \lim_{t\to+\infty} x(t)$ exists.
(ii) The limit point x_∞ verifies

$$-\nabla f(x_\infty) \in \partial\Phi(0), \tag{24.5}$$

i.e. x_∞ is an equilibrium point of (S).

Proof. (i) From Lemma 24.5, the condition $0 \in \operatorname{int} \partial\Phi(0)$ implies the existence of $\alpha > 0$ such that $\Phi \geq \alpha\,|\,.\,|$. On the other hand, from Proposition 24.3 (ii), we have $\Phi(\dot{x}) \in L^1([0, +\infty[, \mathbb{R})$ and it follows that $|\dot{x}| \in L^1([0, +\infty[, \mathbb{R})$. The convergence of the trajectory $x(.)$ immediately results from the equality $x(t) = x(0) + \int_0^t \dot{x}(s)\, ds$.

(ii) Let us argue by contradiction and assume that the set $C := \partial\Phi(0) + \nabla f(x_\infty)$ does not contain 0. It is then possible to strictly separate the convex compact set $\{0\}$ from the closed convex set C. More precisely, there exist $p \in \mathbb{R}^n$ and $m \in \mathbb{R}_+^*$ such that, for every $x \in C$, $\langle x, p\rangle > m$, which amounts to say that C is contained in the open half-space $\mathcal{H}_{p,m}$ defined by

$$\mathcal{H}_{p,m} := \{x \in \mathbb{R}^n, \quad \langle x, p\rangle > m\}.$$

Let us prove that for t large enough, we have

$$\partial\Phi(\dot{x}(t)) + \nabla f(x(t)) \subset \mathcal{H}_{p,m}. \tag{24.6}$$

If this was not true, there would exist a sequence (t_n) tending to $+\infty$ such that $\partial\Phi(\dot{x}(t_n)) + \nabla f(x(t_n)) \not\subset \mathcal{H}_{p,m}$. This means that, for every $n \in \mathbb{N}$, there exists $u_n \in \partial\Phi(\dot{x}(t_n))$ such that

$$\langle u_n + \nabla f(x(t_n)), p\rangle \leq m. \tag{24.7}$$

Since the operator $\partial\Phi$ is bounded on the bounded sets, the sequence (u_n) is bounded. Therefore, there exists $\bar{u} \in \mathbb{R}^n$ and a converging subsequence of (u_n), still denoted by (u_n) such that $\lim_{n\to+\infty} u_n = \bar{u}$. On the other hand, we have $\lim_{n\to+\infty} \dot{x}(t_n) = 0$ and using the graph-closedness property of the operator $\partial\Phi$ in $\mathbb{R}^n \times \mathbb{R}^n$, we conclude that $\bar{u} \in \partial\Phi(0)$. Let us now take the limit in (24.7) when $n \to +\infty$:

$$\langle \bar{u} + \nabla f(x_\infty), p\rangle \leq m,$$

i.e. $\bar{u} + \nabla f(x_\infty) \notin \mathcal{H}_{p,m}$. Since $\bar{u} \in \partial\Phi(0)$, we deduce that $C = \partial\Phi(0) + \nabla f(x_\infty) \not\subset \mathcal{H}_{p,m}$, a contradiction. Hence, we have proved that there exists $t_0 \geq 0$ such that the inclusion (24.6) holds for $t \geq t_0$. Coming back to the (S) system, we infer that, for almost every $t \geq t_0$, $\langle -\ddot{x}(t), p\rangle > m$. Integrating this inequality between t_0 and t immediately yields

$$\langle -\dot{x}(t), p\rangle > \langle -\dot{x}(t_0), p\rangle + m\,(t - t_0)$$

and we deduce that $\lim_{t\to+\infty}\langle \dot{x}(t), p\rangle = -\infty$, a contradiction with the fact that $\dot{x} \in L^\infty([0, +\infty[, \mathbb{R}^n)$. As a consequence, the initial assumption $0 \notin \partial\Phi(0) + \nabla f(x_\infty)$ is false, which ends the proof. □

REMARK 24.7 We notice that in the case of a dry friction, the proof of the convergence of the trajectories $x(.)$ in Theorem 24.6 is elementary. This contrasts with situations involving viscous friction, where the velocity $\dot{x}$ is in L^2 but not in L^1 in general. In the case of a linear damped dynamics, convergence may not hold without further assumptions on f like convexity or analyticity (see for example (Alvarez, 2000, Attouch et al., 2000)). However the convergence of the (S) trajectories under dry friction has a counterpart: in view of (24.5), the limit is just an "approximate" critical point of f. Since our goal is to minimize the function f, we will have to choose a function Φ whose subdifferential set $\partial\Phi(0)$ is "relatively small".

The most remarkable property of the dry friction is the convergence in finite time of the (S) trajectories. The following statement is quite general and there are no further assumptions on the potential f.

THEOREM 24.8 (FINITE TIME CONVERGENCE) *Under the hypotheses of Theorem 24.6, let x be the unique solution of (S) and let $x_\infty \in \mathbb{R}^n$ be defined by $x_\infty := \lim_{t\to+\infty} x(t)$. If $-\nabla f(x_\infty) \notin \mathrm{bd}\,(\partial\Phi(0))$, then there exists $t_1 \geq 0$ such that $x(t) = x_\infty$ for every $t \geq t_1$.*

Proof. From Theorem 24.6, the limit point x_∞ necessarily fulfils $-\nabla f(x_\infty) \in \partial\Phi(0)$, so that the assumption $-\nabla f(x_\infty) \notin \mathrm{bd}\,(\partial\Phi(0))$ is equivalent to $-\nabla f(x_\infty) \in \mathrm{int}\,(\partial\Phi(0))$. This implies the existence of $\varepsilon > 0$ such that

$$-\nabla f(x_\infty) + \mathbb{B}(0, 2\,\varepsilon) \subset \partial\Phi(0).$$

On the other hand, since $\lim_{t\to+\infty} \nabla f(x(t)) = \nabla f(x_\infty)$, there exists $t_0 \geq 0$ such that for every $t \geq t_0$, we have

$$\nabla f(x(t)) \in \nabla f(x_\infty) + \mathbb{B}(0, \varepsilon).$$

Hence,

$$-\nabla f(x(t)) + \mathbb{B}(0, \varepsilon) \subset -\nabla f(x_\infty) + \mathbb{B}(0, 2\,\varepsilon) \subset \partial\Phi(0).$$

This means that, for every $t \geq t_0$ and for every $u \in \mathbb{B}(0, 1)$, we have:

$$-\nabla f(x(t)) + \varepsilon\, u \in \partial\Phi(0).$$

Thus, for every $t \geq t_0$, we deduce

$$\forall u \in \mathbb{B}(0,1), \quad \Phi(\dot{x}(t)) \geq \langle -\nabla f(x(t)) + \varepsilon\, u, \dot{x}(t)\rangle.$$

Taking the supremum over $u \in \mathbb{B}(0, 1)$, we obtain for every $t \geq t_0$,

$$\Phi(\dot{x}(t)) \geq \langle -\nabla f(x(t)), \dot{x}(t)\rangle + \varepsilon\, |\dot{x}(t)|. \tag{24.8}$$

On the other hand, the inequality (24.2) of energy decay can be rewritten as:

$$\frac{1}{2}\frac{d}{dt}|\dot{x}(t)|^2 + \langle \nabla f(x(t)), \dot{x}(t)\rangle + \Phi(\dot{x}(t)) \leq 0 \qquad \text{a.e. on } [0, +\infty[. \tag{24.9}$$

By combining (24.8) and (24.9), we get

$$\frac{1}{2}\frac{d}{dt}|\dot{x}(t)|^2 + \varepsilon\, |\dot{x}(t)| \leq 0. \tag{24.10}$$

By setting $h(t) := |\dot{x}(t)|^2$, it is clear that relation (24.10) can be rewritten as the following differential inequality:

$$\dot{h}(t) + 2\,\varepsilon\,\sqrt{h(t)} \leq 0 \qquad \text{a.e. on } [0, +\infty[. \tag{24.11}$$

Let us prove that there exists $t_1 \geq t_0$ such that $h(t_1) = 0$. Assume, on the contrary that for every $t \geq t_0$, $h(t) > 0$. Dividing (24.11) by $\sqrt{h(t)}$ and integrating on $[t_0, t]$, we obtain

$$\sqrt{h(t)} - \sqrt{h(t_0)} \leq -\varepsilon\,(t - t_0).$$

Letting $t \to +\infty$, the previous inequality leads to $\lim_{t\to+\infty} \sqrt{h(t)} = -\infty$, a contradiction, whence the existence of $t_1 \geq t_0$ satisfying $h(t_1) = 0$.

From (24.11), we deduce that $\dot{h}(t) \leq 0$ almost everywhere and hence $h(t) \leq h(t_1) = 0$, for every $t \geq t_1$. We conclude that $|\dot{x}(t)| = 0$ for every $t \in [t_1, +\infty[$, *i.e.* $x(t) = x_\infty$ for every $t \in [t_1, +\infty[$. □

We may notice that the conclusions of Theorem 24.8 hold under the key assumption $-\nabla f(x_\infty) \notin \mathrm{bd}\,(\partial\Phi(0))$. Since the boundary of the convex set $\partial\Phi(0)$ has an empty interior, it is reasonable to think that the circumstances leading to $-\nabla f(x_\infty) \in \mathrm{bd}\,(\partial\Phi(0))$ are "exceptional". More precisely, we conjecture that generically with respect to the initial data $(x_0, \dot{x}_0) \in \mathbb{R}^n \times \mathbb{R}^n$, the point $x_\infty = \lim_{t\to+\infty} x(t)$ satisfies the condition $-\nabla f(x_\infty) \notin \mathrm{bd}\,(\partial\Phi(0))$.

Let us now give a counterexample to finite time convergence when $-\nabla f(x_\infty) \in \mathrm{bd}\,(\partial\Phi(0))$. For that purpose, take $H = \mathbb{R}$, $\Phi := |\,.\,| + |\,.\,|^2$ (so that $\partial\Phi(0) = [-1, 1]$) and $f := |\,.\,|^2/2$. The differential inclusion (S) then reduces to

$$(S) \qquad \ddot{x}(t) + \mathrm{Sgn}\,(\dot{x}(t)) + 2\,\dot{x}(t) + x(t) \ni 0.$$

Let us choose as initial conditions $x(0) = 2$ and $\dot{x}(0) = -1$. We let the reader check that the unique solution of (S) is given by $x(t) = 1 + e^{-t}$, $t \geq 0$. The trajectory tends toward the value $x_\infty = 1$, which satisfies $-f'(x_\infty) = -1 \in \mathrm{bd}\,(\partial\Phi(0))$. However the convergence does not hold in a finite time.

4. Proof of the existence result

We prove the existence of a function $x : [0, +\infty[\to \mathbb{R}^n$ satisfying points (a), (b) and (c) of Theorem 24.1. For that purpose, let us define for any positive λ, the approximate equation

$$(S_\lambda) \qquad \ddot{x}_\lambda(t) + \nabla\Phi_\lambda(\dot{x}_\lambda(t)) + \nabla f(x_\lambda(t)) = 0,$$

where Φ_λ denotes the Moreau-Yosida approximate of Φ. The general features relative to the Moreau-Yosida approximation can be found in Brezis (Brézis, 1972) or in Rockafellar-Wets (Rockafellar and Wets, 1998). Let us recall that, for any $\lambda > 0$, Φ_λ is a $\mathcal{C}^1$ function from $\mathbb{R}^n$ into $\mathbb{R}$, whose gradient $\nabla\Phi_\lambda$ is Lipschitz continuous. Equation (S_λ) falls into the field of ordinary differential equations. We let the reader check the following lemma.

LEMMA 24.9 *Assume* $(\mathcal{H}_f - i, ii)$ *and* $(\mathcal{H}_\Phi - i, ii)$. *Then, for every* $(x_0, \dot{x}_0) \in \mathbb{R}^n \times \mathbb{R}^n$, *there exists a unique maximal solution* $x_\lambda : [0, +\infty[\to \mathbb{R}^n$ *of* (S_λ) *satisfying* $(x_\lambda(0), \dot{x}_\lambda(0)) = (x_0, \dot{x}_0)$. *Moreover, setting* $E_\lambda(t) := \frac{1}{2} |\dot{x}_\lambda(t)|^2 + f(x_\lambda(t))$, *we have for every* $t \in \mathbb{R}_+$,

$$\dot{E}_\lambda(t) \leq -\Phi_\lambda(\dot{x}_\lambda(t)). \tag{24.12}$$

The proof of the local existence and uniqueness of x_λ relies on the Cauchy-Lipschitz Theorem. Denoting by $[0, T_{\max}[$ the maximal interval on which x_λ is defined, we use an estimate of $\dot{x}_\lambda$ in $L^\infty([0, T_{\max}[, \mathbb{R}^n)$ to prove that $T_{\max} = +\infty$. The decay property (24.12) is trivial.

Coming back to the existence problem in Theorem 24.1, we are going to establish uniform estimates relying on the solutions of (S_λ). Then arguing by compacity, we pass to the limit and exhibit a function x which is proved to fulfil (a), (b) and (c).

Estimations. Since $\min_{\mathbb{R}^n} \Phi_\lambda = \min_{\mathbb{R}^n} \Phi = 0$, we deduce from (24.12) that $\dot{E}_\lambda \leq 0$, *i.e.* E_λ is a non increasing function. Therefore, we have

$$\frac{1}{2} |\dot{x}_\lambda(t)|^2 + f(x_\lambda(t)) = E_\lambda(t) \leq E_\lambda(0) = \frac{1}{2} |\dot{x}_0|^2 + f(x_0)$$

and since f is bounded from below, we immediately infer that

$$(\dot{x}_\lambda) \quad \text{is bounded in } L^\infty([0, +\infty[, \mathbb{R}^n). \tag{24.13}$$

Let us now fix some $T > 0$. From the formula $x_\lambda(t) = x_0 + \int_0^t \dot{x}_\lambda(s)\, ds$, we deduce that

$$\sup_{\substack{\lambda > 0 \\ t \in [0,T]}} |x_\lambda(t)| \leq |x_0| + T \sup_{\lambda > 0} ||\dot{x}_\lambda||_\infty$$

and hence

$$(x_\lambda) \quad \text{is bounded in } L^\infty([0, T], \mathbb{R}^n). \tag{24.14}$$

From the last estimate (24.14), the boundedness of ∇f on the bounded sets implies that

$$(\nabla f(x_\lambda)) \quad \text{is bounded in } L^\infty([0, T], \mathbb{R}^n). \tag{24.15}$$

On the other hand, denoting by $\partial \Phi^0(y)$ the element of minimal norm, it is well-known that $|\nabla \Phi_\lambda(y)| \leq |\partial \Phi^0(y)|$ for every $y \in \mathbb{R}^n$. As a consequence, we have $|\nabla \Phi_\lambda(\dot{x}_\lambda)| \leq |\partial \Phi^0(\dot{x}_\lambda)|$, which combined with estimate (24.13) and the boundedness of $\partial \Phi$ on bounded sets, gives

$$(\nabla \Phi_\lambda(\dot{x}_\lambda)) \quad \text{is bounded in } L^\infty([0, +\infty[, \mathbb{R}^n). \tag{24.16}$$

By taking into account (24.15) and (24.16), we deduce in view of (S_λ) that

$$(\ddot{x}_\lambda) \quad \text{is bounded in } L^\infty([0,T],\mathbb{R}^n). \tag{24.17}$$

Passing to the limit. From the inequality $|x_\lambda(t') - x_\lambda(t)| \leq (\sup_{\lambda>0} ||\dot{x}_\lambda||_\infty)\, |t'-t|$, it ensues that (x_λ) is an equicontinuous bounded sequence in $\mathcal{C}([0,T],\mathbb{R}^n)$ equipped with the supremum norm, and therefore Ascoli Theorem shows the existence of a cluster point $x \in \mathcal{C}([0,T],\mathbb{R}^n)$ to the sequence (x_λ). Then, there exists a subsequence of (x_λ), still denoted by (x_λ) such that

$$x_\lambda \to x \qquad \text{in } \mathcal{C}([0,T],\mathbb{R}^n).$$

In view of estimate (24.17), the same argument applied to the sequence $(\dot{x}_\lambda)$ shows that there exist $u \in \mathbb{R}^n$ and a subsequence of $(\dot{x}_\lambda)$, still denoted by $(\dot{x}_\lambda)$ such that

$$\dot{x}_\lambda \to u \qquad \text{in } \mathcal{C}([0,T],\mathbb{R}^n).$$

Hence we have $\dot{x}_\lambda \to \dot{x}$ and $\dot{x}_\lambda \to u$ in the sense of distributions in $]0,T[$. Identifying both limits, we infer that $\dot{x} \in \mathcal{C}([0,T],\mathbb{R}^n)$, *i.e.* $x \in \mathcal{C}^1([0,T],\mathbb{R}^n)$. From (24.17), the sequence $(\ddot{x}_\lambda)$ is bounded in the space $L^\infty([0,T],\mathbb{R}^n)$, which can be identified with the topological dual of $L^1([0,T],\mathbb{R}^n)$. The Banach-Alaoglu Theorem then shows the existence of $v \in L^\infty([0,T],\mathbb{R}^n)$ and a subsequence of $(\ddot{x}_\lambda)$, still denoted by $(\ddot{x}_\lambda)$ such that

$$\ddot{x}_\lambda \to v \qquad \text{for the topology } \sigma(L^\infty([0,T],\mathbb{R}^n), L^1([0,T],\mathbb{R}^n)).$$

Hence, we have $\ddot{x}_\lambda \to \ddot{x}$ and $\ddot{x}_\lambda \to v$ in the sense of distributions in $]0,T[$. The identification of the limits shows that $\ddot{x} \in L^\infty([0,T],\mathbb{R}^n)$ or equivalently $x \in \mathcal{W}^{2,\infty}([0,T],\mathbb{R}^n)$.

Let us now prove that the map x satisfies (S) almost everywhere on $[0,T]$. Fix $\theta \geq 0$ in $\mathcal{C}_c(]0,T[)$ (the set of continuous functions with compact support included in $]0,T[$). Since Φ_λ is convex, the following inequality holds for every $t \geq 0$:

$$\begin{aligned}\forall \xi \in \mathbb{R}^n, \qquad \Phi_\lambda(\xi) &\geq \Phi_\lambda(\dot{x}_\lambda(t)) + \langle \nabla\Phi_\lambda(\dot{x}_\lambda(t)), \xi - \dot{x}_\lambda(t)\rangle \\ &\geq \Phi_\lambda(\dot{x}_\lambda(t)) + \langle -\ddot{x}_\lambda(t) - \nabla f(x_\lambda(t)), \xi - \dot{x}_\lambda(t)\rangle.\end{aligned}$$

Multiplying both members by θ and integrating on $[0,T]$, we obtain:

$$\int_0^T \theta(t)\,\Phi_\lambda(\xi)\,dt \geq \int_0^T \theta(t)\,\Phi_\lambda(\dot{x}_\lambda(t))\,dt + \int_0^T \theta(t)\,\langle -\ddot{x}_\lambda(t) - \nabla f(x_\lambda(t)), \xi - \dot{x}_\lambda(t)\rangle\,dt.$$

Fix some $\lambda_0 > 0$ and consider $\lambda \in]0, \lambda_0[$. From the monotonicity of the Moreau-Yosida approximation, we have $\Phi_{\lambda_0} \leq \Phi_\lambda \leq \Phi$ so that the last inequality implies

$$\int_0^T \theta(t)\, \Phi(\xi)\, dt \geq \int_0^T \theta(t)\, \Phi_{\lambda_0}(\dot{x}_\lambda(t))\, dt + \int_0^T \theta(t)\, \langle -\ddot{x}_\lambda(t) - \nabla f(x_\lambda(t)), \xi - \dot{x}_\lambda(t) \rangle\, dt. \quad (24.18)$$

Take now the limit when $\lambda \to 0$ in the previous inequality. Since Φ_{λ_0} is Lipschitz continuous on the bounded sets and since $(\dot{x}_\lambda)$ converges uniformly toward $\dot{x}$, we deduce that $\Phi_{\lambda_0}(\dot{x}_\lambda) \to \Phi_{\lambda_0}(\dot{x})$ uniformly in $\mathcal{C}([0,T], \mathbb{R}^n)$, whence

$$\lim_{\lambda \to 0} \int_0^T \theta(t)\, \Phi_{\lambda_0}(\dot{x}_\lambda(t))\, dt = \int_0^T \theta(t)\, \Phi_{\lambda_0}(\dot{x}(t))\, dt. \quad (24.19)$$

Similarly, the uniform convergence of (x_λ) toward x joined with the Lipschitz continuity of ∇f on the bounded sets shows that $\nabla f(x_\lambda) \to \nabla f(x)$ uniformly in $\mathcal{C}([0,T], \mathbb{R}^n)$. In view of the weak convergence of $(\ddot{x}_\lambda)$ toward $\ddot{x}$ in the $\sigma(L^\infty, L^1)$ sense, this implies that

$$\lim_{\lambda \to 0} \int_0^T \theta(t)\, \langle -\ddot{x}_\lambda(t) - \nabla f(x_\lambda(t)), \xi - \dot{x}_\lambda(t) \rangle\, dt = \int_0^T \theta(t)\, \langle -\ddot{x}(t) - \nabla f(x(t)), \xi - \dot{x}(t) \rangle\, dt. \quad (24.20)$$

Let us take the limit when $\lambda \to 0$ in inequality (24.18) by taking into account (24.19) and (24.20):

$$\int_0^T \theta(t)\, \Phi(\xi)\, dt \geq \int_0^T \theta(t)\, \Phi_{\lambda_0}(\dot{x}(t))\, dt + \int_0^T \theta(t)\, \langle -\ddot{x}(t) - \nabla f(x(t)), \xi - \dot{x}(t) \rangle\, dt. \quad (24.21)$$

Taking the limit when $\lambda_0 \to 0$ in the previous inequality and applying the Beppo-Levi Theorem, it is immediate that (24.21) holds with Φ in place of Φ_{λ_0}. The latter being true for all $\theta \geq 0$ in $\mathcal{C}_c(]0,T[)$, it follows that for every $\xi \in \mathbb{R}^n$,

$$\Phi(\xi) \geq \Phi(\dot{x}(t)) + \langle -\ddot{x}(t) - \nabla f(x(t)), \xi - \dot{x}(t) \rangle \qquad \text{ae in } [0,T].$$

From the definition of the subdifferential, this is equivalent to:

$$-\ddot{x}(t) - \nabla f(x(t)) \in \partial \Phi(\dot{x}(t)) \qquad \text{ae in } [0,T],$$

which means that inclusion (S) is satisfied almost everywhere in $[0, T]$.

Let us now summarize our results. For every $T > 0$, we have proved the existence of a function $x \in \mathcal{C}^1([0,T],\mathbb{R}^n) \cap \mathcal{W}^{2,\infty}([0,T],\mathbb{R}^n)$ such that (S) is fulfilled for almost every $t \in [0,T]$. Since $(x_\lambda(0), \dot{x}_\lambda(0)) = (x_0, \dot{x}_0)$ for every $\lambda > 0$, it is immediate that the limit function x also satisfies $(x(0), \dot{x}(0)) = (x_0, \dot{x}_0)$. To avoid confusion, we now denote by x^T the previous function defined on $[0,T]$. For every $T' > T$, we obtain in the same way a function $x^{T'}$ defined on $[0,T']$. From the uniqueness result, it is clear that $x^{T'}_{|[0,T]} = x^T$, so that we can define without any ambiguity a function $x : [0,+\infty[$ in the following manner:

$$\forall t \in [0,+\infty[, \qquad x(t) = x^T(t) \qquad \text{as soon as } T \geq t.$$

It is then immediate that such a function x satisfies items (a), (b) and (c) of Theorem 24.1. □

References

S. Adly, D. Goeleven, A stability theory for second order nonsmooth dynamical systems with application to friction problems, *Journal de Mathématiques Pures et Appliquées*, vol. 83, (2004), 17-51.

F. Alvarez, On the minimizing property of a second order dissipative system in Hilbert space, *SIAM J. on Control and Optimization*, vol. 38, 4 (2000), 1102-1119.

F. Alvarez, H. Attouch, J. Bolte, P. Redont, A second-order gradient-like dissipative dynamical system with Hessian-driven damping. Application to optimization and mechanics, *J. Math. Pures Appl.*, vol. 81, 8 (2002), 747-779.

H. Amann, J. I. Díaz, A note on the dynamics of an oscillator in the presence of strong friction, *Nonlinear Anal.*, vol. 55 (2003), 209-216.

A. Amassad, C. Fabre, Analysis of a viscoelastic unilateral contact problem involving the Coulomb friction law, *J. Optim. Theory Appl.*, vol. 116, 3 (2003), 465-483.

H. Attouch, X. Goudou, P. Redont, The heavy ball with friction method. I The continuous dynamical system, *Communications in Contemporary Math*, vol. 2, 1 (2000), 1-34.

H. Attouch, P. Redont, The second-order in time continuous Newton method. Approximation, optimization and mathematical economics (Pointe--Pitre, 1999), 25-36, Physica, Heidelberg, 2001.

H. Brézis, Opérateurs maximaux monotones dans les espaces de Hilbert et équations d'évolution, Lecture Notes 5, North Holland, 1972.

M. Cadivel, D. Goeleven and M. Shillor, Study of a unilateral oscillator with friction, *Math. Comput. Modelling*, vol. 32, 2/3 (2000), 381-391.

J. Conlisk, Why bounded rationality, *Journal of Economic Literature*, vol. 34 (1996), 669-700.

J. I. Díaz, A. Liñán, On the asymptotic behavior of a damped oscillator under a sublinear friction term, *Rev. R. Acad. Cien. Serie A. Mat.*, vol. 95, 1 (2001), 155-160.

Y. Dumont, D. Goeleven, M. Rochdi and M. Shillor, Frictional contact of a nonlinear spring, *Math. Comput. Modelling*, vol. 31, 2/3 (2000), 83-97.

C. Eck, J. Jarusek, A survey on dynamic contact problems with Coulomb friction, Multifield problems, (2000), 254-261, Springer, Berlin.

W. Han, M. Sofonea, Quasistatic contact problems in viscoelasticity and viscoplasticity, AMS and International Press, 2002.

M. Jean, Unilateral contact with dry friction: time and space discrete variables formulation, *Arch. Mech.*, vol. 40, 5/6 (1988), 677–691.

T. Kato, Accretive operators and nonlinear evolution equations in Banach spaces. Nonlinear functional analysis, *Proc. Symp. Pure Math.*, vol. 18, 1 (1970), 138-161.

M. Mabrouk, Sur un principe variationnel pour un problème d'évolution hyperbolique non linéaire, working paper, Laboratoire de Mécanique Appliquée, Université de Franche-Comté, Besançon.

M. D. P. Monteiro Marques, Differential inclusions in nonsmooth mechanical problems, Progress in nonlinear differential equations and their applications, vol. 9, Birkhauser, 1993.

J. J. Moreau, Evolution problem associated with a moving convex set in a Hilbert space, *J. Diff. Equ.*, (1977), 347-374.

J. J. Moreau, Standard inelastic shocks and the dynamics of unilateral constraints. In: Unilateral Problems in Structural Analysis (ed. by G. Del Piero and F. Maceri), CISM Courses and Lectures, vol. 288, Springer Verlag, Wien, New York 1985, 173-221.

B.T. Polyack, Some methods of speeding up the convergence of iterative methods, *Z. Vylist Math. Fiz.*, vol. 4, (1964), 1-17.

R.T. Rockafellar and R. Wets, Variational analysis, Springer, Berlin, 1998.

A. Rubinstein, Modeling bounded rationality, (1998), MIT Press.

D. A. Russell, The piano hammer as a nonlinear spring, publications of the Science and Mathematics Department, GMI Engineering & Management Institute, Kettering University, 1997.

M. Schatzman, A class of nonlinear differential equations of second order in time, *Nonlinear Analysis*, vol. 2, (1978), 355-373.

J. Sobel, Economists' models of learning, *Journal of Economics Theory*, vol. 94, (2000), 241-261.

Chapter 25

CANONICAL DUALITY IN NONSMOOTH, CONSTRAINED CONCAVE MINIMIZATION

David Y. Gao

Department of Mathematics, Virginia Polytechnic Institute & State University, Blacksburg, VA 24061, USA

gao@vt.edu

Abstract This paper presents a perfect duality theory for solving nonsmooth, concave minimization problems subjected to inequality constraints. By use of the *canonical dual transformation* developed recently, a canonical dual problem is formulated, which is perfectly dual to the primal problem with zero duality gap. It is shown that the global minimizer and local extrema of the nonconvex problem can be identified by the triality theory discovered recently (Gao, 2000). This canonical dual form and the triality theory can be used to develop certain powerful algorithms for solving nonsmooth concave minimization problems.

Keywords: duality theory, concave minimization, canonical dual transformation, quadratic programming.

Dedicated to Professor Jean Jacques Moreau for his 80th birthday

1. Primal Problem and Parametrization

The nonsmooth, concave minimization problem to be discussed in this paper is denoted as the primal problem (($\mathcal{P}$) in short)

$$(\mathcal{P}): \quad \min P(\mathbf{x}) \quad \forall \mathbf{x} \in \mathcal{X}_f, \tag{25.1}$$

where $P(\mathbf{x})$ is a real-valued, nonsmooth, concave function defined on a suitable convex set $\mathcal{X}_a \subset \mathbb{R}^n$, and $\mathcal{X}_f \subset \mathbb{R}^n$ is the feasible space, defined by

$$\mathcal{X}_f = \{\mathbf{x} \in \mathcal{X}_a \subset \mathbb{R}^n | \ B\mathbf{x} \leq \mathbf{b}\}, \tag{25.2}$$

in which, $B \in \mathbb{R}^{m\times n}$ is given matrix such that rank$B = \min\{m, n\}$, and $\mathbf{b} \in \mathbb{R}^m$ is a given vector. The goal in this problem is to find the (global) minimum value that P can achieve in the feasible space and, if this value is not $-\infty$, to find, if it exists, at least one vector $\bar{\mathbf{x}} \in \mathcal{X}_f$ that achieves this value. The concave minimization problem $(\mathcal{P})$ appears in many applications. Actually, even if $P(\mathbf{x})$ is a quadratic function, the problem $(\mathcal{P})$ is NP-hard (see, Horst *et al*, 2000; Floudas and Visweswaran, 1995). Methods and solutions to this very difficult problem are fundamentally important in both mathematics and engineering science.

Mathematically speaking, if the nonconvex function P is continuous on its domain $\mathcal{X}_a$ and the feasible set $\mathcal{X}_f$ is compact, then by the well-known Weierstrass Theorem, the global minimum value is finite, and at least one point in $\mathcal{X}_f$ exists which attains this value. From the point of view of convex analysis, if the convex subset $\mathcal{X}_f \subset \mathbb{R}^n$ is compact, it must be bounded and closed, i.e. the radius r_0 of $\mathcal{X}_f$, defined by $|\mathbf{x}| \leq r_0 \quad \forall \mathbf{x} \in \mathcal{X}_f$, is finite, thus, the problem $(\mathcal{P})$ has at least one solution. However, due to the concavity of the total cost $P(\mathbf{x})$, the problem $(\mathcal{P})$ generally will possess many solutions that are local, but not global, minimizers. The application of standard algorithms designed for solving convex programming problems will generally fail to solve this multi-extremal nonconvex global minimization problem. A detailed discussion and comprehensive review on this topic are given by Benson (1995).

Physically speaking, the primal problem $(\mathcal{P})$ is called *realizable* if there exists a vector $\bar{\mathbf{x}} \in \mathcal{X}_f$ such that $P(\bar{\mathbf{x}}) = \min\{P(\mathbf{x})| \ \forall \mathbf{x} \in \mathcal{X}_f\} > -\infty$. Thus, if $(\mathcal{P})$ is realizable, the total cost P must be bounded below on $\mathcal{X}_f$, i.e. there exists a parameter $\mu > -\infty$ such that $P(\mathbf{x}) \geq \mu \quad \forall \mathbf{x} \in \mathcal{X}_f$. From point view of applications, if a problem is not realizable, then the mathematical modeling of this problem might be not well proposed. By this philosophy, we consider only the realizable primal problem $(\mathcal{P})$ so that the primal problem can be written in the parametrization form $(\mathcal{P}_\mu)$

$$(\mathcal{P}_\mu): \qquad \min_{\mathbf{x}\in\mathbb{R}^n} P(\mathbf{x}) \tag{25.3}$$

$$\text{s.t.} \quad B\mathbf{x} \leq \mathbf{b}, \quad P(\mathbf{x}) \geq \mu. \tag{25.4}$$

LEMMA 25.1 (PARAMETRIZATION) *Suppose that $P : \mathcal{X}_a \to \mathbb{R}$ is a concave function. For a given parameter $\mu \in \mathbb{R}$, the parametrized problem $(\mathcal{P}_\mu)$ has at least one global minimizer $\bar{\mathbf{x}}$ in the parametrical feasible set $\mathcal{X}_\mu \subset \mathbb{R}^n$ defined by*

$$\mathcal{X}_\mu = \{\mathbf{x} \in \mathbb{R}^n | \ B\mathbf{x} \leq \mathbf{b}, \ P(\mathbf{x}) \geq \mu\}. \tag{25.5}$$

Moreover, if the problem $(\mathcal{P})$ is realizable, then there exists a constant $\mu > -\infty$ such that the solution of the parametrized problem $(\mathcal{P}_\mu)$ solves also the primal problem $(\mathcal{P})$.

The main purpose of this paper is to find a canonical dual formulation of the parametric problem $(\mathcal{P}_\mu)$ and extremality conditions for local and global extrema. Duality theory and methods in convex systems have been well established (cf. Moreau, 1966; Ekeland and Temam, 1976). However, many problems are still open in nonconvex systems. It is known that due to the nonconvexity, the well developed Fenchel-Moreau-Rockafellar duality theory usually leads to a duality gap. It turns out that the perfect duality theory and methods have been subjected to an extensively study during the recent years (see, Ekeland, 2003; Goh and Yang, 2002; Robinov and Yang, 2003). Based on the complementary variational principle developed by Gao and Strang in nonconvex mechanics (Gao and Strang, 1989), a so-called canonical dual transformation method and the associated triality theory have been established recently (see Gao, 2000). This method and theory play an important role in global optimization (see Gao, 2003, 2004). In the present paper, a canonical dual problem has been formulated, which is perfectly dual to the parametric problem $(\mathcal{P}_\mu)$ in the sense that they have the same solution set. The global and local extrema are identified by the triality theory in the Section 3.

2. Canonical Dual Problem and Complete Solutions

By the definition introduced in (Gao, 2000), a piecewise Gâteaux differentiable function $P : \mathcal{X}_a \to \mathbb{R}$ is called a *canonical function* on $\mathcal{X}_a \subset \mathbb{R}^n$ if its Gâteaux derivative $\mathbf{x}^* = \delta P(\mathbf{x})$ is piecewise invertible on the range $\mathcal{X}_a^* \subset \mathbb{R}^n$, such that the Legendre conjugate of P can be defined uniquely by

$$P^*(\mathbf{x}^*) = \{\mathbf{x}^T\mathbf{x}^* - P(\mathbf{x})| \ \mathbf{x}^* = \delta P(\mathbf{x}) \ \ \forall \mathbf{x} \in \mathcal{X}_a\}. \tag{25.6}$$

Thus, if $P(\mathbf{x})$ is a canonical function on $\mathcal{X}_a$, then the equivalent relations

$$\mathbf{x}^* = \delta P(\mathbf{x}) \quad \Leftrightarrow \quad \mathbf{x} = \delta P^*(\mathbf{x}^*) \quad \Leftrightarrow \quad P(\mathbf{x}) + P^*(\mathbf{x}^*) = \mathbf{x}^T\mathbf{x}^* \tag{25.7}$$

hold on $\mathcal{X}_a \times \mathcal{X}_a^*$. As it was shown by the author (Gao, 2000) that even if $P(\mathbf{x})$ may be a nonsmooth function, its Legendre conjugate is always smooth (see Fig. 25.1). This simple fact plays an important role in nonsmooth analysis.

Therefore, by the standard procedure of the canonical dual transformation, a dual problem $((\mathcal{P}_\mu{}^d)$ in short) can be formulated as the

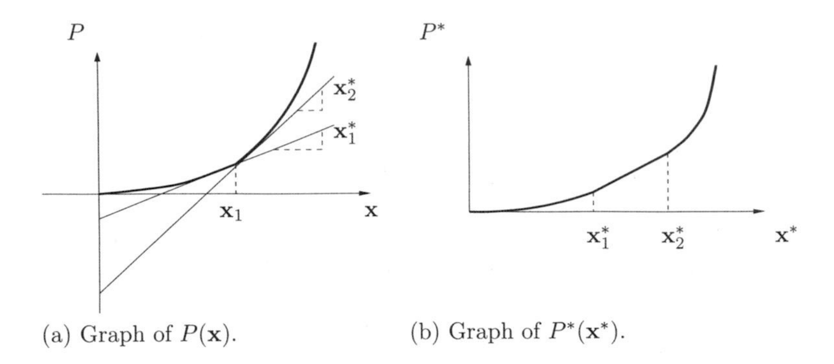

(a) Graph of $P(\mathbf{x})$. (b) Graph of $P^*(\mathbf{x}^*)$.

Figure 25.1. Nonsmooth function P and its smooth Legendre conjugate P^*

following.

$$(\mathcal{P}_\mu{}^d): \quad \max P^d(\boldsymbol{\epsilon}^*, \rho^*) \;\; \forall(\boldsymbol{\epsilon}^*, \rho^*) \in \mathcal{Y}_\mu^*, \tag{25.8}$$

where $P^d : \mathbb{R}^m \times \mathbb{R} \to \mathbb{R}$ is defined by

$$P^d(\boldsymbol{\epsilon}^*, \rho^*) = (\rho^* - 1)P^*\left(B^T\boldsymbol{\epsilon}^*/(\rho^* - 1)\right) + \mu\rho^* - \mathbf{b}^T\boldsymbol{\epsilon}^*, \tag{25.9}$$

in which, $P^*(\mathbf{x}^*)$ is the Legendre conjugate function of $P(\mathbf{x})$ on $\mathcal{X}_a^*$, and the dual feasible space $\mathcal{Y}_\mu^*$ is defined by

$$\mathcal{Y}_\mu^* = \{(\boldsymbol{\epsilon}^*, \rho^*) \in \mathbb{R}^m \times \mathbb{R} | \;\; \boldsymbol{\epsilon}^* \geq 0 \in \mathbb{R}^m, \;\; \rho^* \geq 0, \quad B^T\boldsymbol{\epsilon}^*/(\rho^* - 1) \in \mathcal{X}_a^*\}. \tag{25.10}$$

Since $P(\mathbf{x})$ is a canonical function over $\mathcal{X}_f$, the canonical dual function P^d is well defined on $\mathcal{Y}_\mu^*$.

THEOREM 25.2 (PERFECT DUALITY THEOREM) *Suppose that the concave function $P(\mathbf{x})$ is piecewise Gâteaux differentiable on $\mathcal{X}_a$ such that its Legendre conjugate $P^*(\mathbf{x}^*)$ can be defined by (25.6), then for a given parameter $\mu > -\infty$ and an m-vector $\mathbf{b}$, the problem $(\mathcal{P}^d)$ is canonically (or perfectly) dual to the primal problem $(\mathcal{P})$ in the sense that if $\bar{\mathbf{y}}^* = (\bar{\boldsymbol{\epsilon}}^*, \bar{\rho}^*) \in \mathcal{Y}_\mu^*$ is a KKT point of $(\mathcal{P}_\mu{}^d)$, and $\bar{\mathbf{x}}^* = B^T\bar{\boldsymbol{\epsilon}}^*/(\bar{\rho}^* - 1)$, then the vector $\bar{\mathbf{x}}$ determined by*

$$\bar{\mathbf{x}} = \delta P^*(\bar{\mathbf{x}}^*) \tag{25.11}$$

is a KKT point of $(\mathcal{P}_\mu)$, and

$$P(\bar{\mathbf{x}}) = P^d(\bar{\mathbf{y}}^*). \tag{25.12}$$

Moreover, if the primal problem $(\mathcal{P})$ is realizable, then there exists a finite parameter $\mu > -\infty$ such that the vector $\bar{\mathbf{x}}$ is also a KKT point of the problem $(\mathcal{P})$.

Proof. Suppose that $\bar{\mathbf{y}}^* = (\bar{\boldsymbol{\epsilon}}^*, \bar{\rho}^*) \in \mathcal{Y}_\mu^*$ is a KKT point of $(\mathcal{P}_\mu{}^d)$ and $\bar{\mathbf{x}}^* = B^T\bar{\boldsymbol{\epsilon}}^*/(\bar{\rho}^* - 1)$, then we have

$$0 \leq \bar{\boldsymbol{\epsilon}}^* \ \perp \ B\delta P^*(\bar{\mathbf{x}}^*) - \mathbf{b} \leq 0, \tag{25.13}$$

$$0 \leq \bar{\rho}^* \ \perp \ (\delta P^*)^T \frac{B^T\bar{\boldsymbol{\epsilon}}^*}{\bar{\rho}^* - 1} - P^* - \mu \geq 0. \tag{25.14}$$

Since $(\bar{\mathbf{x}}, \bar{\mathbf{x}}^*)$ is a canonical dual pair on $\mathcal{X}_a \times \mathcal{X}_a^*$, we have

$$\begin{aligned} \bar{\mathbf{x}}^* &= \frac{B^T\bar{\boldsymbol{\epsilon}}^*}{(\bar{\rho}^* - 1)} = \delta P(\bar{\mathbf{x}}) \quad \Leftrightarrow \\ \bar{\mathbf{x}} &= \delta P^*(\bar{\mathbf{x}}^*) \ \Leftrightarrow \\ P(\bar{\mathbf{x}}) &= \bar{\mathbf{x}}^T\bar{\mathbf{x}}^* - P^*(\bar{\mathbf{x}}^*). \end{aligned}$$

Thus in term of $\bar{\mathbf{x}}^* = B^T\bar{\boldsymbol{\epsilon}}^*/(\bar{\rho}^* - 1)$ and $\bar{\mathbf{x}} = \delta P^*(\bar{\mathbf{x}}^*)$, the KKT conditions (25.13) and (25.14) can be written as

$$0 \leq \bar{\boldsymbol{\epsilon}}^* \ \perp \ B\bar{\mathbf{x}} - \mathbf{b} \leq 0, \tag{25.15}$$

$$0 \leq \bar{\rho}^* \ \perp \ P(\bar{\mathbf{x}}) - \mu \geq 0. \tag{25.16}$$

This shows that $\bar{\mathbf{x}} = \delta P^* \left(B^T\bar{\boldsymbol{\epsilon}}^*/(\bar{\rho}^* - 1)\right)$ is indeed a KKT point of the parametric problem $(\mathcal{P}_\mu)$.

By the complementarity conditions (25.15) and (25.16), we have $\mathbf{b}^T\bar{\boldsymbol{\epsilon}}^* = (B\bar{\mathbf{x}})^T\bar{\boldsymbol{\epsilon}}^*$ and $\bar{\rho}^*\mu = \bar{\rho}^*P(\bar{\mathbf{x}})$, respectively. Thus, in term of

$$\bar{\mathbf{x}}^* = B^T\bar{\boldsymbol{\epsilon}}^*/(\bar{\rho}^* - 1), \quad \text{and } \bar{\mathbf{x}} = \delta P^*(\bar{\mathbf{x}}^*),$$

we have

$$P^d(\bar{\boldsymbol{\epsilon}}^*, \bar{\rho}^*) = (\bar{\rho}^* - 1)P^* \left(B^T\bar{\boldsymbol{\epsilon}}^*/(\bar{\rho}^* - 1)\right) - \bar{\mathbf{x}}^T B^T\bar{\boldsymbol{\epsilon}}^* + \bar{\rho}^* P(\bar{\mathbf{x}}) = P(\bar{\mathbf{x}})$$

due to the fact that $\bar{\mathbf{x}}^T\bar{\mathbf{x}}^* - P^*(\bar{\mathbf{x}}^*) = P(\bar{\mathbf{x}})$. □

Theorem 25.2 shows that there is no duality gap between the problems $(\mathcal{P}_\mu)$ and $(\mathcal{P}_\mu{}^d)$. As we know that the KKT stationary conditions are only necessary for the nonconvex optimization problem $(\mathcal{P}_\mu)$. The sufficient condition will be given by the triality theorem in the next section.

3. Triality Theory: Local and Global Minimizers

Let

$$\mathcal{Y}_{\mu+}^* = \{(\boldsymbol{\epsilon}^*, \rho^*) \in \mathcal{Y}_\mu^* | \ \rho^* > 1\}, \tag{25.17}$$

$$\mathcal{Y}_{\mu-}^* = \{(\boldsymbol{\epsilon}^*, \rho^*) \in \mathcal{Y}_\mu^* | \ 0 \leq \rho^* < 1\}, \tag{25.18}$$

and

$$\mathcal{X}_{\mu b} = \{\mathbf{x} \in \mathcal{X}_f | \quad P(\mathbf{x}) = \mu\}. \tag{25.19}$$

THEOREM 25.3 (TRIALITY THEOREM) *Suppose that for a given finite parameter $\mu > -\infty$, the vector $\bar{\mathbf{y}}^* = (\bar{\boldsymbol{\epsilon}}^*, \bar{\rho}^*)$ is a KKT point of the problem $(\mathcal{P}^d_\mu)$, and $\bar{\mathbf{x}}^* = B^T \bar{\boldsymbol{\epsilon}}^* / (\bar{\rho}^* - 1)$.*

If $\bar{\rho}^* > 1$, then $\bar{\mathbf{x}} = \delta P^*(\bar{\mathbf{x}}^*)$ is a global minimizer of $P(\mathbf{x})$ over $\mathcal{X}_{\mu b}$, and

$$P(\bar{\mathbf{x}}) = \min_{\mathbf{x} \in \mathcal{X}_{\mu b}} P(\mathbf{x}) = \max_{(\boldsymbol{\epsilon}^*, \rho^*) \in \mathcal{Y}^*_{\mu+}} P^d(\boldsymbol{\epsilon}^*, \rho^*) = P^d(\bar{\mathbf{y}}^*). \tag{25.20}$$

If $0 \le \bar{\rho}^* < 1$, then $\bar{\mathbf{x}} = \delta P^*(\bar{\mathbf{x}}^*)$ is a global maximizer of $P(\mathbf{x})$ over $\mathcal{X}_\mu$ if and only if $\bar{\mathbf{y}}^* = (\bar{\boldsymbol{\epsilon}}^*, \bar{\rho}^*)$ is a maximizer of P^d over $\mathcal{Y}^*_{\mu-}$ and

$$P(\bar{\mathbf{x}}) = \max_{\mathbf{x} \in \mathcal{X}_{\mu}} P(\mathbf{x}) = \max_{(\boldsymbol{\epsilon}^*, \rho^*) \in \mathcal{Y}^*_{\mu-}} P^d(\boldsymbol{\epsilon}^*, \rho^*) = P^d(\bar{\mathbf{y}}^*). \tag{25.21}$$

If $0 < \bar{\rho}^* < 1$, then $\bar{\mathbf{x}} = \delta P^*(\bar{\mathbf{x}}^*)$ is a global minimizer of $P(\mathbf{x})$ over $\mathcal{X}_{\mu b}$ if and only if $\bar{\mathbf{y}}^* = (\bar{\boldsymbol{\epsilon}}^*, \bar{\rho}^*)$ is a global minimizer of P^d over $\mathcal{Y}^*_{\mu-} \cap \{\rho^* \neq 0\}$ and

$$P(\bar{\mathbf{x}}) = \min_{\mathbf{x} \in \mathcal{X}_{\mu b}} P(\mathbf{x}) = \min_{(\boldsymbol{\epsilon}^*, \rho^*) \in \mathcal{Y}^*_{\mu-}} P^d(\boldsymbol{\epsilon}^*, \rho^*) = P^d(\bar{\mathbf{y}}^*). \tag{25.22}$$

Proof. Since $P^*(\mathbf{x}^*)$ is concave on $\mathcal{X}^*_a$, then for a given duality pair $(\bar{\mathbf{x}}, \bar{\mathbf{x}}^*)$, the inequality

$$P^*(\mathbf{x}^*) \le \bar{\mathbf{x}}^T(\mathbf{x}^* - \bar{\mathbf{x}}^*) + P^*(\bar{\mathbf{x}}^*) \ \forall \mathbf{x}^* \in \mathcal{X}^*_a \tag{25.23}$$

holds on $\mathcal{X}^*_a$. In terms of $\mathbf{y}^* = (\boldsymbol{\epsilon}^*, \rho^*)$, and $\bar{\mathbf{y}}^*(\bar{\boldsymbol{\epsilon}}^*, \bar{\rho}^*) \in \mathcal{Y}^*_\mu$, the inequality (25.23) also holds for

$$\mathbf{x}^* = \frac{B^T \boldsymbol{\epsilon}^*}{\rho^* - 1}, \quad \bar{\mathbf{x}}^* = \frac{B^T \bar{\boldsymbol{\epsilon}}^*}{\bar{\rho}^* - 1}.$$

Particularly, we let $\bar{\mathbf{y}}^* = (\bar{\boldsymbol{\epsilon}}^*, \bar{\rho}^*) \in \mathcal{Y}^*_\mu$ be a KKT point of the problem $(\mathcal{P}_\mu{}^d)$ such that $\bar{\mathbf{x}}^* = B^T \bar{\boldsymbol{\epsilon}}^* / (\bar{\rho}^* - 1)$. By Theorem refthm:cavedual we know that the vector $\bar{\mathbf{x}} = \delta P^*(\bar{\mathbf{x}}^*)$ is a KKT point of the problem $(\mathcal{P}_\mu)$. Thus, if $\rho^* > 1$, the inequality (25.23) leads to

$$\begin{aligned}(\rho^* - 1) P^* \left(\frac{B^T \boldsymbol{\epsilon}^*}{\rho^* - 1} \right) \ &\le \ \bar{\mathbf{x}}^T B^T \boldsymbol{\epsilon}^* - \bar{\mathbf{x}}^T \frac{B^T \boldsymbol{\epsilon}^*}{\bar{\rho}^* - 1} (\rho^* - 1) \\ &+ \ (\rho^* - 1) P^* \left(\frac{B^T \boldsymbol{\epsilon}^*}{\rho^* - 1} \right) \ \forall (\boldsymbol{\epsilon}^*, \rho^*) \in \mathcal{Y}^*_\mu.\end{aligned}$$

By the complementarity conditions $\bar{\mathbf{x}}^T B^T \bar{\boldsymbol{\epsilon}}^* = \bar{\mathbf{x}}^T \mathbf{b}$, and $\bar{\rho}^*(\mu - P(\bar{\mathbf{x}})) = 0$, as well as the equality $P(\bar{\mathbf{x}}) = \bar{\mathbf{x}}^T \bar{\mathbf{x}}^* - P^*(\bar{\mathbf{x}}^*)$, we have

$$\begin{aligned} P^d(\mathbf{y}^*) &\leq P^d(\bar{\mathbf{y}}^*) + (B\bar{\mathbf{x}} - \mathbf{b})^T \boldsymbol{\epsilon}^* + \rho^*(\mu - P(\bar{\mathbf{x}})) \\ &\leq P^d(\bar{\mathbf{y}}^*) \quad \forall (\boldsymbol{\epsilon}^*, \rho^*) \in \mathcal{Y}_\mu^*, \quad \rho^* > 1. \end{aligned}$$

This shows that if $\bar{\rho}^* > 1$, the KKT point $\bar{\mathbf{y}}^* = (\bar{\boldsymbol{\epsilon}}^*, \bar{\rho}^*)$ maximizes P^d on $\mathcal{Y}_{\mu+}^*$.

We now need to prove that the vector $\bar{\mathbf{x}}$ associated with the KKT point $\bar{\mathbf{y}}^*$ of $(\mathcal{P}_\mu{}^d)$ minimizes P on $\mathcal{X}_\mu$. By the concavity of the canonical function P on $\mathcal{X}_a$, the Legendre conjugate

$$P^*(\mathbf{x}^*) = \min_{\mathbf{x} \in \mathcal{X}_a} \{\mathbf{x}^T \mathbf{x}^* - P(\mathbf{x})\}$$

is uniquely defined on $\mathcal{X}_a^*$. Particularly, for a given $\mathbf{y}^* = (\boldsymbol{\epsilon}^*, \rho^*) \in \mathcal{Y}_\mu^*$, we let $\mathbf{x}^* = B^T \boldsymbol{\epsilon}^* / (\rho^* - 1)$. Thus, we have

$$P^d(\boldsymbol{\epsilon}^*, \rho^*) = \min_{\mathbf{x} \in \mathcal{X}_a} \Xi(\mathbf{x}, \boldsymbol{\epsilon}^*, \rho^*) \tag{25.24}$$

where

$$\Xi(\mathbf{x}, \boldsymbol{\epsilon}^*, \rho^*) = \mathbf{x}^T B^T \boldsymbol{\epsilon}^* - (\rho^* - 1)P(\mathbf{x}) + \mu\rho^* - \mathbf{b}^T \boldsymbol{\epsilon}^* \tag{25.25}$$

is the so-called *total complementary function*, or the *extended Lagrangian*, which can be obtained by the standard canonical dual transformation (see (Gao, 2000)). Clearly, for any given $\rho^* > 1$, $\Xi(\mathbf{x}, \boldsymbol{\epsilon}^*, \rho^*)$ is convex in $\mathbf{x}$ and concave (linear) in $\boldsymbol{\epsilon}^*$ and ρ^*. Thus, by the classical saddle-minimax theory, we have

$$\begin{aligned} P^d(\bar{\boldsymbol{\epsilon}}^*, \bar{\rho}^*) &= \max_{\rho^* > 1} \max_{\boldsymbol{\epsilon}^* \geq 0} \min_{\mathbf{x} \in \mathcal{X}_a} \Xi(\mathbf{x}, \boldsymbol{\epsilon}^*, \rho^*) \\ &= \max_{\rho^* > 1} \min_{\mathbf{x} \in \mathcal{X}_a} \max_{\boldsymbol{\epsilon}^* \geq 0} \Xi(\mathbf{x}, \boldsymbol{\epsilon}^*, \rho^*) \\ &= \min_{\mathbf{x} \in \mathcal{X}_a} \max_{\rho^* > 1} \{(1 - \rho^*)P(\mathbf{x}) + \mu\rho^*\} \quad s.t. \;\; B\mathbf{x} \leq \mathbf{b} \\ &= \min_{\mathbf{x} \in \mathcal{X}_f} \left\{ P(\mathbf{x}) - \min_{\rho^* > 1} \rho^*(P(\mathbf{x}) - \mu) \right\} \\ &= \min_{\mathbf{x} \in \mathcal{X}_f} P(\mathbf{x}) \quad s.t. \;\; P(\mathbf{x}) = \mu, \end{aligned}$$

since the linear programming

$$\theta_1 = \min_{\rho^* > 1} \rho^*(P(\mathbf{x}) - \mu)$$

has a solution in the open domain $(1, +\infty)$ if and only if $P(\mathbf{x}) = \mu$. This shows that the KKT point $\bar{\mathbf{y}}^* = (\bar{\boldsymbol{\epsilon}}^*, \bar{\rho}^*)$ maximize P^d on $\mathcal{Y}_{\mu+}^*$ if and only if $\bar{\mathbf{x}}$ is a global minimizer of $P(\mathbf{x})$ on $\mathcal{X}_{\mu b}$.

In the case that $0 < \bar{\rho}^* < 1$, the total complementary function $\Xi(\mathbf{x}, \boldsymbol{\epsilon}^*, \rho^*)$ is concave in $\mathbf{x} \in \mathbb{R}^n$ and concave in both $\boldsymbol{\epsilon}^* \in \mathbb{R}^m_+$ and $\rho^* \in (0,1)$. Thus, if $(\bar{\boldsymbol{\epsilon}}^*, \bar{\rho}^*)$ is a global minimizer of P^d on $\mathcal{Y}^*_{\mu-}$, then by the so-called *bi-duality theory* developed in (Gao, 2000), we have either

$$\max_{\mathbf{x}\in\mathbb{R}^n} \max_{(\boldsymbol{\epsilon}^*,\rho^*)\in\mathcal{Y}^*_{\mu-}} \Xi(\mathbf{x}, \boldsymbol{\epsilon}^*, \rho^*) = \max_{(\boldsymbol{\epsilon}^*,\rho^*)\in\mathcal{Y}^*_{\mu-}} \max_{\mathbf{x}\in\mathbb{R}^n} \Xi(\mathbf{x}, \boldsymbol{\epsilon}^*, \rho^*) \tag{25.26}$$

or

$$\min_{\mathbf{x}\in\mathbb{R}^n} \max_{(\boldsymbol{\epsilon}^*,\rho^*)\in\mathcal{Y}^*_{\mu-}} \Xi(\mathbf{x}, \boldsymbol{\epsilon}^*, \rho^*) = \min_{(\boldsymbol{\epsilon}^*,\rho^*)\in\mathcal{Y}^*_{\mu-}} \max_{\mathbf{x}\in\mathbb{R}^n} \Xi(\mathbf{x}, \boldsymbol{\epsilon}^*, \rho^*). \tag{25.27}$$

Thus, if $\bar{\mathbf{y}}^* = (\bar{\boldsymbol{\epsilon}}^*, \bar{\rho}^*)$ is a maximizer of P^d over $\mathcal{Y}^*_{\mu-}$ then we have

$$\begin{aligned}
P^d(\bar{\boldsymbol{\epsilon}}^*, \bar{\rho}^*) &= \max_{(\boldsymbol{\epsilon}^*,\rho^*)\in\mathcal{Y}^*_{\mu-}} P^d(\boldsymbol{\epsilon}^*, \rho^*) \\
&= \max_{\boldsymbol{\epsilon}^*\geq 0,\ \rho^*\in[0,1)} \max_{\mathbf{x}\in\mathbb{R}^n} \Xi(\mathbf{x}, \boldsymbol{\epsilon}^*, \rho^*) \\
&= \max_{\rho^*\in(0,1)} \max_{\mathbf{x}\in\mathbb{R}^n} \max_{\boldsymbol{\epsilon}^*\geq 0} \Xi(\mathbf{x}, \boldsymbol{\epsilon}^*, \rho^*) \\
&= \max_{\rho^*\in(0,1)} \max_{\mathbf{x}\in\mathbb{R}^n} \left\{(1-\rho^*)P(\mathbf{x}) + \mu\rho^* - \mathbf{x}^T\mathbf{b}\right\} \quad s.t.\ \ B\mathbf{x} \leq \mathbf{b} \\
&= \max_{\mathbf{x}\in\mathcal{X}_f} \left\{ P(\mathbf{x}) - \min_{\rho^*\in[0,1)} \rho^*(P(\mathbf{x}) - \mu) \right\} \\
&= \max_{\mathbf{x}\in\mathcal{X}_f} P(\mathbf{x}) \ \ s.t.\ \ P(\mathbf{x}) - \mu \geq 0,
\end{aligned}$$

by the fact that the domain $[0,1)$ is closed on the lower bound and open on the upper bound, the problem

$$\theta_2 = \min_{\rho^*\in[0,1)} \{\rho^*(P(\mathbf{x}) - \mu)\} \tag{25.28}$$

has a solution if and only if $P(\mathbf{x}) \geq \mu$, and for this solution, $\theta_2 = 0$.

On the other hand, if $\bar{\mathbf{y}}^* = (\bar{\boldsymbol{\epsilon}}^*, \bar{\rho}^*)$ is a minimizer of P^d over $\mathcal{Y}^*_{\mu-} \cap \{\rho^* \neq 0\}$ then we have

$$\begin{aligned}
P^d(\bar{\boldsymbol{\epsilon}}^*, \bar{\rho}^*) &= \min_{(\boldsymbol{\epsilon}^*,\rho^*)\in\mathcal{Y}^*_{\mu-}} P^d(\boldsymbol{\epsilon}^*, \rho^*) \\
&= \min_{\boldsymbol{\epsilon}^*\geq 0} \min_{\rho^*\in(0,1)} \max_{\mathbf{x}\in\mathbb{R}^n} \Xi(\mathbf{x}, \boldsymbol{\epsilon}^*, \rho^*) \\
&= \min_{\rho^*\in(0,1)} \min_{\mathbf{x}\in\mathbb{R}^n} \max_{\boldsymbol{\epsilon}^*\geq 0} \Xi(\mathbf{x}, \boldsymbol{\epsilon}^*, \rho^*) \\
&= \min_{\rho^*\in(0,1)} \min_{\mathbf{x}\in\mathbb{R}^n} \{(1-\rho^*)P(\mathbf{x}) + \mu\rho^*\} \quad s.t.\ \ B\mathbf{x} \leq \mathbf{b} \\
&= \min_{\mathbf{x}\in\mathcal{X}_f} \left\{ P(\mathbf{x}) - \max_{\rho^*\in(0,1)} \rho^*(P(\mathbf{x}) - \mu) \right\} \\
&= \min_{\mathbf{x}\in\mathcal{X}_f} P(\mathbf{x}) \ \ s.t.\ \ P(\mathbf{x}) = \mu,
\end{aligned}$$

since the linear minimization

$$\theta_3 = \max_{\rho^* \in (0,1)} \rho^*(P(\mathbf{x}) - \mu)$$

has a solution on the open domain $(0, 1)$ if and only $P(\mathbf{x}) - \mu = 0$. By the fact that $P^d(\bar{\boldsymbol{\epsilon}}^*, \bar{\rho}^*) = P(\bar{\mathbf{x}})$ for all KKT points of $(\mathcal{P}_\mu)$, the theorem is proved. □

Theorem 25.3 shows that if the KKT point $\bar{\rho}^*$ is in the open set $(0, 1)$, then $\bar{\mathbf{x}}$ is a minimizer of P only if $\bar{\mathbf{x}}$ is located on the boundary of the feasible set $\mathcal{X}_\mu$, i.e. $P(\bar{\mathbf{x}}) = \mu$, which can not be located, generally speaking, by standard algorithms designed for convex minimization problems. This is the main reason that why the primal problem $(\mathcal{P})$ is NP-hard. However, the dual problem in this case is a convex minimization over the open set $\mathcal{Y}^*_{\mu-} \cap \{\rho^* \neq 0\}$, which is much easier than the primal one. The triality theorem can be used to develop certain powerful algorithms for solving this concave minimization problem with inequality constraints.

The canonical dual transformation method and triality theory were originally developed from nonsmooth and nonconvex mechanics (see Gao, 2000). The key idea of this method is to choose certain geometrically admissible measure $\Lambda : \mathcal{X}_a \to \mathcal{Y}_a$ such that the Legendre-Fenchel-Moreau transformation holds on the canonical dual spaces $\mathcal{Y}_a \times \mathcal{Y}^*_a$ (see Gao, 2000, 2003a,b). In the present paper, since the concave function P is a canonical function, this geometrical measure is simply chosen to be $\Lambda(\mathbf{x}) = P(\mathbf{x})$. In the case that P is a quadratic function, the constraint $P(\mathbf{x}) \leq \mu$ can be simply replaced by the normality condition $\frac{1}{2}|\mathbf{x}|^2 \leq \mu$, which has been studied very recently by the author (see Gao, 2004). The results presented in this paper show again that the canonical dual transformation and associated triality theory may possess important computational impacts on global optimization.

References

Benson, H (1995), Concave minimization: theory, applications and algorithms, in *Handbook of Global Optimization*, eds. R. Horst and P. Pardalos, Kluwer Academic Publishers, 43-148.

Ekeland, I. (2003). Nonconvex duality, in *Proceedings of IUTAM Symposium on Duality, Complementarity and Symmetry in Nonlinear Mechanics*, D.Y. Gao (Edited), Kluwer Academic Publishers, Dordrecht / Boston / London, 13-22.

Ekeland, I. and Temam, R. (1976),*Convex Analysis and Variational Problems,* North-Holland, and SIAM, Philadelphia, 1999, 402pp.

Floudas, C.A and Visweswaran, V. (1995). Quadratic optimization, in *Handbook of Global Optimization*, R. Horst and P.M. Pardalos (eds), Kluwer Academic Publ., Dordrecht / Boston / London, pp. 217-270.

Gao, D.Y. (2000). *Duality Principles in Nonconvex Systems: Theory, Methods and Applications*, Kluwer Academic Publishers, Dordrecht / Boston / London, xviii+454pp.

Gao, D.Y. (2003). Perfect duality theory and complete solutions to a class of global optimization problems, *Optimisation*, Vol. **52**, 4-5, 467-493.

Gao, D.Y. (2003). Nonconvex semi-linear problems and canonical duality solutions, in *Advances in Mechanics and Mathematics*, Vol. II, edited by D.Y. Gao and R.W. Ogden, Kluwer Academic Publishers, Dordrecht / Boston / London, pp. 261-312..

Gao, D.Y. (2004). Canonical duality theory and solutions to constrained nonconvex quadratic programming, *J. Global Optimization*, to appear in a special issue on duality edited by D.Y. Gao and KL Teo.

Gao, D.Y. and Strang, G. (1989), Geometric nonlinearity: Potential energy, complementary energy, and the gap function, *Quart. Appl. Math.*, **47**(3), 487-504, 1989.

Goh, C.J. and Yang, X.Q. (2002). *Duality in Optimization and Variational Inequalities,* Taylor and Francis, 329pp.

Horst, R., Panos M. Pardalos, Nguyen Van Thoai (2000). *Introduction to Global Optimization*, Kluwer Academic Publishers.

Moreau, J.J. (1966). *Fonctionelles Convexes*, Collège de France, 1966, and Instituto Poligrafico E Zecca Dello Stato, S.p.A, Roma, 2003.

Rubinov, A.M. and Yang X.Q. (2003). *Lagrange-Type Functions in Constrained Non-Convex Optimization.* Kluwer Academic Publishers, Boston / Dordrecht / London, 285 pp.

Appendix: Theoretical and Numerical Nonsmooth Mechanics International Colloquium

From 17 to 19 November 2003, the *Laboratoire de Mécanique et Génie Civil* in Montpellier, France, has organized an international colloquium in honour of the 80th birthday of Jean Jacques Moreau, Emeritus Professor at University Montpellier II. The colloquium was devoted to the presentation of recent works in the fields where the contribution of Jean Jacques Moreau was original and significant.

Some of the participants were able to prepare the content of this book, in the same spirit, but all participate to the success of the event. Many thanks to all of them. They were:

- Hedy Attouch, Université Montpellier II, France,
- Jean-Pierre Aubin, Université Paris-Dauphine, France,
- John Ball, University of Oxford, United Kingdom,
- Bernard Brogliato, Institut National de Recherche en Informatique et en Automatique, France,
- Giuseppe Buttazzo, Università di Pisa, Italy,
- Francis Clarke, Université Claude Bernard Lyon I, France,
- Louis Marie Cléon, Société Nationale des Chemins de Fer, France,
- Bernard Dacoragna, École Polytechnique Fédérale de Lausanne, Switzerland,
- David Dureisseix, Université Montpellier II, France,
- Gianpietro Del Piero, University of Ferrara and Accademia Nazionale dei Lincei, Italy,
- Michel Frémond, Laboratoire Central des Ponts et Chaussées, France,
- David Gao, Virginia Polytechnic Institute & State University, USA,
- Christoph Glocker, Swiss Federal Institute of Technology Zurich, Switzerland,
- Jean-Baptiste Hiriart-Urruty, Université Paul Sabatier, France,
- Michel Jean, Laboratoire de Mécanique et d'Acoustique, CNRS, France,
- Anders Klarbring, University of Linköping, Sweden,
- Franco Maceri, University of Roma II Tor Vergata, Italy,
- Manuel Marques, Universidade de Lisboa, Portugal,
- Henry Keith Moffatt, University of Cambridge, United Kingdom,
- Bernard Nayroles, Institut National des Sciences Appliquées de Rouen, France,
- Quoc Son Nguyen, École Polytechnique, France,
- Alain Pumir, Institut Non Linéaire de Nice, CNRS, France,
- David Quéré, Laboratoire de Physique de la Matière Condensée, CNRS, France,
- Farhang Radjaï, Laboratoire de Mécanique et Génie Civil, CNRS, France,
- Michel Raous, Laboratoire de Mécanique et d'Acoustique, CNRS, France,
- Renzo Ricca, Università di Milano, Italy, and University College, United Kingdom,
- Ralph Tyrrell Rockafellar, University of Washington, USA,
- Michelle Schatzman, Laboratoire de Mathématiques Appliquées de Lyon, CNRS, France,

- Georgios E. Stavroulakis, University of Ioannina, Greece,
- Pierre Suquet, Laboratoire de Mécanique et d'Acoustique, CNRS, France,
- Michel Valadier, Université Montpellier II, France,
- Jean-Pierre Vilotte, Institut de Physique du Globe de Paris, France.

This event was also made possible thanks to the sponsorship of:

- Ministère de la Jeunesse, de l'Éducation Nationale et de la Recherche,
- Centre National de la Recherche Scientifique,
- Région Languedoc-Roussillon,
- Conseil Général de l'Hérault,
- Communauté d'Agglomération de Montpellier,
- Université Montpellier II,
- Association Française de Mécanique,
- Société de Mathématiques Appliquées et Industrielles,
- Société Nationale des Chemins de Fer.

Index